CABLES, CRISES, and the PRESS

NORTH
CANADA
Montreal
Washington
Wilmington
Bermudas
THE AZORES
ATLANTIC
Madeira
Bahama Is
WEST INDIA ISLANDS
Jamaica
Kingstown
CARRIBBEAN SEA
OCEA
CAPE VERD Is
Panama
VENEZUELA
Georgetown
Paramaribo
Guiana
Buenaventura
COLOMBIA
ECUADOR
Para
Maranham
Fortalezza
Pernambuco
PERU
Lima
BRAZIL
Bahia
Mollendo
BOLIVIA
SOU
Rio Janeiro
Santos
Antofagasta
PARAGUAY
Desterro
ATLAN
URUGUAY
Rio Grande
Valparaiso
Santiago
Buenos Ayres
Concepcion
OCEA

CABLES, CRISES, and the PRESS

THE GEOPOLITICS OF THE NEW INTERNATIONAL INFORMATION SYSTEM IN THE AMERICAS, 1866–1903

John A. Britton

University of New Mexico Press | Albuquerque

Printed in the United States of America
18 17 16 15 14 13 1 2 3 4 5 6

Library of Congress Cataloging-in-Publication Data
Britton, John A.
Cables, crises, and the press : the geopolitics of the new international information system in the Americas, 1866–1903 / John A. Britton.
pages cm
Includes bibliographical references and index.
ISBN 978-0-8263-5397-9 (hardback) — ISBN 978-0-8263-5398-6 (electronic)
1. Journalism—History. 2. Journalism—Technological innovations.
3. Online journalism. 4. Digital media. I. Title.
PN4801.B74 2013
070.4—dc23
2013030754

Designed and typeset by Karen Mazur
Text composed in ScalaOT 10/13.6
Display type is News 701 BT and ScalaSansOT

Contents

Acknowledgments

THIS STUDY BEGAN AS what appeared to be a brief inquiry into how journalist Carleton Beals sent his telegraphed reports on his 1928 interview with Nicaraguan rebel Augusto Sandino to the editorial office of the *Nation* in New York. I wrote Daniel Headrick who directed me to Mary Godwin and the Cable and Wireless Archive (then in London now in Porthcurno, Cornwall). Archivist Godwin provided me with detailed information on the extensive holdings of the Cable and Wireless Archive. The Francis Marion University Sabbatical and Release Time Committee granted me sabbatical leave for the spring of 1999, and I divided my time between the Cable and Wireless Archive in England and the Library of Congress and the National Archives in Washington, D.C., to explore the building and operations of the undersea cable network that linked Latin America with the United States and Europe. Mary Godwin soon led me to Jorma Ahvanainen. His correspondence was of great help in understanding the large scope and importance of this cable system. Eventually Jorma and I met at the Dibner Conference at the Massachusetts Institute of Technology in 2002 where I also benefitted from discussions with Menahem Blondheim, Bernard Finn, Peter Hugill, and other scholars. By this time it had become evident that I was dealing with far more than a background chapter on the history of journalistic communications in the Americas. The cable system and its use by journalists and government officials had a history of its own.

My research came to concentrate on several collections of papers held by the Manuscript Division of the Library of Congress and the National Archives and also the Public Records Office in England as well as the Cable and Wireless Archive. The archivists at these institutions were of immense help in my work over the years. A large part of this study is based on newspapers. Because the great bulk of this research was completed before digitalization, I relied on the microfilm holdings in the Government Documents section of the Cooper Library of the University of South Carolina in Columbia and the Newspaper and Current Periodical Reading Room of the Library of Congress. The staff members at both institutions were consistent in their patience and helpfulness. My research into the Latin American aspects of this study was concentrated in the Hispanic Division of the Library of Congress where Katherine McCann gave generous and thoughtful support. The reference and interlibrary loan specialists at the Rogers Library of Francis Marion University were also of much importance in my work, and Nathan Flowers led me into the new world of digitalized information. The Francis Marion University Board of Trustees' support of my work as an FMU Research Scholar made possible the completion of this long-term project. The FMU Professional Development committee also provided essential support.

This project at one point had as a main theme the involvement of Latin American governments, political movements, and individual activists, but research in the technological and business aspects of the new communications system and mass-circulation newspapers led to a focus on the United States. The author hopes to expand upon the roles of Latin American and other non-U.S. participants in the global communications system in later work.

A large number of individuals contributed to this project through commentary at professional meetings, correspondence, archival and reference assistance, and informal conversations. They include Ralph Lee Woodward, Lester Langley, Robert Greenhill, Alan McPherson, Jurgen Buchenau, Barbara Tennenbaum, Georgette Dorn, Dolores Martin, Helen Delpar, Marshall Eakin, Rhoda Desbordes, Dwayne Winseck, Larry Clayton, Richard V. Salisbury, John Orbell, Don Stewart, G. Wayne King, Scott Kaufman, and Buddy Bilbry. I owe a special acknowledgement to David Paull Nickles whose excellent pioneering study and sage advice over the years helped guide me through the nuances and complexities of communications history.

My greatest debt is owed to my lovely wife Kathy Smith Britton who gave of her time, energy, intelligence, and knowledge of standard written English to help me in the organization and composition of this manuscript.

PERMISSIONS

The following gave permission to publish illustrations in this book:

The Cable and Wireless Archive of the Porthcurno Telegraph Museum granted permission to use the portrait of John Pender.

The Cable and Wireless Archive of the Porthcurno Telegraph Museum granted permission to use the picture of the cable ship *Dacia*.

The Library of Congress, Prints and Photographs Division granted permission to use the photograph of Theodore Roosevelt and Richard Harding Davis, Reproduction Number: LC-DIG-stereo-1s01905 (digital file from original) LC-USZ62–23023 (black-and-white film copy negative).

The National Anthropological Archives of the Smithsonian Institution granted permission to use the photograph of the execution of Pedro Prestán, photo lot 81–54.

The following articles originally appeared in the journals named below, which gave permission for them to be published in a revised form here:

"'The Confusion Provoked by Instantaneous Discussion': The New International Communications Network and the Chilean Crisis of 1891–1892 in the United States." *Technology and Culture: The International Quarterly of the Society for the History of Technology* 48.4 (2007): 729–57.

"Communications and Military Intervention in Historical Perspective: The United States and Latin America." *Forum on Public Policy: A Publication of the Oxford Round Table* (Summer 2007): forumonpublicpolicy.com.

"International Communications and International Crises in Latin America, 1867–1881." *Latin Americanist: SECOLAS Annals* 52.1 (2008): 131–54.

Introduction

Main Themes and Organization of the Book

THE FLOOD OF INFORMATION that has inundated the internet in the last few decades has played a large role in both internal and international crises on what may seem to be an unprecedented scale. A close examination of events in the late nineteenth century, however, reveals an interesting parallel. The new communications system that emerged in the 1860s and 1870s inspired hopes of peace and progress but also brought about stress and instability. Advances in submarine-cable technology made possible the creation of a network of insulated copper wires that tied together North America, the Caribbean, South America, and Europe. The telegraph was too expensive for the average citizen, but the link between electronic communication and the front page brought a mass of readers into the information system. Telegraphed news reached front pages with unprecedented speed. The main themes in this book concern the rapid acceleration of information moving through the system of cables and printing presses that soon acquired a dynamic of its own beyond the control of presidents, prime ministers, secretaries of state, and even the newspaper editors who promoted and exploited the improvements in telegraphic communication. At the same time, newspapers changed their front-page formats, used larger and more frequent illustrations, and enhanced printing speed and circulation. The acceleration of the messages in this information system reached millions of newspaper readers and affected the work of government leaders, diplomats, and journalists.

The first four chapters examine the establishment of the new information system. Chapter 1 surveys the technological innovations and business operations of newspapers in the United States. This chapter also explores the historical context in Latin America with emphasis on the influences of the United States and European nations in the region. Chapters 2 and 3 discuss the building of the international network and the initial enthusiasm for these electronic means of communication to resolve international conflicts. Chapter 4 examines the early attempts by U.S. and British newspapers to cover the War of the Pacific fought between Chile and the alliance of Peru and Bolivia. Chilean Marcial Martínez was one of the first officials to grasp the opportunities in the new information system to benefit government policy. The flawed press coverage of the war and the subsequent conflicts between the United States and Chile indicated gaps and weaknesses in the new information system.

The next four chapters trace the escalation of international conflicts involving the new information system. Chapter 5 describes the political turmoil and European intrusions in Central America that set the stage for the U.S. intervention in Panama in 1885. Washington responded to an outburst of violence on the isthmus with an armed intervention that made extensive use of cable communications for the first time in such an operation. Press coverage of this event was limited, but Commander John Grimes Walker anticipated the potential problems that front-page newspaper reports could generate. Chapter 6 focuses on the dramatic confrontation between Chile and the United states in 1891–1892 that saw press coverage in mass-circulation newspapers in the United States emerge as a factor in the crisis. President Benjamin Harrison used the cable system to win concessions from Chile. International news became popular with the public, and chapter 7 explains how Richard Harding Davis and other writers exploited themes that centered on the quest for wealth and empire in books and magazines as well as newspapers. Chapter 8 examines the confrontation between Britain and the United States in the Venezuelan boundary controversy of 1895–1896. The outburst of patriotism and the threat of war in defense of the Monroe Doctrine in the U.S. newspaper press surprised and distressed leaders on both sides of the Atlantic.

Chapters 9–11 focus on the role of cables and the press in the Spanish-American War. Sylvester Scovel, Richard Harding Davis, and dozens of young male reporters fed stories of Spanish atrocities and Cuban heroism into the mass-circulation press of the United States from 1895 to 1898.

Chapter 10 studies the pressures created by the anti-Spanish slant of this coverage among diplomats and government leaders in the United States and Spain. The sensationalism associated with Hearst and Pulitzer spread across the press landscape in early 1898 and amplified the impact of the Dupuy de Lôme scandal and the explosion of the *Maine*. Nearly unanimous expressions of outrage on front pages and in editorial columns doomed President William McKinley's efforts at a peaceful solution. The war itself received patriotic coverage highlighted by cabled reports from Cuba and the circum-Caribbean as analyzed in chapter 11.

Theodore Roosevelt collaborated with reporter Davis in Cuba to project his image as a war hero, which constituted one step on his way to the White House. Chapter 12 concentrates on Roosevelt's determination to build a canal through Panama. He took advantage of the machinations of engineer-turned-promoter Philippe Bunau-Varilla to seize the Canal Zone. Bunau-Varilla proved to be the master of the new information system by arranging for Panamanian secession from Colombia through the use of secret cables and then by outwitting his Panamanian allies in the signing of the treaty with the United States that secured his financial future and U.S. domination of Panama. Chapter 13 explains the role of newspapers, magazines, and books in the celebrations of the U.S. victory in the Spanish-American War, the construction of the Panama Canal, and the establishment of the U.S. empire in the Caribbean and the Pacific.

The conclusion expands on this celebratory theme with an analysis of the roles of the electric cable system and the press that interfaced with it. This new information system had broad ramifications in the politics and popular culture of 1866–1903. The interface of the new technologies of cables and newspapers made possible outbursts of belligerent patriotism that undercut customary diplomacy and cast an ominous cloud over the geopolitical environment. This study suggests that technological change in communications can produce unanticipated consequences that challenge established political traditions and institutions.

CHAPTER ONE

Introduction to the New International Information System

THE USE OF THE TERM "information system" suggests recent developments in computers and fiber-optic cables that allow the movement, storage, and analysis of large quantities of information from stock market prices to breaking news on political crises. The focus of this study is not on the first years of the twenty-first century, however, but the last decades of the nineteenth century. From 1866 through the early 1900s a network of submarine cables and land telegraph lines tied the six populated continents of the world together through a new means of communication. This new electric system was one of the great innovations of the era. Politicians, corporate executives, and the general public regarded it as a manifestation of material progress and a triumph over time and space. This nineteenth-century information system did not accommodate the storage and analysis of information in the manner of the twenty-first century system, but it did increase the speed and reach of telegraphic messages to unprecedented degrees. This system brought about extensive change to the ways in which governments, businesses, and newspapers carried out their routine tasks. One of the most obvious involved velocity—diplomats, business managers, and journalists saw an acceleration in the movement of information that was crucial to their work. The second type of change concerned reach: government officials, business leaders, and newspaper editors saw their areas of responsibility increased from cities, regions, and/or nations to a larger and more complex international arena.

These increases in speed and range were accompanied by a growth in competitiveness among institutions and individuals to gain the maximum benefit from the movement of information. This potential for competition in the use of the system is exemplified in the contrasting and often conflicting work of diplomats and journalists in times of international crisis. Diplomats and other government officials attempted to resolve the disagreements that caused the crisis, while journalists often sought to publicize those very same points of controversy to capture the attention of their readers. These conflicts lead to some important questions to be addressed in this study: did the new information system bring improvements to the relations between nations, in the formulation of government policies, and in the public understanding of current events?

In 1895 the ramifications of an obscure event in a little-known part of South America provided evidence of the types of stress and strain under study here. A decades-old boundary dispute near the north coast of that continent suddenly became a major international controversy that drew Venezuela, the United States, and Great Britain into a series of widely publicized confrontations. Outspoken politicians and leading newspaper editors in the United States called for war. This crisis soon achieved a global dimension that affected the rivalry between Germany and Great Britain in southern Africa. The disputed area along the boundary of Venezuela and British Guiana was reputed to have considerable economic potential that seemed to be on the verge of realization with the discovery of gold there in 1856, which was followed by a gold rush of modest proportions in the 1880s and early 1890s. A few mining companies and an unnumbered army of prospectors roamed the region while the governments in Caracas and London sparred over the issue of where to draw the boundary.[1] Even after the onset of the gold rush, the area was difficult to reach. Steamships from New York continued their slow-paced voyages to the region, arriving in nearby Ciudad Bolívar on the Orinoco River every six weeks and in Georgetown, the capital of British Guiana, every three weeks.[2] Yet the Venezuelan government was disturbed by what it saw as an encroachment on this territory by the world's greatest imperial power. This dispute drew little attention in London or Washington, however, until 1895, when politicians on both sides of the Atlantic decided that this heretofore ignored affair required prompt resolution.

How did this marginal dispute become an important international issue? Poor leadership and diplomatic misjudgments played large parts, as historians have noted, but it is the contention of this study that the use and misuse of transoceanic cables and land-based telegraph lines and the connection of these technologies with the mass-circulation press deserve more attention as factors in this episode in the history of international misunderstandings.[3]

Transatlantic cables were available at the time, but the principals chose not to use them in the first phase of the controversy. Secretary of State Richard Olney and President Grover Cleveland wrote an abrasive message that reached London on August 7, 1895, via transoceanic steamer and the U.S. embassy in London, a route that took eighteen days. Olney told the British to back away from their territorial claims in the disputed area. He insisted that the United States was the appropriate mediator of any disagreements over territory in this area because it was "practically sovereign" over the Western Hemisphere as established under the Monroe Doctrine. President James Monroe made this unique policy statement in 1823, warning the European powers to stay out of the Western Hemisphere in order to protect the newly independent nations of Latin America as well as the United States. British Prime Minister Lord Salisbury waited nearly five months to reply and, when he did, sent his rejection of the Monroe Doctrine and Washington's current assertions of hegemony via ship mail. Salisbury's response was of course unacceptable to Olney and Cleveland, and the two administrations became entangled in a game of bluff and blunder that carried their countries in the direction of war.[4]

These diplomatic communications moved by ocean transport, but news coverage of these events moved by electric cable. This dichotomy—the slow movement of diplomacy contrasted with the rapidity of journalism—created an information imbalance into which reporters and editors injected facts, fragments of facts, and forceful editorial opinions to hold the attention of readers. This type of journalism was especially prevalent in the United States. Cables from London, Caracas, and Washington in October 1895 confirmed that the Venezuelan government had rejected a British ultimatum to evacuate the disputed territory and that both governments were preparing for war, a conflict that many U.S. newspapers concluded would draw the United States into hostilities with Britain. This October war scare soon passed, but in December tension again rose to the crisis level when Salisbury's long-awaited message arrived in Washington. Months of frustration and press speculation created an ambience into which President Cleveland, usually a calm and careful speaker, thrust angry prose written largely by Olney. As the nation's chief executive, Cleveland defended the Monroe

Doctrine in unequivocal language and then applied it to the dispute by calling for a commission to be appointed by his administration to determine the boundary. The U.S. government would then enforce this ruling, presumably by military means if necessary.[5] Newspapers from New York to San Francisco and from Chicago to New Orleans echoed the president's belligerent tone. The *San Francisco Examiner*'s coverage was representative: it pointed out the superiority of the Royal Navy over the U.S. Navy and then recklessly urged a strong stand against London and for war if necessary.[6]

The patriotic excitement among those who used the U.S. information system found stimulation and sustenance in a pamphlet written by diplomat-turned-publicist William L. Scruggs with the support of Venezuela's representative in Washington, José Andrade. First published in October 1894, *British Aggressions in Venezuela; or, The Monroe Doctrine on Trial* circulated widely in the United States, with two more editions printed in 1895. Its central message resonated through the telegraph network and onto newspaper editorial and front pages cross the country: the defense of the Venezuelan boundary was, in effect, a defense of the Monroe Doctrine. The Scruggs-Andrade argument supported the policy of Cleveland and Olney.[7]

News of the Cleveland-Olney retort to Salisbury reached Caracas by the cable and elicited an enthusiastic response. Printed copies of the statement circulated through the city and telegraphed versions made their way via telegraph into provincial cities. The public reaction was overwhelming. The U.S. diplomatic representative reported that a crowd of twenty thousand assembled in front of the legation in Caracas to express their gratitude and, in the excitement, nearly destroyed the legation's headquarters. A more organized demonstration occurred on Christmas Day: a parade led by the flags of Venezuela and the United States flying was held in a show of binational unity.[8]

Patriotic demonstrations in Venezuela and jingoistic rhetoric in the United States threw the British off balance. By late December both British and U.S. officials had begun to seek a compromise, but another international crisis soon emerged that placed the Venezuela boundary dispute in a larger context. Prime Minister Salisbury, who also served as his own foreign minister, and his cabinet—especially the head of the Colonial Office, Joseph Chamberlain—were surprised by cabled news of the December 29 raid on the Boer republic of Transvaal in southern Africa led by renegade British colonial official Leander Starr Jameson. The Transvaal authorities defeated and captured Jameson and his raiders. Embarrassed by these events, Chamberlain and Salisbury received more unsettling news. Germany's Kaiser

Wilhelm II sent President Paul Kruger of Transvaal a congratulatory telegram. This communication implied a close relationship between Transvaal, a troublesome rival to the British in southern Africa, and Germany, a major player in European politics. The British had already decided to deemphasize their differences with the United States over Venezuela. Business leaders on both sides of the Atlantic favored a peaceful solution. After the Kruger telegram this reconciliation became even more important in Britain's global policy.[9] Over the next few months Washington and London resolved their differences through diplomatic channels, although jingoistic newspaper editors and politicians in the United States continued their patriotic outbursts. The eventual compromise was not what Venezuela had hoped for, but the two major powers had decided that a mutually satisfactory solution suited their needs. After all of the turmoil, Caracas remained on the sidelines.[10]

The connection of a regional boundary dispute in northern South America with a bungled invasion in southern Africa through the Northern Hemisphere capitals of Washington, London, and Berlin was made possible by the international cable system that linked the four continents bordering on the Atlantic Ocean. The electrically transmitted messages that combined and fueled these crises traversed over ten thousand miles of submerged insulated copper wire and overland telegraphic lines.[11] The Venezuelan boundary dispute appears, in retrospect, to have been largely a media event. The press—especially the excitable newspapers in the United States—chose to publish the most dramatic information, which it reinforced with jingoistic editorials and belligerent quotations from politicians such as Senator Henry Cabot Lodge, who regarded any policy of a European power that contradicted the Monroe Doctrine as "an act of hostility towards the United States."[12] British ambassador Julian Pauncefote reported to Salisbury on the close connection between press and politics in Washington. Members of the U.S. Congress picked up copies of the afternoon editions of newspapers that printed Cleveland's sharply worded message on their way to the House chamber *before* they heard the president's presentation. Historian H. W. Brands points to Salisbury's judgment that about 80 percent of Olney's purpose was to arouse an audience of potential voters in the United States. Newspapers were the most convenient medium through which to reach these potential voters.[13]

The patriotic response in the press to this war scare represented a departure from the way it had reacted to earlier ones. Two examples from the expansionist administration of James K. Polk are relevant. The issue of war with Mexico over the disputed Texas border loomed in 1845 and early 1846 as the U.S. Congress and the nation's newspapers also debated the threat of hostilities with Great Britain over the Oregon territory. In both cases the possibility of war stimulated heated disagreements (the war with Mexico began in May 1846). Since most newspapers were partisan publications in the 1840s, discord prevailed. There was nothing that resembled the widespread support for war with the British in the 1890s.[14] In marked contrast, by the last decade of the nineteenth century the national telegraph network and the undersea cables supplied newspaper readers across the country with a consistent message that transcended partisan and regional differences.

THE INFORMATION SYSTEM

The ten thousand miles of undersea cables and telegraph landlines that were conduits in this dispute were but a fraction of the total mileage of insulated copper wire used in international communications in the 1890s. The U.S. Navy's Bureau of Navigation published its survey of the world's submarine cables in 1892 that counted approximately 155,000 miles in this new electric message system. A handful of British private companies owned and operated about 97,000 miles, or 63 percent of the world's total. The first submarine cable ran between Dover, England, and Calais, France, and began operations in 1851. After four failures, the Anglo-American Telegraph Company succeeded in spanning the North Atlantic from Ireland to Nova Scotia in 1866. With that signal accomplishment, transoceanic cable construction entered a boom period, with the British at the forefront. British companies extended their electric-communications network to India, Hong Kong, Australia, and South Africa. British technicians also placed cables in the Caribbean and along the Atlantic coast of South America, which is of importance to this study. Thus, by the 1870s Latin America had cable ties to Europe, the United States, and, through European connections, Africa and Asia.[15] The creation of this triangular network of cables beneath the Atlantic Ocean linking the Americas and Europe is examined in more detail in chapter 2.

Once in place, these cables accelerated the speed of communication between and among cities such as Buenos Aires, Rio de Janeiro, Havana, New York, Washington, London, and Berlin. There were numerous delays and breakdowns in the 1870s and 1880s, but by the 1890s the system had become fairly reliable owing to improvements in technology and the lessons of experience. As the system evolved and improved over three decades, governments and businesses started to place more confidence in cables to carry important messages. The expense of long-distance cables meant that few individuals sent personal messages, but diplomats, other government officials, and businessmen found the rapidity of communications to be an advantage—especially in times of international crisis. Mail sent by ships crossing the Atlantic from New York to London in the 1890s took seven days or more to reach its recipient, but a cabled message took only a few hours. An ocean voyage between Buenos Aires and London usually lasted three weeks, but a message sent by electric wire reached its destination in a day or two. Merchants, bankers, and business managers found the speedup in communications to be of value in their quest for competitive advantage. The cost of long-distance cables declined after 1890, when a series of technical changes were made that improved speed and reliability. For example, the duplex system doubled the capacity of most cables by making it possible to send two messages in opposite directions at the same time. Automatic retransmission equipment reduced the chances of human error in the movement of messages through the network. As a result of these and other changes, the 1890s saw a revolution in international reporting. Journalists made more frequent use of the cable network in one of the most highly competitive businesses of the late nineteenth century: mass-circulation newspapers.[16]

The utilization of this new means of communication by two institutions—governments and the press—is the focal point of this study. There are several well-researched and thoughtful studies of international communications during this period that are relevant here. They generally fall into three categories: studies that explore the development of the cable network itself, those that look at the use of the communications system by diplomats and other government officials, and those that investigate the use of the communications system by businesses, including the wire services and newspapers that gathered and distributed international news. Each one of these approaches a useful perspective on the arrival of the new technology. This study, however, intends to take a more inclusive, holistic approach and examine this network of submarine cables, the landlines that they joined, and the movement of messages in these electric conduits as parts of an information system. These electric conduits formed an interface with newspapers and other print media that, in turn, completed the communications process by helping these information outlets reach their large readerships.[17] This information system made possible in unprecedented ways the rapid transmission of diplomatic notes and journalistic scoops in patterns that, in times of crisis, often pitted not only nation against nation but also institution against institution, or, to be more explicit, governments against the press.

The second focal point of this study is the interaction between popular culture and this new information system. Governments initially greeted the cables as conveyances of progress, harmony, and efficiency but soon began trying to restrict the flow of information and to prevent newspapers from printing certain stories. These press organs reached millions of readers, who reacted to international crises in terms of the values and practices embedded in the nation's popular culture and traditional politics (the Monroe Doctrine for example). Public attention focused on government policy sometimes exacerbated the crisis. It is also essential to note that this information system was not monolithic in structure but rather was a loose amalgam of several institutions. The electric portion of the system was operated by the cable companies, which were regulated by governments and international agreements and pressured by other private corporations to do their bidding. In spite of these restrictions and hurdles put in place by powerful institutions, there were openings for messages to get through from a variety of sources (both electric and print) on the fringes of the new international system. At the same time, newspapers in the United States functioned with a sense of independence often expressed in the phrase "freedom of the press." In this largely unregulated information flow, messages appeared that sometimes offended powerful institutions and provoked public debate. Such disputes had the potential to energize the general public of the United States with unpredictable consequences.

Before we continue the exploration of the twists and turns of the movement of information in this system, it will be helpful to survey publications in the three main areas of communications history. First, there are several impressive studies of the building of the cable system. Bernard Finn and Vary Coates pioneered historical research on the establishment of the system and its early use in their *A Retrospective Technology Assessment: Submarine Telegraphy*. Ken Beauchamp's *A History of Telegraphy* is a masterful study of the historical evolution of the system's technological base. Daniel Headrick's *The Invisible Weapon: Telecommunications and International Politics, 1851–1945* not only provides a comprehensive survey of the development of the international cable system but also examines its role in diplomacy and war. Peter Hugill's *Global Communications since 1844* places the cables in their broad geographical and technological contexts.[18] Jill Hills covers the political, legal, and managerial aspects of the cable system.[19] Dwayne Winseck and Robert Pike make effective use of the concepts of globalization and empire to examine the complex interplay between governments and private corporations in the creation of the system.[20] Tom Standage presents an entertaining narrative of the work of scientists, entrepreneurs, and engineers in the formative years of cable technology.[21] Of special importance to this study, Jorma Ahvenainen's two volumes on the history of submarine cables in the Caribbean and South America combine in-depth research with extensive coverage. These books give exceptional detail on the founding and early operations of British and other European companies.[22]

Compared to the number of works on the building of the system, there are relatively few scholarly studies of how the system was put to use. Headrick's insightful *Invisible Weapon* stresses the importance of the cable system in diplomatic and military history. David Paull Nickles scores a historiographical breakthrough in his monograph entitled *Under the Wire*, in which he uses archival documentation to show how diplomats struggled with the new acceleration of information flow in times of international crises.[23] Other studies in diplomatic history mention the role of telegraphic communications in particular episodes, but, in general, the focus of these works is on diplomacy and not the employment of the new technology. For example, the publications by Richard Bradford, Marshall Bertram, William Harris, and John Offner provide this sort of useful groundwork in understanding the impact of cables in international relations in the late nineteenth century.[24]

The impact of international cables on business in this period is the subject of some valuable scholarship, but much work remains to be done. Two pathbreaking studies appeared in the edited volume *A Nation Transformed by Information*: Richard R. John's "Recasting the Information Infrastructure for the Industrial Age" and JoAnne Yates's "Business Use of Information Technology during the Industrial Age." Although very helpful, these articles concentrate on the domestic U.S. economy.[25] One of the most challenging analyses of the use of the international cable system is Alex Nalbach's "'Poisoned at the Source'? Telegraphic News and Big Business in the Nineteenth Century." Two older studies contain additional helpful research on the use of cable communications by British entrepreneurs in South America. H. S. Ferns details the activities of Barings Bank in Argentina, and Harold Blakemore follows the career of John Thomas North in Chilean nitrates.[26]

Communications through electric wires had an especially large effect on journalism, a story that has been covered in a variety of studies by both scholarly and popular writers. The history of press associations or wire services as news-gathering agencies spanning the large geographical expanse of the United States is the primary focus of Menahem Blondheim's deeply researched book *News over the Wires*.[27] Historians Richard Schwarzlose and Ted Curtis Smythe have written comprehensive surveys that include examinations of the press's adoption of new technologies in this period.[28] The rise and consolidation of the three major international agencies, Reuters, Havas, and Wolff, is a complex story that remains the subject of ongoing research. By the 1890s the Associated Press wire service of the United States had also established an international news branch. These four organizations attempted to cover the world, each within their own designated region. Donald Reed's *The Power of News* concentrates on Reuters but also gives attention to Havas and Wolff. Simon Potter concentrates on the newspaper press in the context of the British Empire. Oliver Boyd-Barrett and Michael Palmer offer broad institutional surveys of the agencies, and Winseck and Pike connect those agencies to the framework built by the submarine-cable companies. Rhoda Desbordes focuses on Havas in her thoroughly documented work.[29] Nalbach's nuanced article encompasses all four of the world's major international agencies.[30]

The emergence of the international news agencies was accompanied by rival coverage by some of the larger newspapers with enough revenue to afford their own representatives overseas. The London *Times*, the *New York Tribune*, and the *New York Times* led the way in this practice in the 1880s and 1890s. One result of the growth of foreign reportage was the rise to prominence of a handful of daring foreign correspondents whose personalities sometimes became as important as the events they covered. William Howard Russell, Archibald Forbes, Nellie Bly, Richard Harding Davis, and Sylvester Scovel became the star reporters of this generation.[31]

These works explore in detail the role of international cables in the actions of diplomats, political leaders, and reporters. Their findings tend to concentrate on one or two of these professions and their arenas of activity. The purpose of this study is to examine the places where these arenas of activity overlapped in the use of cable communications. Diplomats used the cable as a presumably secure point-to-point means of communication. They sent messages (usually in code) on the assumption that their colleagues hundreds or thousands miles away would read the decoded message and respond to their queries or demands in a helpful way that would contribute to the resolution of the current crisis, whether the crisis be in intergovernmental relations or international business. However, through the same cables used by agents of diplomacy flowed messages written by reporters whose intent was to see that their words (or edited versions thereof) would show up on the front pages of newspapers. These two types of communicators—those who sent ostensibly secure messages for their own internal institutional purposes and those who sent presumably secure messages in order to be the first to reach a great mass of readers—relied on the same network of cables and were, therefore, contributors to the same information system.

During crises such as the Venezuela boundary dispute of 1895–1896 these professionals often worked against each other. President Cleveland and Secretary Olney had their disagreements with Prime Minister Salisbury and Ambassador Pauncefote, but the coverage of these unfriendly exchanges became an issue of public interest in newspaper editorials and political speeches across the United States. These four protagonists in the controversy were surprised and dismayed by the extent of the public reaction to this issue, a reaction that included loud proclamations of patriotic excess. This new information system that transmitted diplomatic notes and reportorial scoops was not always a means to the peaceful world that pundits anticipated in the 1860s and 1870s.

Given the differing motivations and attitudes of those who used the new information system, an obvious question is who made the best use of the system. Both Cleveland and Salisbury were disturbed by the outburst of jingoism in the U.S. press, an apparent reflection of the public mood in the United States that ran contrary to the policies that these leaders pursued. As archival research in the papers of Cleveland, Salisbury, and other officials reveals, diplomats and politicians were frequently frustrated by the published revelations in the press. This study includes correspondence and unpublished reports from the files of four U.S. presidents and the papers and memoirs of several diplomats as well as editors, journalists, and publicists involved in the coverage of these events. Because of the speed of cable communications, the front page of the newspaper would have a version of the diplomatic note or government report on the same day that the diplomat or government official read what was supposed to be a secure, presumably secret message. In some cases the newspaper coverage affected the attitudes and actions of diplomats and businessmen. Did this interaction, largely generated through the information system, have an impact on government policy decisions? The cases studied here indicate that governments often couldn't keep up with the movement of information in the mass press and, indeed, had to adjust their policies.

In order to understand the difficulties experienced by diplomats, it is helpful to review the standard modes of communication before the advent of the telegraph. The practices in place in the early and middle years of the nineteenth century relied on the movement of messages by sailing ship, stagecoach, and railroad. Diplomats usually had several weeks to deal with overseas crises before they would receive a reply from their nation's capital after sending out an initial message. Even after submarine cables crossed the Atlantic, diplomats retained their sense of "diplomatic time," which meant a rather leisurely schedule. David Nickles's instructive comparison of the U.S. diplomatic response to the French revolution of 1848 (before the cable) with Washington's reaction to the Franco-Prussian War of 1870 indicates that there were few changes in the speed of policy implementation.[32]

The U.S. government used the preelectric means of communication to deal with a series of crises in its relations with Mexico and France in the 1860s. France intervened in Mexico in 1862 and placed Emperor Maximilian at the head of its colonial enterprise in that nation. The United States, embroiled in its own civil war, protested and eventually began to issue pointed diplomatic warnings. The French soon withdrew their forces from Mexico, and Maximilian's regime collapsed in 1867, as the United States threw its support to the government of Benito Juárez. The Lincoln and Johnson administrations handled these difficulties under circumstances that included the possibility of war with France. Diplomats on both sides of the Atlantic used the older modes of communications well enough to bring a peaceful outcome to this three-cornered entanglement.[33]

Diplomats tended to cling to the old means of communication. The formulation and implementation of policy in the pretelegraphic world was characterized by languidness into the 1870s, but in the last two decades of the century the pace and complexity of diplomacy began to intensify under pressure from the new information system. Viewed from the perspective of government officials, this information system became a potent but unpredictable tool. The content and tone reflected the interests of the powerful institutions, but there were also opportunities for less powerful governments, political movements, and individuals to insert their version of events. For example, much of the U.S. public's interest in the Venezuela–British Guiana dispute came from the publication of *British Aggression in Venezuela*. Neither Scruggs, Andrade, nor the administration in Caracas had special connections with the Associated Press or major newspapers in the United States, but their pamphlet, prepared and distributed as an old-fashioned hard-copy publication, soon entered the domain of cabled news in the United States, traveled across the Atlantic in Britain, and eventually was disseminated throughout Latin America. The information system distributed the Scruggs-Andrade alarmist warning that the Monroe Doctrine and the security and honor of the United States were at stake. As their correspondence and unpublished reports reveal, many diplomats and political leaders were surprised by the extent of the public uproar exacerbated by the medium of electricity.

Another conceptual issue concerns the question of whether this information system served mainly as a conveyor of selected and manipulated images and ideas, in other words, as a propaganda device, or was rather simply the reflection of the mood and values of the public and the people who consumed the information. History is complex, and this dichotomy is to some extent an artificial simplification. The information system played both roles at given times. It is, therefore, difficult if not impossible to determine which was the dominant function in the case of a given crisis. Instead of dwelling on the issues surrounding the dichotomy between the information system as opinion maker or indicator of the public mind-set, much of the following analysis relies on a different perspective, generally viewing the movement of information within the system as an interlocutory function. The information system, from this perspective, carried out both functions almost simultaneously. It was both a source of compelling words and illustrations that conveyed the attitudes and ideas of politicians, editors, and reporters and an indication of the public mood (or a rough estimate of public opinion) that newspapers and the profit-seeking policies of editors and publishers sought to give voice to in order to boost advertising revenues.

In other words, newspapers reflected the nation's popular culture. By the 1890s, that popular culture encompassed a heady mixture of traditional values and policies such as the Monroe Doctrine and was sustained by an aggressive, heavily masculine attitude that placed its priorities on honor, action, weaponry, and combat. Editors and journalists were aware of this popular culture and its role in newspaper circulation. Since most of them were men, they felt at ease with this approach to the role of the United States in world affairs and often became its advocates.[34]

AN OVERVIEW OF THE INTERNATIONAL HISTORY OF THE ATLANTIC WORLD IN THE LATE NINETEENTH CENTURY

The stories of the U.S. victory over Spain in 1898 and its taking of the Panama Canal Zone in 1903 give the impression that the decision makers in Washington controlled the international affairs of the Western Hemisphere. Images of cheering crowds and patriotic celebrations in the United States obscure the fact that diplomacy, politics, and business activity in Latin America were often influenced and, at times, dominated by the European powers from the 1860s through the early 1900s. The intrusive undertakings of Spain, France, Germany, and especially Great Britain frequently preoccupied the nations of Latin America as well as the United States.[35]

The Latin American nations themselves and political movements within those nations played a large role in the international history of the Americas. The Latin American countries came into existence in the 1810s and 1820s as the result of several movements for independence within the Spanish Empire and the Portuguese Empire. Following the expulsion of these European colonial masters, the newly independent countries began the difficult task of nation building. The history of each of the eighteen Latin American nations in the 1800s is unique, but it is helpful to note that, in general, the establishment of stable governments and prosperous economies eluded most of them. Mexico was one of the largest countries in the world at the time of independence in 1821, stretching from the ill-defined Oregon territory and Texas on the north to the isthmus of Central America on the south, but by the late 1840s it had lost half of its national territory to revolutions and its neighbor the United States. Argentina, Chile, and Peru struggled with instability and internal strife, and the political entity known as Gran Colombia in northwestern South America split into three countries: Venezuela, Colombia, and Ecuador. Likewise the isthmian nation of Central America (after breaking away from Mexico) divided into five countries: Guatemala, El Salvador, Honduras, Nicaragua, and Costa Rica. Political dismemberment was often accompanied by economic disarray. The bright spots in business were often limited to exports of primary products such as Argentine wheat, Peruvian guano, or Brazilian coffee. Brazil, the sole descendant of the Portuguese Empire, held together politically in spite of strong regionalist tendencies but experienced slow growth in its economy.[36]

Given the problems of nation building in Latin America, the European powers and the United States saw opportunities to exploit resources in the region. Many Latin American nations lacked the political unity and material infrastructure to expand their own economies. Foreign entrepreneurs with credit and potential markets in Europe and the United States took advantage of opportunities in mining, agriculture, and transportation. The Latin American nations had achieved political independence but had to rely on external investments and trade for economic development.

This reliance on Europe and the United States for economic development created an asymmetrical or unbalanced power relationship that is often termed "informal imperialism." In these circumstances, the more advanced industrial nations had several advantages over the less-developed countries. For example, Chile and Mexico typically exported agricultural products and minerals and, in turn, depended on the banks, shipping companies, technical experts, and factories of the United States and Europe. The industrial powers generally did not annex territory but used their advantages in trade, finance, resource ownership, and the threat of military intervention to dominate the region. Informal imperialism contrasts with formal imperialism, in which the more powerful nation takes over territory, as the United States did in its wars with Mexico (1846–1848) and Spain (1898) and as the British did in the Falkland Islands (1830s) and British Honduras (1860s). Formal imperial expansion was unusual in Latin America in the last half of the nineteenth century. Informal imperialism is more important to this study.[37]

The British were the most active of the European powers. They retained their old colonial possessions in Jamaica, the smaller islands of the West Indies, and British Guiana, and they added new ones in the Falkland Islands and British Honduras.[38] These colonial bastions joined Great Britain's global empire. The sheer size of this formal empire coupled with the industrial, commercial, and financial strengths of Britain impressed the new nations of Latin America and the United States. The presence of a small British force on the disputed border between Venezuela and British Guiana provoked fears of invasion not only in that South American country but also in the United States. The reputation of the British navy combined with the activities of British merchants, bankers, railroad and harbor construction engineers, and telegraph-cable technicians and operators created the impression that wherever British trade and investment penetrated Latin America, the might of the formal empire was not far behind. The presence of a few mining companies and an undetermined number of gold prospectors near the mouth of the Orinoco River and the determination of London to secure the westernmost boundary between Venezuela and British Guiana led to discussions in the U.S. press of the striking power of the Royal Navy in the Western Hemisphere.

The image of an aggressive British Empire put forth by Venezuelan and U.S. spokesmen and members of the press in the United States did not match the reality of London's circumspect and generally limited policies in the Western Hemisphere. The British government had supported that nation's merchants in the early decades of the nineteenth century in order to find markets for the export of manufactured goods, but by the last third of the century the Foreign Office was less inclined to intervene to help merchants and investors who had become entangled in difficulties in the region. British diplomats and naval officers stepped back to leave businessmen to deal with whatever issues they encountered. As both Rory Miller and Joseph Smith have confirmed, the government generally sought to avoid conflicts with the United States, Venezuela, Argentina, and the other American governments. The initiative was in the hands of the overseas merchants and bankers, and they continued to be active.[39]

The reluctance of the British Empire to exert its imperial weight in the Americas was not understood in either Washington or Caracas, in other Latin American capitals, or in newspaper offices in the United States. The separation of the British government from Britain's financial institutions must have seemed to be a very fine and even artificial point, as illustrated in the widely publicized Argentine financial crisis of the early 1890s. The negotiations involved the Argentine government and Barings Bank and other financial institutions, with British diplomats as interested observers, but on the large geopolitical stage, Barings and Whitehall seemed to be components of the immense entity—the British Empire.[40]

French activism in the Western Hemisphere was extensive in the last half of the nineteenth century, but its most ambitious projects turned into failures. During the early 1860s, with the United States weakened by its civil war, Napoleon III attempted to exploit Mexico's problems by placing an emperor on a French-imposed throne in Mexico City. Maximilian von Hapsburg of Austria accepted this throne only to discover that Benito Juárez had the support of much of his nation and some aid from the United States. French rule in Mexico collapsed in 1867, and Emperor Maximilian met his death before a firing squad.[41] In the 1880s a French company headed by Ferdinand de Lesseps turned to informal imperialism in its attempt to build a transoceanic canal across Panama. By 1889 a combination of landslides, torrential rains, tropical diseases, and unexpected expenses ended this project.[42] If the French had achieved their aims in either of these efforts, their nation would have become one of the major powers of the Americas. Even though they failed, the French continued to hold a unique place in Latin America through the enduring influence of French enterprise and culture. Napoleon III and business promoter Felix Belly publicized the expression "Latin America" in the 1860s to emphasize the shared Latin background of France with the former colonies of Spain and Portugal. French merchant vessels plied the coastal waters of South America and the Caribbean.[43] A French submarine-cable company linked northern South America with Haiti and New York, thereby competing with British and U.S. cable enterprises. French traders and bankers were busy from Buenos Aires to Mexico City, and French novelists (especially Victor Hugo) and philosophers (especially Auguste Comte) had large readerships in Latin America. In addition, Paris retained the mainland colony of French Guiana as well as its colonial bases on the islands of Martinique and Guadalupe.[44]

In spite of the dismemberment of its American empire in the early 1800s, Spain retained two colonies: Cuba and Puerto Rico. Both of these islands exported sugar and other tropical products. Spain, like France, took advantage of the United States' absorption in its civil war in early 1860s, sending naval and military expeditions to the American to try to regain some of its colonial presence. An invasion of Santo Domingo and blockades and bombardments of key ports in Peru and Chile brought no lasting gains. Spain's forces withdrew after 1866. Although the Spanish held Cuba and Puerto Rico, the dislocations brought about the emancipation of slaves in the 1880s and the growing activism among Cubans for independence resulted in two major uprisings on that island (1868–1878 and 1895–1898). In spite of its unsuccessful efforts at rebuilding parts of its empire, Spain retained influence in the culture, language, and political traditions of its former colonies. Spain was an important player in the Americas, but its position was weakening.[45]

In the worldwide arena of imperial expansion, Germany was a latecomer. The unification of Prussia and the other Germanic states was completed only in 1871. Supported by a prosperous economy and spurred by a strong sense of nationalism, Germany wanted to compete with the other imperial powers but faced at least two disadvantages in Latin America. First, Germany, unlike Spain, France, and Britain, had no formal colonies in the region. And second, German businessmen had something of a late start compared with their colleagues from the other European countries. In spite of these disadvantages, Germany made rapid strides from the 1870s to 1914. These strides covered so much ground that they inspired anxiety among its European rivals and especially the United States. German trade made inroads in Mexico, Guatemala, Nicaragua, Venezuela, Colombia, Brazil, Argentina, and Chile, where German merchants opened stores and warehouses. German immigrants settled in these nations and gained respect among local peoples for efficient business methods and at the same time retained a sense of identity with the fatherland. Added to this commercial success was the presence of German naval vessels in American waters. A small German squadron made an impressive appearance on the coast of Nicaragua in the 1870s. By the 1890s U.S. admiral George Dewey (hero of the Battle of Manila Bay in 1898) voiced suspicions regarding the intentions of Germany in the Western Hemisphere.[46]

The foregoing indicates that the United States was not the hegemon of the Western Hemisphere in the late nineteenth century. If we take 1882 as a chronological point of reference, it is easy to understand why government and business leaders in the Americas and Europe saw most of Latin America as a field overrun by enterprises with corporate roots in London, Paris, or Berlin. In that year Ferdinand de Lesseps's project in Panama seemed to be making headway, British investors in Argentine railways and Chilean nitrates were reporting profits, and German merchants were optimistic about their future in Venezuela and several other countries.

The submarine cable was vital in all of these businesses. De Lesseps's company received cabled reports from Panama.[47] British businessmen also relied on cables for information regarding the financial status of the Argentine government and nitrate mining in Chile.[48] German businessmen communicated with Hamburg and Berlin regarding political conditions in Venezuela.[49] Consuls and diplomats reported regularly to their home governments on the general health of the regional economy and their enterprising nationals who sought profits in their areas of activity. If cable use had been confined to the transmission of these point-to-point, presumably secure, usually coded messages, the long copper strands that stretched along the ocean bottoms would have been mainly conduits of helpful information for making business decisions and formulating government policy. But these cables also had a much different use. Increasingly in the 1870s and 1880s, enterprising journalists began to rely on them to supply reports to their headquarters with a slant intended to capture the attention of the readers of newspapers, especially in the United States. These reporters were not content to describe the routine operations of a bank or the opening of a new mining enterprise. They sought news of the exceptional and the exciting. By the 1890s stories of conflict and confusion, rumors of wars and civil strife, and tales of political conspiracies and barroom brawls were finding their way into the electric information system alongside the cautious diplomatic and business reports.

PENETRATION OF AN IMAGINED COMMUNITY: LATIN AMERICA IN THE U.S. PRINT MEDIA

In this fluid information system, nationalism often played a prominent role. The pronouncements of national leaders and heads of revolutionary movements frequently contained appeals to the need to remain unified in the face of outside threats, whether they emanated from the British Empire along Venezuela's eastern border or the Spanish Empire on the island of Cuba. This type of rhetoric served to create an "imagined community" in many countries. Such was the case in the United States when wire services carried essentially the same story across the expanse from New York to San Francisco. The establishment of such imagined communities is one of the main themes of this study. It is necessary to emphasize, however, these imagined communities were in the process of formation throughout the Americas from Venezuela and Cuba to Argentina and Chile.[50]

In fact, Chile, the seemingly isolated South American nation with the three-thousand-mile coastline facing the vast southern Pacific Ocean, proved to be the most capable competitor of the United States in securing geopolitical advantage in that part of the hemisphere and also in the use of the new information system. Chilean diplomats and publicists were among the first to grasp the opportunities available in the U.S. mass press and to exploit them by injecting information into that system to the benefit of their country. Diplomatic conflicts between Washington and Santiago, therefore, featured the movement and manipulation of information that often reached the mass press. Chileans were not the only Latin Americans with expertise in this area. Argentines, Venezuelans, Cubans, Mexicans, and Panamanians used similar techniques. These Latin American diplomats and propagandists spoke for their respective national interests, defending their imagined communities in the internal information system of their powerful neighbor to the north.

The new information system, under these conditions, made possible a new dimension in geopolitics. Clearly the system promoted trade and investment across national boundaries. There is a tendency to assume that the nations and companies that built and operated new technology largely controlled the use of this technology.[51] This generalization is reasonably consistent when it comes to the use of submarine cables by corporate managers, diplomats, and military officers from Britain, France, Germany, and the United States. It does not, however, take into consideration those Latin Americans who, with an understanding of how this system worked and, in particular, its impact on the press, public, and politics of the United States, attempted with some success to defend and promote their nations (and in the cases of Cuba and Panama, potential nations) at the expense of the great powers. Venezuela and Chile were the most effective of the countries featured in this study in injecting their views and information of their choosing into the new communications network. The newspapers in the United States and, to some extent, Europe accepted these prepared write-ups and placed them on or near their front pages. Likewise, revolutionary groups such as the Cuban Junta in New York supplied the receptive metropolitan press with stories that justified their movement for independence from Spain. Philippe Bunau-Varilla and his Panamanian coconspirators also plied the New York press with information aimed at promoting the idea of locating the transoceanic canal in their part of the Central American isthmus. The information system in the United States was porous, that is, open to the influx of deliberately structured messages from outside.

Argentina and Mexico used a variant of this technique. The political leaders of these two nations used the information system to paint a favorable picture of the commercial potential of their two nations. The administrations of Julio Roca of Argentina and Porfirio Díaz of Mexico tracked the images of their respective nations in the foreign press. The Argentines had close ties with British financial interests, and Mexico was sensitive to Wall Street and press coverage in the United States. Both Buenos Aires and Mexico City attempted to assimilate the international business ethos, but both experienced difficulties in meeting the demands of this early phase of globalization. They both struggled with international convergence, a process that was plagued with uncertainties, as revealed in the same international information system that these two nations attempted to master.

In summary, the international information system introduced more uncertainty and, to many participants and observers, instability to the imperial geopolitics of the period. What had been a routine, predictable, somewhat plodding series of arrangements involving high-ranking government officials and diplomats became a much faster-paced set of meetings, missives, and reluctant public announcements, giving rise to a scenario in which a curious public, aroused by patriotic and/or monetary concerns, was able to read about these events in the newspapers at about the same time or soon after government officials received their heretofore privileged communications. Such unpredictability combined with public exposure to intensify tensions and make the geopolitics of imperialism more difficult, often disruptive, and, at times, explosive.

THE INTERNATIONAL INFORMATION SYSTEM IN THE UNITED STATES

The intensification of tensions at the international level was, in part, a reflection of technological and political trends at the national level. It is helpful at this point to examine the evolution of the internal or domestic information system in the United States. There is a paradox here. The United States was among the cutting-edge innovators in technology, but its political leadership was poorly informed about international affairs. Communications pioneers such as Samuel F. B. Morse and Cyrus Field and newspaper editors such as Whitelaw Reid and Joseph Pulitzer incorporated innovations in technology into their businesses, placing the United States alongside Britain, France, and Germany at the forefront in this area. The United States serves as the focal point in this study because by the 1890s leaders in Washington and newspaper editors across the nation had adopted an assertive approach in foreign policy based to a large extent on the popularity of this approach with newspaper readers who were also voters. In general, this aggressiveness was connected to efforts to build national unity in the decades after the Civil War. The U.S. information system was an essential link in the policy formation and politics of this era; close scrutiny reveals the part that it played as a conveyor of images and ideas to the public and, from another perspective, as a reflection of the public mood.

This period in the history of the United States was marked by struggles to adjust to the consequences of the Civil War (1861–1865) that had preserved national unity and ended slavery but left the general population divided along sectional and racial lines.[52] Under these circumstances, the internal communications system acquired a special importance. In the pre–Civil War era, the national government had devoted much attention to the Post Office and the reliable movement of the mail. The postal service delivered newspapers as well as letters, and so up-to-date news often appeared first at the local post office in smaller cities and towns a day or two after it broke in big-city newspapers.[53] The building of Morse's intercity telegraph line between Washington and Baltimore ushered in the age of electric communication in 1844, and by 1848 Washington was tied to New York, Chicago, and New Orleans.[54] The war between the United States and Mexico received coverage via telegraphy at least part of the way: the news came from northern Mexico and Mexico City by horseback and by ship to New Orleans and then the rest of the way by a combination of railroad, pony express, and telegraph.[55]

The Associated Press partnered with the rapidly expanding Western Union Telegraph Company in the 1840s and 1850s to become the nation's primary corporation in the gathering and distributing information for newspapers. The AP's hold on this function solidified during the Civil War in the states outside the Confederacy. In the late 1860s it became, in effect, a national news monopoly that concentrated on domestic news.[56] This wire association carried the news of Abraham Lincoln's assassination on April 14, 1865, creating thereby what amounted to an electric community encompassing Washington, New York, Chicago, the small cities and towns of the Great Plains, and even the South. In the words of historian Richard D. Brown, "It was the telegraph in conjunction with the newspaper that employed this electronic marvel that exercised the central, dominant role in conveying the calamitous news to the public."[57] Six years later, the devastating Chicago fire also received coast-to-coast coverage in newspapers thanks again to the telegraph. In the aftermath of the burning of Chicago, the *New York Post* editorialized on the telegraph, calling it "the nerve of the continent."[58] Newspapers printed stories from the smoldering city that, in spite of the destruction, remained a national communications center: "For two days daily newspapers printed round-the-clock dispatches from the fire and crowds mulled about the telegraphic offices in Boston and New York awaiting the latest word."[59]

Telegraphic coverage of Lincoln's assassination and the Chicago fire served to unify the nation during times of tragedy, but few U.S. citizens had an interest in international news in the 1860s and 1870s. There was a limited precedent for telegraphed news from foreign countries in the U.S.-Mexican War. The British put together electric connections with the battlefield during the Crimean War of 1854–1856, but it had marginal impact in the United States. The breakthrough in international news coverage in the United States occurred in the Franco-Prussian War of 1871. George W. Smalley, a lawyer-turned-journalist with experience as a battlefield reporter in the

Civil War, wrought a major innovation in foreign correspondence. Working for the *New York Tribune,* Smalley convinced his editor in New York to establish an office in London from which he—Smalley—supervised the assignments and collected the reports from the front in France. The cable was crucial. Reporters made their way from the scene of the fighting (or close to it) to London, where Smalley cajoled the exhausted journalists to compose their accounts. He then sent them via cable to the *Tribune* at great expense. Smalley's *Tribune* office in London coordinated with the *London Daily News,* and the two newspapers combined their resources to produce a series of scoops from France well ahead of the *New York Times* and other U.S. newspapers. Cyrus Field's transoceanic cables were essential in this movement of information across the Atlantic. The *New York Times* was at first critical of the *Tribune*'s accelerated flow of information but eventually admitted that its rival had scored an important journalistic triumph.[60]

Within weeks another major story appeared in France. Following the defeat of the French army, the city of Paris was torn apart by an internal uprising known as the Paris Commune. The radicals rejected the surrender to Germany and, in the parts of Paris under their control, took steps to create not only a new government but also a new social order. The Communards and their opponents resorted to violence that swept across the national capital. The radical movement soon collapsed, but, in the meantime, it inspired another surge of newspaper coverage in the United States that again relied on submarine cables. The journalists on the scene in Paris wrote their accounts, which then traversed the international information system into the United States. The widespread press coverage in the United States stimulated a conservative reaction that spilled over to the national scene, leading to condemnations of local popular movements such as urban labor unions and the agrarian Grangers that amounted to an early version of the red scare.[61] Apparently the newspaper account of the Paris Commune was the first instance of cabled news from a foreign locale inspiring an intense public reaction within the United States.

Smalley's proficient reporting from Europe came at a time when newspapers in the United States had become committed to the maximization of profits. Organized as private corporations, newspapers by the 1870s had moved away from their earlier role as trumpets for a particular faction or party to become profit-seeking businesses in search of news items that would boost circulation and advertising revenue. The Associated Press increased its coverage of foreign news in the 1880s, a trend that expanded in the next decade.[62] The public response to the Franco-Prussian War and the Paris Commune showed that foreign news could help profitability.

The profit motive also drove editors and owners to install new technology to increase productivity. The information system in the United States had already received a boost through the use of the on-land telegraph and the undersea cable. In the 1880s new designs in printing presses made it possible to turn out larger numbers of newspapers at a faster pace. The Hoe press was the best of its kind. By 1883 the Hoe Company was selling equipment that could print both sides of a sheet that was part of a continuous roll of paper that would then be folded and cut into what readers recognized as the finished product—a newspaper. In the 1890s spot-color printing was added to the repertoire. Hoe's main customers were big-city newspapers with their large readership and lucrative advertising revenues. At the same time, the cost of newsprint (the paper used in newspapers) fell considerably.[63]

Setting the type for each page of the newspaper had been a time-consuming process that required skill and patience. Whitelaw Reid, the editor of the *New York Tribune*, was searching for a labor-saving technology and worked closely with Ottmar Mergenthaler, a German immigrant who had experience in the minute internal operations of clocks. Mergenthaler designed a device that used a keyboard for the typesetting process. First employed by the *Tribune* on an experimental basis in March 1885, the Mergenthaler system, or Linotype, was soon a success. In 1890 Mergenthaler began to market his innovation to other newspapers. Annual sales grew rapidly in spite of the depressed state of the economy. He sold 288 Linotype machines in 1892, 568 in 1893, 890 in 1894, and 1,076 in 1895.[64]

These innovations in technology soon reached newspapers in larger cities, where income from advertising and large circulation provided a substantial operating budget. This study uses ten daily newspapers from eight major cities across the United States to monitor the transmission of international stories through the telegraph lines and to document the content of foreign news coverage. These newspapers are the *New York Tribune*, the *New York Times*, the *Boston Globe*, the *Philadelphia Inquirer*, the *Washington Post*, the *Chicago Tribune*, the *Atlanta Constitution*, the *New Orleans Picayune* (*Times-Picayune* after 1915), and the *Examiner* and *Chronicle* of San Francisco. These were among the nation's leading newspapers, with readerships that extended outside their home cities. While any selected group of newspapers may reflect particular ideological and/or political leanings, these ten provide a fairly representative cross section of U.S.-press opinion in this period.

Henry Grady's strong editorial influence dominated the *Atlanta Constitution*, and this paper along with the *New Orleans Picayune* and most newspapers in the states of the former Confederacy was in the Democratic Party's camp. The *San Francisco Examiner* was also inclined toward the Democratic Party, especially after its purchase by William Randolph Hearst in 1887. The *Boston Globe* shifted to the Democratic Party in the 1870s, when it came under the leadership of editor Charles Taylor. On the Republican side were Whitelaw Reid's *New York Tribune*, Joseph Medill's *Chicago Tribune*, and the *San Francisco Chronicle*, run by the de Young brothers, Charles and Michael Henry. Loyal Republican James Elverson was editor of the *Philadelphia Inquirer*. The *New York Times* usually steered an independent course, and when it allied with a faction or party, it was usually the reform-minded Republicans or the Democrats. The *Washington Post* also followed an independent track with occasional ventures in the direction of the Republicans.[65]

In addition to political and ideological affiliations, the nation's geography and demography were reflected in these ten metropolitan newspapers. These newspapers tested the capacity of Western Union, the Associated Press, and other press services to distribute the news across the nation. These eight cities in 1890 had a combined population of 4,959,527, constituting 7.9 percent of the nation's total of 62,622,250. With circulations that reached into surrounding communities, these newspapers were often primary sources of information on national and international issues in eleven states, with a combined population of 24,961,282, or 39.8 percent of the of the U.S. total.[66]

Europe was the traditional center of attention for foreign reporting in the United States, but in the last third of the nineteenth century Latin America began to take on a larger importance. Cuba, Mexico, Central America, and Venezuela were closer to the United States than Britain or France, and New York investors were increasingly involved in railroads, sugar mills, and silver mines in Mexico and the circum-Caribbean. Another stream entered the mix of facts and opinions that flowed through the new international system. The United States was undergoing a revival of nationalism after the Civil War. This imagined community often found expression on the front pages of newspapers. One important component of this print version of the new nationalism was the role of the United States in world affairs as a manifestation of the country's strength. The information system carried stories of the activities of British, French, and German diplomats, businessmen, and naval officers in Latin America and also the lingering presence of the Spanish Empire in the Caribbean. In effect, the actions of these foreign rivals of the United States in the Western Hemisphere and the often exaggerated perception of such rivals fueled the concern, the indignation, and even the outrage of the U.S. public. European impingements in Latin America, whether real or concocted, became important challenges to the self-image of the United States. The British, the Spanish, and other Europeans became increasingly disturbed as this tendency—admittedly somewhat haphazard in the 1870s and 1880s—found rallying points in the 1890s in the revived Monroe Doctrine and a concomitant nationalism. The Monroe Doctrine's status had become uncertain in the middle years of the nineteenth century, but by 1895 the policy precept of 1823 reemerged to occupy center stage in the national discussion of what had been a minor dispute about a remote boundary in northern South America.[67]

The explosion of interest in the Monroe Doctrine and the position of the United States in the international power structure took place at a time when public officials were at work on a closely related project—the promotion of national unity through public education and popular books and magazine articles. As historian Michael Kammen observes, respect for national as opposed to European culture and traditions constituted a "renaissance of patriotism that began to be manifest late in the 1880s and absolutely surged during the following decade."[68] Public schools became centers for the dissemination of patriotic propaganda. The campaign to sell flags to schools to support subscriptions of *Youth's Companion,* an American children's magazine, that began in 1888 became a nationwide crusade to place a U.S. flag in every schoolhouse. The Pledge of Allegiance soon became a national ritual for school children. Cecilia O'Leary's observation is especially important for this study: this effort to publicize patriotism soon shifted its emphasis "from individual speakers addressing local communities to mass newspapers and magazines."[69] The mass press needed a mass audience, and the United States had such an audience. The literacy rate for the native white population in 1880 was 92.3 percent. While literacy for African Americans and Native Americans was lower, about 90 percent of the recently arrived European immigrant population could read and write. Native whites and recent immigrants were receptive audiences for patriotic indoctrination.[70]

The Linotype, the Hoe press, cheaper newsprint, the lower cost for sending international telegrams, and technical innovations in submarine cables combined with Smalley's organization of foreign reporting to create an information system that enabled news from distant lands to reach the front pages of newspapers in the United States within a day or two. That the newspapers were able to cover stories from abroad so quickly created a sense of immediacy that gave foreign news an air not only of freshness but also of excitement. As editors discovered the public's appetite for this type of journalism in the 1870s and 1880s, newspapers began to give more space to foreign news. At the same time a new type of reporting appeared. Sometimes referred to as the "new journalism," critics usually termed it "yellow journalism." The question of the extent of its sensationalism continues to provoke arguments among historians and media commentators. Some of this news coverage, however, detailed political and corporate malfeasance and was truthful in certain respects. The most renowned practitioners of this form of journalism were Joseph Pulitzer and William Randolph Hearst, whose respective publications—the *World* and the *Journal*, both of New York—inspired the ire of cultural analysts and the interest of a wide spectrum of newspaper readers. Media historians John Maxwell Hamilton, David Spenser, Joseph Campbell, John Stevens, Kevin Barnhurst, and John Nerone point out that one of the main goals of the yellow press was to stimulate reader interest by presenting the news in an entertaining fashion. Banner headlines, bold-faced type, and dramatic illustrations in the form of political cartoons filled the front pages. Only one of the ten newspapers in this study falls into the category of the yellow press (the *San Francisco Examiner* was owned by Hearst), but several began to use the techniques of the "new journalism" in the 1890s.[71]

Somewhat less titillating but also of widespread public interest were events in faraway places. The newspaper-reading public shared common ground in nationalistic values, but, at the same time, these readers were interested in news items datelined from the Caribbean, South America, and Europe. Stories from abroad accorded with the yellow press's emphasis on variety and excitement. The yellow press and more traditional newspapers added coverage of Monroe Doctrine disputes, diplomatic discord, threats of war, and military battles. The larger goals of the expanding urban press found informational sustenance through the undersea cables that made it possible to print stories from Havana, Cuba; Santiago, Chile; London, England; and Madrid, Spain, the day after their datelines indicated they had been written. No less a figure than Hearst wrote his own account of the destruction of the Spanish naval squadron off Santiago, Cuba, introduced by a boastful eighty-six-word dateline that began: "On the quarterdeck of the Spanish cruiser *Vizcaya,* seen by many New Yorkers during her visit to their harbor and lying beached on the southern coast of Cuba."[72]

Of particular importance were the convergence of innovations in both the operations of the submarine-cable network and the printing of newspapers. The front pages of major metropolitan newspapers underwent a transformation from eight-column, small-type, all-print formats to a more dramatic presentation that combined larger type, banner headlines, and large illustrations buttressed by telegraphed news from distant sources. At the same time the cable network experienced a series of technical advances that made it faster, more reliable, and less costly. The convergence of these new technologies (seldom discussed by historians) helps to explain the excitement in news coverage as much as the slanted, hyperbolic journalism of Joseph Pulitzer and William Randolph Hearst.

In the closing decades of the century, the technological, political, and cultural components were in place for an eruption of patriotic responses to international issues. The submarine cables injected fast-breaking accounts of foreign crises into the information system with unprecedented speed. The newspapers, tied together by Western Union, the AP, and other wire services, broadcast these stories from coast to coast. The Hoe press, the Linotype, and inexpensive newsprint made the printing process faster and cheaper. And, just as important, the schoolhouse-flag movement and spontaneous celebrations of national traditions and newly found rituals of patriotism created a receptive audience of readers. No matter what party or ideology a newspaper advocated, the press in general was deeply involved in politics and had an impact on discussions of foreign-policy issues in Washington. The extent of public involvement grew from brief controversies involving the United States and Cuba and Spain, Chile and Peru, and Panama and Colombia in the 1870s and 1880s to intense confrontations with Chile, Britain, and Spain in the 1890s. These discussions and debates ebbed and flowed through the electric information system that came to feature day-to-day coverage of events that tied together South America, the Caribbean, and Europe with Washington, New York, and other cities and towns across the United States.

CHAPTER TWO

Building the International Cable System

CYRUS FIELD GAINED WIDE ACCLAIM in 1866 as the entrepreneur who spanned the North Atlantic with a functioning submarine cable. His earlier failures were overshadowed by this technical triumph that linked Britain and continental Europe with North America. Field's cable formed the northernmost connection in the information system under study here. The remaining electrical links came into being between 1867 and 1882. In spite of their essential contributions to this unique three-cornered transoceanic network, they gained little public attention outside the Latin American ports and capitals that benefited from these connections. Instead of being overseen by a single corporate entity, this system was managed by ten different companies headquartered in three countries. The British, world leaders in submarine-cable technology, operated six of these firms. U.S. entrepreneur James Scrymser founded three, and a joint venture that originated in Chile with Argentine support established the tenth.[1]

There was no central plan for this system. Instead, these companies assembled the network in bits and pieces. In technical and operational terms, these companies had to collaborate in order to move messages from one city to another across international boundaries. As Winseck and Pike emphasize, these companies entered into numerous agreements in order to accomplish this basic function. At the same time, these companies maintained separate management structures, issued their own financial reports, and had their own stock offerings. The cooperation among the firms created an information system that made possible the flow of telegraphed messages over great distances, but at the same time, managers and investors watched carefully the activity and profitability of their particular firms. This system was a mix of technical teamwork and capitalist aggrandizement.[2]

The focal point of this study, the international communications system, was dependent on the new transoceanic cable network. The establishment of this network was a major event in the history of the Atlantic Ocean. These cables and the messages that moved through them are important parts of the relatively new field of Atlantic history. Students in this field consider the Atlantic Ocean to be a useful unit for historical analysis. Submarine cables on the floor of the Atlantic formed the electric nerve system for governments and corporations. The Atlantic was far ahead of the Pacific and Indian oceans with respect to the communications revolution of the late nineteenth century. These lines tied the Latin American countries to Britain, Spain, France, and Germany as well as the United States. By the turn of the century, the new information system was bringing an unprecedented flow of information into and out of the region.

THE EXTANT SYSTEM: COMMUNICATION BY SHIP

Submarine cables entered a world in which information could move from continent to continent only as fast as the fastest ship. Sailing vessels were susceptible to the vagaries of ocean currents, winds, and storms. They were the mainstays of this traffic until the 1850s, when heavier and larger steamships began to take on a significant role. Transoceanic international telegraphic connections came into being in the 1860s and 1870s, but well into the 1880s most corporations and governments depended on mail and merchant ships for the movement of time-sensitive information.

The transoceanic shipping lines between Europe and North America were more extensive than those that moved north and south, but significant traffic also flowed between Europe and South America. The steam engine–powered ships on this six-thousand-mile trade route represent one of the developments in the communications system that spanned the Atlantic from north to south. William Hadfield's experiences over five decades of travel between England and South America document these innovations. He stood on the deck of the *Brazileira*, a vessel that combined paddle-wheel steam power and sail, as it departed Liverpool for Rio de Janeiro in 1853. A businessman who served as secretary of the Buenos Aires Great Southern Railway before becoming the founding editor of the biweekly *South*

American Journal in 1863, in which capacity he served until his death in 1887, Hadfield had a vital interest in the promotion of British business enterprise in the Western Hemisphere. Of his 1853 voyage aboard the *Brazileira*, Hadfield noted with enthusiasm that the trip to Rio and back had been completed in about three months, a considerable reduction from the five-month transit for a similar voyage by sailing ship in the 1820s. He made this trip several times in the 1860s and in 1870 noted that, in spite of the fact that the Royal Mail Steam Packet Company employed somewhat antiquated paddle-wheel steamers (in contrast to the faster, more modern screw-propeller craft of other lines), the passage to Rio was reduced to twenty-two days and, by implication, the round trip to less than two months.[3]

Hadfield's account is consistent with the findings of historian Robert Greenhill, who concludes that Royal Mail Steam Packet plodded into the Caribbean and along the east coast of South America providing regular service using older ships that were generally slower than those of the British-owned Pacific Steam Navigation Company, which held the Royal Mail contract for the west coast of South America, and than those of Germany's Hamburg-Amerika and France's Chargeurs Reunis. Newer ships made the run from Europe to Rio in three weeks (as Hadfield had experienced in 1870), which meant that in the last decades of the century messages could be sent and replies received in two months or less if storms or mechanical breakdowns did not intervene.[4]

Hadfield's publication, originally known as the *Brazil and River Plate Mail*, was directly tied to the arrivals and departures of the mail ships from and to South America. The editor's plan was to publish the journal on the seventh and twenty-first of each month in order to furnish readers with recent information from Rio, Montevideo, and Buenos Aires. Hadfield's boosterism was evident in this explanation of his editorial plans: "The commercial transactions" of England "with Brazil and the River Plate are yearly increasing in magnitude, and promise to become second only in importance to those with our Indian Empire."[5] Boosterism had its limits, however. In 1870 he complained about Royal Mail Steam Packet's failure to expand its service to provide a second monthly voyage to Brazil and Argentina.[6] Like most newspaper editors, Hadfield focused on speed. He wanted South American "news" that was more recent than the four-to-eight-week-old material that usually appeared in his publication at that time.

While the British, French, and German lines dominated the international mail, passenger, and freight service in South America and the

Caribbean, the vessels of some Western Hemisphere nations offered competition. On the west coast of South America, the Compañía chileana de vapores was a successful rival of Pacific Steam Navigation, especially during and after the War of the Pacific (1879–1883), when this Chilean enterprise enjoyed a special relationship with the government in Santiago.[7] U.S. shippers connected ports such as New York, Baltimore, and New Orleans to Latin America, but they generally lagged behind their British counterparts. John Roach's United States and Brazil Steamship Company attempted to find profitable routes but, lacking the reliable mail contracts and expanding commercial ties of the British companies, the ventures struggled and eventually failed.[8]

These companies formed the basic channels for the flow of information to and from the Latin American nations and Europe and the United States into the 1880s. Hadfield and other businessmen as well as diplomats and politicians were encouraged by the increased speed and reliability of correspondence delivered by steamship as compared with the uncertainties of the sailing ships, but the potential of the telegraph inspired their expectations for even more rapid communication.

INITIAL TELEGRAPHIC LINKS

The most radical change in communications in the nineteenth century came with the introduction of transoceanic submarine telegraphy. The center stage for this dramatic innovation was the deep waters of the North Atlantic Ocean, where Cyrus Field's cable linked Britain with the United States. Field's firm, Anglo-American Telegraph, relied heavily on British investors for its capital, and prominent among them was John Pender, who had established himself in the business through the Telegraph Construction and Maintenance Company. In the 1860s Pender's firm and other British companies placed submerged lines in the Mediterranean and on to India. By the end of that decade, the British had established their leadership in the building of the world's submarine-cable network.[9]

Before we consider the larger enterprises in the Western Hemisphere, it is necessary to examine two small pathbreaking submarine telegraph lines that eventually formed vital parts of the electronic network that was to connect Latin America with Europe and the United States. These initial connecting points involved submerged cables across the Río de la Plata estuary between Argentina and Uruguay and the Florida Straits between Cuba and the United States.

The River Plate Telegraph Company was organized by Liverpool merchant John Proudfoot and Glasgow engineer Matthew Gray. A specialist in submarine telegraph cables, Gray also ran the India Rubber, Gutta Percha, and Telegraph Works Company for several years. River Plate Telegraph obtained the necessary concessions (the formal permission granted by a government to bring submarine cables on shore at specified points) from the Argentine and Uruguayan governments and completed construction in late 1867, approximately a year and a half after the project began. The submerged cable ran from Punta Lara on the Argentine side to Colonia, Uruguay. Landlines completed the connections with Buenos Aires and Montevideo. Although the company's operations were located in the Río de la Plata region, the headquarters were in Scotland.[10]

The first exchange of telegraph messages between Buenos Aires and Montevideo was the occasion for official celebration on both sides of the Río de la Plata. General Venacio Flores, head of state in Uruguay, led in salutes to Argentina and "our foreign residents in general, and the Electric Company in particular." Governor Adolfo Alsina of Buenos Aires Province (in the absence of President Bartolome Mitre, who was personally involved in the Paraguayan War) echoed the praise of "the wonderful inventions of Englishmen and Yankees. . . . What a change since the time of Rosas!" (Rosas was the former dictator of Argentina). Vice President Marcos Paz summed up the sentiments of the officials present at the ceremony: "Who can forget how much we owe to England and Englishmen." On both sides the ceremonies included the playing of their respective national anthems as well as further veneration of the British with a respectful rendition of "God Save the Queen."[11]

River Plate Telegraph's second annual report, dated November 1867, revealed that there was very little message traffic in the first months of operation and, therefore, the company decided to reduce the amount charged for telegrams by about half. In March 1869 the directors of the company reported "a steady increase over the year preceding" that "may be attributed to the permanent growth of telegraphic intercourse." By 1873 it had duplicated its Uruguayan line with an undersea cable along that nation's coastline to avoid interruptions in service resulting from outbursts of civil strife that damaged the landline. In spite of this turmoil, the directors were looking forward to increased business with the extension of submarine telegraphy to southern South America.[12]

Eager promoters in the United States shared the same vision and were willing to venture into the field dominated by British technology and investments. Concerted efforts to connect the state of Florida with the Spanish colony of Cuba by submarine cable began as early as 1857, when businessman and sometime diplomat Horatio J. Perry devised a scheme to bring together the governments of Spain and Great Britain in the construction of such a link. Historian Jorma Ahvenainen explains in detail the work of Perry, whose plans were delayed by the Civil War. James Scrymser entered the scene in 1866, having purchased Perry's Spanish concession for the construction of a Cuba-Florida connection. This transaction was essential to Scrymser's International Ocean Telegraph Company in its campaign to fend off a rival company based in Spain. With Perry on board as a shareholder, International Ocean obtained the necessary permissions from the U.S. government and contracted with the India Rubber, Gutta Percha, and Telegraph Works for the laying of the cable. The problems of submarine cables in the rough bottom terrain of the Caribbean (to be discussed in more depth later in this chapter) became evident in the form of cable breaks and consequent delays, which meant that the Havana–Key West–Punta Rassa connection did not open until early September 1867.[13]

Scrymser's ambitions extended far beyond spanning the Florida-Cuba channel. A veteran of the Civil War and a New York–based entrepreneur, Scrymser consulted with U.S. secretary of state William Seward, West Indies merchant Moses Taylor, and French canal builder Ferdinand de Lesseps and obtained financial backing from Alfred Pell, Alexander Hamilton Jr., and Oliver King. Well connected on Wall Street and in Washington, Scrymser made public his intentions of extending International Ocean's lines to Mexico and Panama to the west and south and then onward through the West Indies to the coast of South America and eventually to Brazil. He obtained several of the necessary concessions and also created a list of which governments would be willing to subsidize the project. For example, the governor of the British colony of Demerara expressed his eagerness for the system that would bring "daily communication with the United States as to the markets and other matters of importance."[14]

Meanwhile, International Ocean showed profits from its Cuba-Florida operations from the outset. A protracted and ultimately unsuccessful revolt for Cuban independence (1868–1878) created substantial telegraph traffic between Spanish officials in strife-torn Cuba and the government in Madrid. These telegrams moved through International Ocean's lines, the U.S. telegraph network, and the North Atlantic cables. However, Scrymser and Perry's plans for a Florida–Cuba–West Indies connection to South

America were not realized. Instead, Scrymser accepted a buyout by Western Union in 1873 after wrestling with the problems of broken cables in the waters off Florida and political unrest in Cuba.[15] Nevertheless, Scrymser's International Ocean had tied Cuba into the United States' communications system just before the outbreak of the revolt for independence. International Ocean had established a strategically important connection between a temporarily disrupted but potentially productive Cuban economy that, after the end of the fighting, would enjoy a period of growth based largely on the export of sugar and tobacco to the United States.[16]

These small links across the Río de la Plata and the Florida Straits constituted pioneering endeavors in what was to become a network that would eventually include New York, London, Rio de Janeiro, Santiago, Lima, and Mexico City as well as Buenos Aires, Montevideo, and Havana. The *Banker's Magazine* of July 1870 listed the world's sixty-four operational submarine cables. Most European nations, Egypt, and India were mentioned, but South America was not. Only because of its three-thousand-mile connection with Great Britain and its much shorter link with Cuba was the United States listed.[17] As of 1870 the Western Hemisphere was far behind Europe and its colonies, still relying on sail and steam to convey most of its transoceanic messages.

THE WEST INDIES: DISAPPOINTED EXPECTATIONS

Even before the Argentines and Uruguayans celebrated the opening of their new telegraphic line, the British had established themselves as the world's premier submarine-telegraph specialists.[18] Perhaps Scrymser's decision to sell International Ocean to Western Union was due in part to his recognition of the superiority of the British in this field and his awareness of their interests in the Caribbean.[19] With their successes in the North Atlantic, the Mediterranean, and the Persian Gulf, British submarine-cable companies enjoyed widespread respect on the London stock market, and it was to this community of investors that entrepreneurs turned with their plans for a West Indian enterprise. Sir Charles Bright, a recognized cable expert, assumed leadership in the founding of West India and Panama in July 1869. The £650,000 of shares sold quickly enough, but the original enthusiasm for the company diminished when Bright had difficulty in finding prominent citizens to serve on the company's board of directors. These uncertainties seemed to be allayed when the fledgling company secured the concessions held by International Ocean in the trans-Caribbean and worked out a "joint traffic agreement" with that company to gain access to Florida and the busy Western Union lines. Many of the local colonial governments in the Caribbean reaffirmed their intentions of providing subsidies to West India and Panama upon the completion of the construction of the submerged system that would tie their port cities to Europe and the United States.[20]

Cuba was crucial for the enterprise and, therefore, the ten-year war for independence became a serious problem in the construction process. This conflict pitted Cuban guerrilla forces against Spain's imperial army in a violent fight that continued until 1878, at the conclusion of which Madrid still retained a hold on its most valuable colony. The original plans called for a landline along the length of the island between Santiago and Havana, but heavy fighting—especially in eastern Cuba—made immediate implementation impossible. Experts advised laying a submerged cable from Santiago along the south coast of the island that could connect with a landline at Batabanó that was already connected to Havana only thirty-five miles to the north in the relatively secure western part of the island. In order to deal with this circumstance, the West India and Panama backers established a separate London-based company. Capitalized with only £160,000, the Cuba Submarine Telegraph Company experienced a bustling subscription for its stock and, with leadership tied to the parent company, quickly set out on its appointed task. Joint traffic agreements enabled the new company to cooperate with West India and Panama and also International Ocean. Organized in January 1870, Cuba Submarine was operating a submerged cable between Santiago and Havana within eight months.[21]

West India and Panama now faced the daunting task of placing cables from Santiago south to Jamaica and then east and south to Puerto Rico, St. Thomas, Barbados, Trinidad, and British Guiana as well as to several islands in between. A second line was to run south from Kingston, Jamaica, to Panama. Although estimates for the length of cable needed for this project varied, there was general agreement that the company would have to place over two thousand miles of cable in a sea known for difficult currents and sharp, cable-cutting coral. The records of West India and Panama reveal an unfortunate story of cable breaks, futile grappling (attempts to recover broken cable at the sea bottom), lost lines, and, as a consequence, unanticipated expenses. In 1871 the cable lay from Jamaica to Panama came to a halt when the cable broke and grappling efforts failed. A similar plight delayed the connection from Puerto Rico to Jamaica. In January 1873, the Dominica-Martinique line failed, functioned intermittently, and then fell completely silent (or "dumb" in the language of the engineers) on September 2, 1873.[22]

The early economic performance of West India and Panama fell far below expectations. Ahvenainen summarizes the misfortunes of its operations in financial terms: the original prospectus anticipated annual revenues of £170,000, but from 1872 to 1875 the actual annual income averaged only £30,000. The central problems were easy to identify. Cable traffic was light in this period because the cable line to Panama was not in working order, and the connection to South America simply did not exist. Also the frequent interruptions caused by broken lines cut into income.[23] While the smaller and less ambitious Cuba Submarine had a more respectable record, the overall British experience with the initial phase of cables in the Caribbean was not promising.

ACROSS THE ANDES

While British submarine-cable experts met their match in the depths of the Caribbean, an entirely separate project surmounted the intimidating Andes Mountains in southern South America. Merchants in Chile were watching the progress of the international telegraphic system being built by Europeans. Juan Clark, a hardware and machinery dealer in Valparaíso, surmised in 1868 that a submarine telegraph cable would soon reach down the east coast of South America to Buenos Aires. With its harbor on the lightly populated south Pacific coast, Chile might have to wait many years for a submarine cable to venture around the forbiddingly stormy Cape Horn. Clark joined with Argentine businessman Mariano E. de Sarrata to form a corporation with the purpose of tying Buenos Aires to Valparaíso with a telegraph line across the Andes. Their company was known as the Compañía del telégrafo transandino and was financed by local investors: two-thirds were Chilean and one-third were Argentine. Company headquarters were in Valparaíso, with branch offices in Buenos Aires and Liverpool. Clark obtained the necessary concessions from the two governments and began the construction job during the early part of 1870 to take advantage of the summer and fall months in the Southern Hemisphere. This Chilean-Argentine venture filled one of the most challenging gaps in the international communications network.[24]

Transandino contracted with its founder's construction company to carry out the actual building of the line. Juan Clark secured the services of engineer Javier Villanueva y García who led the construction crew on the Argentine side, while Juan Clark and his brother Mateo supervised construction in Chile. Clark and company placed orders for the cable with British firms and estimated the total cost of the project at Ch$335,000. Given the task in front of them and the newness of the technology, it is not surprising that this estimate proved to be too low. The total cost was actually Ch$467,220, about 40 percent above the original estimate.[25]

The construction crews encountered the curiosity and even the hostility of local people as they moved across the plains and foothills in their approach to the Andes. Marcial Ocampo, a wealthy landowner on the Chilean side, demanded Ch$10,000 from Clark in exchange for permission to run the telegraph line across his property. Clark refused, and Ocampo went to court and eventually to the national congress where, much to his chagrin, the verdict favored Clark by requiring a payment of only Ch$47.[26]

On the Argentine side, local irritation manifested itself in the form of raids by gauchos and Native Americans. As the work crews moved through frontier zones, bands of roving locals pillaged their base camps for food and supplies, sometimes resorting to violence. Clark and company had difficulty in keeping work crews at sufficient strength in these areas. As historian John J. Johnson has noted, the attacks made it necessary "for the workmen to keep their guns at their sides at all times." Hostile local communities aside, topography and altitude posed large challenges. In November 1879, work crews reached the higher Andes—ten to twelve thousand feet above sea level. Fast-moving mountain streams, steep, rocky terrain, and the thin, cold air made work difficult. The cables, heavy and cumbersome to transport under any conditions, had to be cut into short lengths in order to move them into the higher altitudes. Once at the proper site, engineers had to splice them together, a job that required the skills of a specialist brought from England especially for this task.[27]

Rock slides, gaucho raids, and irate hacendados delayed but did not stop Clark and his crews. On July 26, 1872, telegraphic communication opened over the eight hundred miles from Valparaíso to Buenos Aires. Transandino took over the operations amid celebrations at both ends of the line. The transmission of information had dropped from two weeks on board ship around Cape Horn to two hours by telegraph.[28] The erection of an up-to-date line of communication across some of the most difficult terrain in the Western Hemisphere was a major triumph for this Chilean corporation, local managers, and investors, and Chilean and Argentine workers.

In order to move messages from Valparaíso and Santiago into the soon-to-be-created international network's station in Buenos Aires, the Clark brothers' Transandino had to rely on the Argentine national network. The Administración central de los telégrafos led by Carlos Burton was an agency in the Ministry of Interior and operated under the scrutiny of ministry head Dalmacio Vélez Sarsfield. Like Transandino, the Administración central was new, having been established in October 1869 by the Argentine national congress at the enthusiastic urging of President Domingo Sarmiento, one of the most ardent promoters of international cable communications on the continent of South America.[29] Vélez Sarsfield and Burton shared the president's enthusiasm, but they faced a difficult job. They had to coordinate a patchwork of telegraph lines, many of which had originated with the building of the nation's railroads. The Administración central imposed a degree of organization on these telegraph lines, the facilities built by provincial governments, and the telegraphs set up by the Argentine military. Such an improvised network had its difficulties, including defective, outdated equipment and poorly maintained wires. There were frequent breakdowns and numerous complaints, but the linkage from the Transandino to Buenos Aires provided adequate service, and so business firms and government agencies used it with increasing frequency in the 1880s and 1890s.[30]

SPANNING THE ATLANTIC: NORTH TO SOUTH

Chile's Clark brothers had correctly anticipated the extension of the European communications system into the southwest Atlantic—first to Rio de Janeiro and then to Buenos Aires. The magnitude of this cable lay was obvious—especially to the engineers, technicians, and entrepreneurs who were cognizant of the problems that dogged West India and Panama in the tropical waters on the north coast of South America. One of the two major corporations involved in this project, the Brazilian Submarine Telegraph Company, placed the transoceanic cable in three sections: the first comprised 653 miles from Lisbon to Madeira, the second 1,260 miles from Madeira to St. Vincent, and the third 1,953 miles from St. Vincent to Pernambuco. The Western and Brazilian Telegraph Company (originally known as Great Western Telegraph Company) on the Brazilian coast carried the line from Pernambuco to Bahia and on to Rio de Janeiro. The line reached Buenos Aires, completing the connection with River Plate Telegraph, through arrangements with a smaller corporation, the Compañía telégrafica platino-brasileira (which was bought out by the London Platino Brazilian Company in 1877). The total distance of the cable linkages from Lisbon to Rio was around 5,800 miles, comparable to some of the longest cable connections in the world at the time.[31]

The improvement in the information flow between the expanding economies of Brazil, Argentina, and Uruguay and the commercial centers of Great Britain that would be secured by the conquest of such an imposing distance made it a necessity as well as an opportunity, at least from the perspective of investors in South American securities on the London stock exchange. William Hadfield's *South American Journal* had advocated such a project as early as November 1871.[32] An irate observer of international markets agreed with Hadfield in a letter of complaint that appeared in the London *Times* of July 19, 1872. The writer noted that the recent "depreciation in South American stocks would have been arrested" in a more timely fashion if a modern communications link existed between southern South America and England. Rumors of a recent dispute between Argentina and Brazil had resulted in a downturn in some securities on the London market. With some dismay the writer concluded that the news from "South America has still no better means of intelligence than that rendered by the postal service."[33]

As the economic potential of southern South America attracted the attention of London investors, the projected transoceanic cable drew the interest of several cable entrepreneurs. The businessman who had the upper hand in this competition was John Pender, a Scottish merchant who rose into the upper echelon in Manchester textiles before shifting his capital to submarine telegraphy in the 1850s. Pender was one of 345 British investors in Cyrus Field's cable adventures. Pender took over the directorship of the Atlantic Telegraph Company in December 1856 and was a central figure in guiding this company through frustration and failure to success in the next decade. An astute businessman, Pender quickly grasped the advantages of the division of risky cable ventures among different companies. In 1864 he took the initiative in the formation of Telegraph Construction and Maintenance, which manufactured and submerged the cable between England and North America in 1866. As a powerful presence in both North Atlantic cable and Telegraph Construction and Maintenance, Pender was easily one of the commanding leaders in the new industry.[34]

Pender was most likely a participant in discussions concerning the viability of a transatlantic cable to South America in 1871, if not earlier.[35] This large undertaking faced two major problems. First, the technological challenge, great distance, and the economic potential of the South American project meant that much more capital was needed than had been for past ventures. Brazilian Submarine, for example, was capitalized at twice the amount as West India and Panama had been. Second, competition for concessions was more intense. For example, a European group headed by a "Signor Balestrini" claimed a concession for a submarine cable from Portugal to Brazil based on an 1864 collaborative agreement approved by the governments of France, Italy, and Haiti in addition to Portugal and Brazil. The Portuguese, however, decided to allow this agreement to lapse in early 1872, which voided the concession for all parties.[36]

John Pender did not acknowledge any direct role in voiding the Balestrini concession, but he was openly a central player in the formation of the two large submarine-cable companies and in obtaining the necessary concessions for the Lisbon to Rio de Janeiro route. Brazilian Submarine had a capital investment of £1,446,400 and held the exclusive concession for the submarine cable across the Atlantic from Lisbon to Pernambuco.[37] Western and Brazilian, the second company, had in its incarnation as Great Western intended to run a cable from Britain to New York but had decided to shift from the increasingly competitive North Atlantic to coastal Brazil, where it would hold an exclusive concession from the government of Emperor Pedro II. This Pender company was capitalized at £1,350,000.[38]

Pender's Telegraph Construction and Maintenance contracted to lay the cables. Although all of these companies were essentially a part of Pender's cable empire, they had separate boards of directors, issued their own stock offerings on the London stock exchange, and signed contracts with each other for the completion of the work.[39] Therefore, the cable connection between Europe and South America was a typical Pender arrangement in which three different companies with their own managements, boards, and stockholders built and operated what would eventually become an interconnected communications system. These separate companies, however, acted to compartmentalize the risks in submarine-cable construction and operation, a prudent arrangement after the difficulties of the North Atlantic in the 1850s and the Caribbean in the early 1870s.

There remained one important missing link in telegraphic communications between London and Buenos Aires: the 1,200 miles from Rio de Janeiro to Montevideo (already connected to Buenos Aires by River Plate Telegraph). An expensive rivalry between Western and Brazilian and Platino-brasileira to construct it, however, was a distinct possibility. Andres Lamas, a Uruguayan entrepreneur who had envisioned a telegraphic connection between Rio and Buenos Aires via Montevideo as early as the 1840s, held the appropriate concessions from the three governments and had also raised most of the required capital in Brazil and Argentina. The point of rivalry involved the possible duplication of lines from Rio to the Brazil-Uruguay border, which was permitted under overlapping concessions granted to Platino-brasileira and Western and Brazilian. From March to August 1873 representatives of the two companies entered negotiations that resulted in a compromise solution, avoiding "a ruinous competition," as Major I. F. Rickard explained to the stockholders of Platino-brasileira.[40] In essence, Western and Brazilian would operate the cable, collect the revenues, and turn over to Platino-brasileira £115,000 in stocks and, on an annual basis, 20 percent of the gross receipts. With competition and duplication thus eliminated, direct communication between Buenos Aires and London became practicable, and also the Chilean cities of Santiago and Valparaíso were brought into the network through the facilities of Transandino.[41]

The deployment of the cable was relatively uneventful in contrast to the difficulties encountered by the West India and Panama. Operations proceeded simultaneously across the Atlantic and along the Brazilian coast from January to June 1874. However, Telegraph Construction and Maintenance did experience short-term troubles: a break in the cable between Lisbon and Madeira required grappling and repair, and the line from Madeira and Pernambuco had to be placed in very deep water that put pressure on the available supply of cable.[42] Telegraph Construction and Maintenance remedied these problems and on June 23, Viscount Rio Branco of Brazil and Viscount Monck, chairman of the board of directors of Brazilian Submarine, exchanged electronic greetings to mark the opening of the connection between their two countries. On August 7 President Domingo Sarmiento of Argentina sent celebratory telegrams to Queen Victoria and the pope and then addressed a large gathering of government and business officials in the nation's capital. The Argentine president referred to his "old friend and colleague, Don Andres Lamas" with whom he had shared "dreams of thirty years ago" of connecting the cities of the Río de la Plata with Europe by telegraphic communication. Lamas also spoke and reinforced Sarmiento's recognition of investors in Argentina, Brazil, and Uruguay who had supplied most of the capital for Platino-brasileira.[43]

The early economic performances of Brazilian Submarine and Western and Brazilian offered a substantial contrast. With its monopoly on telegraphic communications between southern South America and Europe, Brazilian Submarine showed a record a steady revenue and comfortable profit margins averaging over £100,000 per year in both income and profit for the first five years of operation.[44] Western and Brazilian was not so fortunate. While revenue from messages sent was fairly consistent at about £100,000 per year, company expenses fluctuated considerably, exceeding £120,000 in 1878 because of extensive damage to the coastal cables.[45] Because these two companies maintained separate financial records, the impressive profits of Brazilian Submarine did nothing to alleviate the troubles of Western and Brazilian.

PENDER'S CABLE EMPIRE AND SCRYMSER'S CHALLENGE

Separate accounting systems, however, did not disguise the fact that by 1875, one group of companies dominated telegraphic communications in the Western Hemisphere south of Florida, the Gulf of Mexico, and the Rio Grande. John Pender's presence on the boards of directors, his ownership of stock, and his influence in British business and politics cast him as the central figure in these two communications networks.[46] The first network included Spanish Cuba, the British West Indies, the French islands of Guadeloupe and Martinique, British Guiana, and Colombia's province of Panama. The second network extended from London via Portugal to Brazil and then to south to Uruguay and Argentina and, by arrangement with Transandino, to Chile. Pender and his allies built these networks from 1869 to 1874. Given the contrasts in climate and ocean bed topography and the uncertainties in politics and commerce in this large region, Pender's accomplishments were impressive.

The main weakness in Pender's two networks was that they did not connect and therefore left gaps for competitors to fill. There were two likely lines of connection between the South American companies and their Caribbean counterparts, and, as Pender and his investors hesitated and/or met delays, two non-British operations moved in. A French company, the Société française de télégraphs sous marins, began work in the late 1880s and by 1891 had cables running from Cuba to Haiti and then to the French islands of Guadeloupe and Martinique and on to the Venezuelan capital of Caracas by way of the Dutch island of Curaçao. Later in the 1890s French interests extended out from this Caribbean base, reaching New York in the north and Brazil to the south.[47]

John Pender, British cable magnate who led the way in the construction of the cable network in the Americas (Cable and Wireless Archive, Porthcurno Telegraph Museum, Cornwall, UK).

James Scrymser, head of three U.S. cable companies that operated in Latin America (Alfred K. Fricke, ed., *A Half Century of Service to the Three Americas* [New York: All-America Cables, 1928], 6).

Cable-laying ship Dacia, which placed cables for both British and U.S. cable companies off the west coast of South America (Cable and Wireless Archive, Porthcurno Telegraph Museum, Cornwall, UK).

Cable landing sites often were located at isolated points on the coast (Herbert Laws Webb, "With a Cable Expedition," *Scribner's Magazine*, October 1890, 401).

Scrymser's Central and South America Telegraph Company's cable station in Salina Cruz, Mexico, in 1882 (Alfred K. Fricke, ed., *A Half-Century of Service* [New York: All-America Cables, 1928], 20).

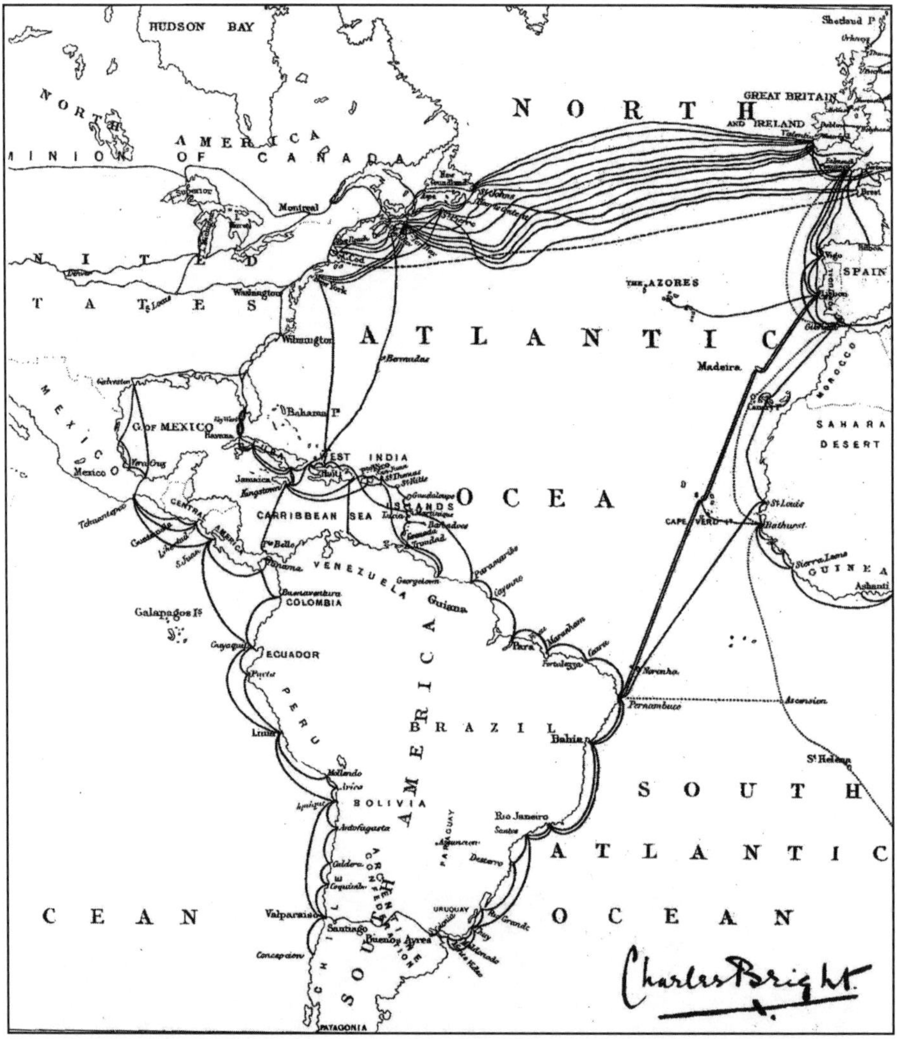

Cable map for 1898 showing the expansion of the network in thirty-two years (Charles Bright, *Submarine Telegraphs, Their History, Construction, and Working* [London: Lockwood, 1898], facing page 208).

The larger concern for Pender, however, was the Pacific coast of South America, where British and U.S. companies became involved in a competition for control of or, at least, primary access to potential profits from the telegraph traffic between the nations of the temperate climes of southern South America on one hand and the industrial and commercial centers of Europe and North America on the other. In a story to be discussed in more depth in the next chapter, British interests established the West Coast of America Telegraph Company to run lines north from Valparaíso and managed to reach Lima by 1876. The War of the Pacific that pitted Chile against Peru and Bolivia delayed further work from 1879 to 1882, but the main problem for the British was the decision of the prime mover of International Ocean, James Scrymser, to return to international telegraphy. Scrymser established two companies, the Mexican Telegraph Company and the Central and South American Telegraph Company, that by 1882 had stretched their cables from Galveston, Texas, to Lima, and in 1891 they reached Buenos Aires. Scrymser and the French made their contributions to a network that by the last decade of the century drew South America, Central America, Mexico, and the Caribbean further into the world's telegraph system. It must be emphasized, however, that even with these competitive incursions from the United States and France, the international telegraph system in Latin America remained largely in British hands, with London-based companies in charge of approximately two-thirds of the submarine cables and virtually all of the cable repair services in the region in 1892. In terms of day-to-day operations, the movement of messages through the system required the various companies to cooperate in their transmission. Nevertheless, the technological and business foundations were set for a quiet but often bruising power struggle between Britain and the United States.[48]

MAGNETS, MIRRORS, AND LIGHT

This international corporate competition involved three continents, the Atlantic and Pacific Oceans, and the Caribbean Sea, but the movement of the messages through the network of submerged cables depended on slender copper wires insulated by a heretofore little-used substance called gutta-percha (or India rubber). The gutta-percha originated in the jungles of Malaya and the East Indies and proved to be an excellent source of protection from seawater. The strength of the cables came from the iron (or, subsequently, steel) wire, which was wrapped around the gutta-percha, the copper lines being at the

The mirror galvanometer received the electronic signal from the cable and translated it into a focused beam of light that highlighted the appropriate letter on the horizontal alphabet (Norman H. Schneider, *Electrical Instruments and Testing* [New York: Spon and Chamberlain, 1913, 13]; Sylvanus P. Thompson, *Elementary Lessons in Electricity and Magnetism* [London: Macmillan, 1881], 169).

center. Although engineers later made improvements, the basic components remained essentially the same into the twentieth century.[49]

The transmission of messages over hundreds and even thousands of miles was a testimony of the strength of the cables themselves, which held up under the weight of water at the bottom of the ocean floor, but completion of the process relied on a delicate sensitivity in receiving the messages calling for visual perception and mental coordination that most humans found extremely challenging if not impossible. The trials and errors in the first transatlantic cables made clear that too much strength in the electronic signal would burn out the copper core, from which the pioneering cable experts concluded that a relatively weak electronic impulse was the only way to avoid "frying" the cable. The main problem, therefore, was the creation of a receiving apparatus that was sufficiently sensitive to register the low level of electric current that the copper could handle. A Scottish scientist, William Thomson, devised the appropriate apparatus by modifying work in earlier experiments. Thomson patented the mirror galvanometer in 1858, and it furnished the standard means of receiving submarine-cable messages for several decades. Thomson's invention solved the problem of how to receive weak electronic signals, but it imposed great demands on the operators.[50]

An essential part of the mirror galvanometer was a tiny concave mirror suspended by a thread (usually silk). The mirror was located inside a wire coil that became, in effect, a magnet when an electric current passed through it. The telegraphed message that arrived via the submarine cable was the source of the electric current. Each impulse activated the magnetized wire coil, which caused the mirror to move to the left or the right. After this initial movement another small magnet moved the mirror back to its centered position. Then there quickly followed another electric current that magnetized the coil and turned the mirror again. The sensitivity of the magnet and the mirror to the weak, long-distance signals made the system work. The mirror concentrated a small beam of light from a nearby kerosene lantern onto a horizontal strip of letters. As the signals arrived at varying strengths, the electromagnet moved the mirror so that its beam of light spelled out the message. The light itself was simple in its origin. It came from a kerosene lantern placed behind an opaque shield with an optical lens in a small opening or aperture to concentrate the light so that only a small stream came to rest on the mirror.[51]

The delicate nature of this apparatus had to be matched by sensitivity in the operator. No sound was involved—only the movement of the beam of light across the horizontal strip (a few inches in length). The receiving room for messages was dark, so that the receiving operator would not be distracted. This operator usually called out the letters to a partner who wrote them down, allowing the receiver to concentrate on the movement of the light along the strip. Most of the messages were sent by businesses and governments; these were in coded form, and so an intuitive, common sense understanding of the message was not possible. Therefore, every letter was important. Obviously both operators had to maintain full concentration and be able to launch into reading and writing at the moment the mirror began to move.[52]

Herbert Laws Webb, a reporter for *Electrical World* (an engineering periodical), described the experience of a veteran "mirror man" in an article that illustrated the nuances in the work of these operators: "A friend of mine, who had been an operator on the west coast of South America, and was a first-class mirror man, on coming home to England applied for a position in one of the Atlantic cable companies. He was asked if he could read mirror, and replied: 'Oh yes; thirty words a minute.'" The company hired him for work in a cable office on the Irish coast, but this "first-class mirror man" soon discovered that the mirror galvanometer on the North Atlantic functioned differently from the apparatus on the west coast of South America. He told Webb that "instead of the movements being clear and well defined, the spot was wandering continually all over the scale," and so, Webb concludes, "my friend had to confess himself beaten (although he had been reading mirror for five years) until he had mastered the eccentricities of that diabolical spot."[53] Although a new system that created a print record of received messages emerged in the 1880s (the siphon recorder to be discussed in chapter 6), most of the South American, circum-Caribbean, and Mexican stations continued to use the mirror galvanometer into the 1890s.[54]

The British dominated the field of mirror galvanometer operations largely because John Pender established a school for the training of operator in this specialized skill at Porthcurno on the coast of Cornwall in England in the 1870s. The vast majority if not all of these Porthcurno-trained telegraphers were young, single British men.[55] The school and its graduates gave the British companies an advantage in the submarine-cable business, even if some of the telegraphers were recruited to work for Scrymser. The Pender-Scrymser rivalry obscured the fact that the basic communications technology was in place and, in spite of its flaws, tied North and South America and Europe into a functional system. Who would use the system and to what effect are questions that would be answered over the next few decades.

CHAPTER THREE

Raising False Hopes

International Communications and International Crises in Latin America, 1866–1881

THE YOUNG TELEGRAPHER STOOD with his back to the electric key as he faced the Spanish soldiers who guarded the cable office. With his hand concealed behind his back, he quietly tapped out his message of distress: "There's the devil to pay in Santiago. They're butchering men of all nations." This startling alert would have traveled over the submarine cable from the south coast of Cuba to the British colony of Jamaica. The butchery in Cuba had taken the lives of fifty-three people from the cargo vessel *Virginius* and another fifty lives hung in the balance. This electric communication alerted the British gunboat *Niobe*, which reached the port of Santiago in time to save the remainder of the crew and passengers from the clutches of the deranged Spanish general Juan Burriel. This story of the heroic young telegrapher is apocryphal. There is no solid historical documentation to support it. However, the existence of this concocted version of events among the English-speaking residents of Santiago in 1885—some twelve years after the executions—provides an indication of the capacity of the telegraph and its potential for rapid communication over large distances to generate a kind of popular iconography. An unnamed reporter for the *New York Times* recorded this story when he visited the Spanish colonial city of Santiago, where the memory of the horrific events remained strong. In fact, telegraphic communications had saved the lives of fifty captives. The main embellishments were the "behind the back" skills and personal audacity of the operator, along with the presence of the Spanish soldiers.[1]

Four thousand miles to the south and nineteen years earlier, Adolfo Alsina, governor of Buenos Aires province, hailed the opening of the cable connection between his capital city and Montevideo, Uruguay. On November 29, 1866, Governor Alsina spoke to banquet hosting guests from business and government, who joined to celebrate the arrival of cable technology to their city. Alsina welcomed the telegraph as a device with the power to conquer both distance and time. His remarks coincided with similar comments being made by Governor Venancio Flores in Montevideo in front of another festive gathering. The cable lines of River Plate Telegraph carried the exchange of reciprocal toasts beneath the Río de la Plata. Business and civic leaders on both sides of the estuary were confident that a new day had dawned in international relations.[2]

The international telegraph network introduced a new means of communication through submerged cables in the 1860s and 1870s that tied together Latin America, the United States, and Europe. There was widespread confidence in the potential of fast-moving messages to improve the international environment for diplomacy and business by resolving disputes and promoting good will across national boundaries. Studies of international crises in Latin America from the end of the Civil War in the United States into the 1880s have examined their diplomatic, political, and economic dimensions in depth.[3] A few scholars have given serious attention to the role of telegraphic communications in these crises. Richard Bradford and Jay Sexton have examined the *Virginius* affair, and Andres Cisneros and Carlos Escudé have studied the so-called "wire treaty" between Argentina and Chile. All take into account the role of international telegraphy.[4] Comprehensive in their detail on particular events, they do not look into the functioning of the communications network as a whole. The main purposes of this chapter are to discuss the operation of this network and to evaluate its use by diplomats, politicians, naval officers, business leaders, and journalists in three crises that occurred between 1873 and 1881. What role did international telegraphy play in the international power structure of this era? Did the new system of communications help to establish stability and peace? The findings of Bradford, Cisneros, and Escudé indicate that governments made good use of the telegraph in the settlement of international crises. A broader, more comprehensive study of the functioning of the new communications system, however, may lead to other conclusions. It is on the margins and beneath the surface of these crises that the potential for the new system to generate false impressions, tension, and disruption became evident.

THE *VIRGINIUS* AFFAIR

Cuba's abortive revolution for independence (1868–1878) spawned a crisis that tested not only the diplomatic skills of the U.S. and Spanish governments but also the communications potential of the new submarine telegraph system. The *Virginius* incident was one of many in which Spain, its colony Cuba, and the United States were embroiled in the last decades of the nineteenth century. Discontented Cuban criollos and mulattos took on Spain's colonial army in a bloody struggle, depending for their success on men and supplies shipped from the United States. The *Virginius*, an overworked, somewhat dilapidated merchant ship, had participated in the transfer of war materials to the Cuban insurgents. Owned by Cuban rebel sympathizers but captained by Joseph Fry, a U.S. Naval Academy graduate and a Confederate Civil War veteran, the *Virginius* flew the U.S. flag and had several U.S. citizens aboard when the Spanish Navy captured this vessel on the night of October 31–November 1, 1873, as it approached Cuban waters. This episode would have been a minor event in the uneasy relations between Spain and the United States except that on November 4 the frustrated and emotionally unstable General Juan Burriel, the Spanish military commander of the port of Santiago, ordered the execution of four of the men captured on the *Virginius*, one of whom, George Washington Ryan, was a U.S. citizen. The other three were Cuban insurgents: Pedro Céspedes, Jesús del Sol, and Bernabé Varona. Three days later Burriel ordered and oversaw the execution of forty-nine more prisoners, including Captain Fry. The Spanish general defended his action on the grounds that the *Virginius* was a pirate ship with no protected status even though it flew the U.S. flag (with dubious justification, at best) and had a crew and passengers from several nations, including the United States and Great Britain.[5]

Given the pace of communications by oceangoing ships in the 1870s, Burriel would have had control of the situation for a week or more after the first four executions if communications had been restricted to this means of conveyance, allowing him to escape not only the authority of his Spanish commander but also the power of the United States and the British. However, the recently installed facilities of the International Ocean, the Cuba Submarine, and its sister firm, West India and Panama, soon brought Santiago to Washington's attention.[6] Two of these three companies had experienced problems in recent months. International Ocean's lines had been out of service for the early part of 1873, and West India and Panama reported interruptions between Jamaica and Puerto Rico from January to August, but by November 1873 all of these cables were operational.[7] Furthermore, the transatlantic cable, which had been working since 1866 through Anglo-American Telegraph, connected Washington with London and, by way of a European extension, also Madrid.[8] This communications network played a vital role in the naval and diplomatic responses to Burriel's actions.

As early as November 2 a suspicious and somewhat alarmed U.S. vice-consul, Emil Schmitt in Santiago, attempted to reach Washington via Cuba Submarine and its International Ocean connection but was denied access to the telegraph office on orders of General Burriel. British acting vice-consul in Santiago, Theodore Brooks, knew that British citizens were among those designated for execution, and on November 6 he managed to send a telegraphic warning to the British governor of Jamaica. For two days Brooks had worked feverishly to contact Burriel in order to prevent further executions. The challenge of maintaining telegraphic communications between Kingston and Santiago was taken up in Jamaica by Cuban merchant and revolutionary sympathizer Altamont de Córdova. Aware of the content of Brooks's telegram, de Córdova convinced the manager of the West India and Panama office in Kingston to remain open around the clock and, furthermore, persuaded the Kingston manager to order the Santiago office to do likewise. Instead of the courageous and dexterous young operator of the concocted story, it was the Cuban revolutionary activist who kept open the lines of communication between Santiago and the outside world. Kingston-based British officials, including navy commander A. F. R. Horsey, sent telegrams of protest to Burriel. De Córdova was a human whirlwind of activity. He cajoled the U.S. consul in Kingston into cabling an order to the USS *Wyoming* in Aspinwall to start for Santiago and also convinced Horsey in Kingston to send the British gunboat *Niobe* under the command of Sir Lambert Lorraine to Santiago. It was the *Niobe* that arrived in Santiago at 1:00 p.m. on November 7. The death toll in executions had reached fifty-three, but fifty prisoners were spared the same fate with the arrival of the British gunboat.[9]

Vice-consuls, a governor, and a navy captain had acted on the basis of information transmitted to them within the Caribbean, but news of the executions soon reached Washington, London, and Madrid as well. On November 7, 1873, Hamilton Fish, the U.S. secretary of state, received an urgent telegram from A. L. Cobright, a journalist working for the Associated Press in Havana, that contained news of the first four executions, and by November 12 Fish had received another telegram, this one from a U.S. consular official in Havana who reported "the stunning news of the execution of Captain Fry and 36 (actually 48) crew members."[10] From the receipt of this telegram until early December, the "*Virginius* affair" remained a large-scale diplomatic crisis whose main players were Washington and Madrid, although London offered considerable input as well. This crisis included a nasty quarrel between the belligerent, undiplomatic U.S. minister to Spain, Dan Sickles, and Spain's foreign minister José de Carvajal; owing to this quarrel, most of the negotiations ended up taking place in Washington, with Fish and Spanish minister José Polo de Barnabé calmly searching for mutually acceptable solutions. Polo would conclude the first phase of negotiations by telegraphing Madrid for confirmation. The use of the transatlantic cable proved essential to the relatively quick solution to the crisis that played out against a growing drumbeat for war in the United States. As a result of the harmonious exchanges between Fish and Polo from November 18 to 28, the government in Madrid agreed to settle the dispute. Spain stopped Burriel from carrying out any further executions, extended a formal apology to the United States, and eventually paid an indemnity of $80,000.[11]

By the end of 1873, Hamilton Fish was one of the world's most experienced diplomats in the use of submarine telegraphy in negotiations during a time of crisis. Obviously pleased with the peaceful solution, Fish nevertheless expressed a lack of confidence in diplomacy through telegraphy because "our information came to us by telegraph where frequently words and sentences are unintelligible and have to be guessed at, and . . . [i]t would not do to communicate to Congress in this imperfect form."[12]

The excitability of the press and public was an important factor in the *Virginius* crisis, but it did not drive events beyond the control of diplomats. The use of telegraphic "scoops" by journalists to report the latest twists and turns in the story before diplomats had the opportunity to find grounds for negotiation was not yet common. The *New York Times*, the *New York Tribune*, and other newspapers provided substantial coverage of events in Cuba and their ramifications in the United States, Spain, and Great Britain. Both the *Times* and the *Tribune* frequently used brief telegraphic accounts from Havana, Madrid, London, and Washington to piece together the story as it evolved on these four fronts. For example, the November 8, 1873, issue of the *Times* carried a telegraphic report datelined November 7 from Havana via International Ocean on the first four executions that had taken place on November 4. Since Cobright had given Secretary Fish essentially the same message on November 7, Fish and the Grant cabinet were able to discuss this unexpected crisis a day before it appeared on page 3 of the *New York*

Times.[13] During this crisis, many journalists cooperated with Secretary Fish by supplying him with new information, just as Cobright had done.[14]

Nevertheless, the public mood and press coverage became excited and even warlike as the crisis ran its course. At least two newspapers in New York, the *Herald* and the *Sun*, engaged in sensationalist coverage and editorial commentary. The *New York Times* chose not to follow this jingoistic line in its editorials, but several of its reports detailed news of bellicose gatherings of large numbers of people in the streets of New York. The Cuban population of the city, already disposed to criticize the Spanish colonial government of the island, organized several public gatherings to express their outrage at the deaths of these champions of Cuban independence.[15] The *Times* also reported via telegraph on public assemblages and rousing speeches in Albany, New York, Raleigh, North Carolina, and St. Louis, Missouri, and the *New York Tribune* provided coverage of a mass meeting in Baltimore.[16] One advocate of this wave of patriotism, an unnamed "gentleman holding high official position," claimed that a war with Spain would unite the states of the former Confederacy with the nation, thereby giving international hostilities an especially nationalistic purpose. There was additional evidence of an aggressive public temper in the United States. President Ulysses S. Grant received letters from Civil War veterans offering to raise volunteer regiments for combat, and the *New York Tribune* reported on public meetings in which Spain was subjected to derogatory insults. A headline for one story read "Madness and Cruelty in Havana."[17] The *Boston Globe* also used provocative language, printing headlines such as "Cuba: Another Horrible Tale," "The Massacre," and "More Slaughter!" The *Globe* also gave supportive coverage to preparations for war.[18] Likewise, the Spanish press and public seemed eager for military action with the United States in November and early December.[19]

Some newspapers in the United States, however, did not engage in a clamor for war, and, in fact, the *New York Times* excoriated the *Herald* for its excesses: "The persistent effort to create public excitement, and to incur all the consequences of an international war through raising false issues and appealing to violent passions is nothing less than a crime." There were other expressions of editorial opinion that tended to be moderate in tone. For example, the *San Francisco Chronicle* proclaimed its sympathy for the Cuban revolutionaries but did not make heated condemnations of Spain. Likewise, the *Chicago Tribune* advocated a cautious policy. These two papers, as well as the *Philadelphia Inquirer*, the *Atlanta Constitution*, the *New Orleans*

Picayune, the *San Francisco Examiner*, and the *New York Tribune*, covered the story fairly closely through mid- to late November. Most accounts were datelined from Washington or New York and a few cabled were from Cuba. These papers carried some reports on outbursts of public agitation and provided coverage of war preparations, but intense belligerence similar to that expressed in the *New York Herald* and the *Boston Globe* was momentary.[20] By the end of December sensationalism in both U.S. and Spanish newspapers had diminished, and the public agitation for war died down.

Technically, the cable network had performed admirably in this crisis. De Córdova and the British gunboat had brought a halt to the executions, and Fish and Polo had found a diplomatic remedy to avoid further conflict. Yet on the margin of these events, a potential problem had appeared: the interaction of sensationalistic press fed by cable reports with an excitable public. Diplomats held the advantage in the *Virginius* affair, but that amorphous entity—public opinion—goaded by a jingoistic press had put in an appearance.

GREAT BRITAIN AND THE TURMOIL IN ARGENTINA

Argentina and Great Britain were the main actors in another crisis, or more precisely a series of crises, in which telegraphic communications played a role. President Domingo Sarmiento, who had hailed the arrival of the cable connection with Europe with great fanfare, stepped down from the presidency of Argentina in 1874. His successor, Nicolás Avellaneda, followed Sarmiento in promoting economic development throughout the 1870s. As his term came to an end, a dynamic figure moved to the forefront of national politics as the leading candidate for the presidency in 1880. He was Julio Roca, an outspoken champion of commercial expansion and the military leader who directed the campaign that pushed the Native Americans off the grasslands known as the Pampas. Roca also advocated the federalization of Buenos Aires that, in essence, would have placed the city and surrounding territory under the control of the national government. This issue divided the nation and also split the *porteño* population. His drive for the presidency became embroiled in a sharp and ultimately bloody controversy. The early months of 1880 were fraught with tension arising from the discord about federalization. The precipitating issue was the election of Roca to the presidency in April, an event that marked a defeat for the province of Buenos Aires. Leaders of this large province were determined to prevent Roca from taking charge of the nation. Avellaneda, the incumbent president, was equally determined to defend the national government and the outcome of the election. Both sides armed themselves for the reappearance of the civil strife that had characterized so much of Argentina's first half century of independence.[21]

The anti-Roca forces began to import weapons from abroad. In these circumstances the Argentine Navy seized two British-owned merchant steamers, the *Plato* and the *Bessel*, in early May and searched them for war material. News of this action reached London in hours by way of the Western Brazilian-Brazilian Submarine cables and caused public outcries of protest as well as expressions of concern in Parliament. The substantial British investment in Argentina and the British population there both seemed to be at risk, at least, from the perspective of London. In reality, the Argentine search and seizure did little if any harm to either ship. The *Plato* and *Bessel* incidents made clear, however, that telegrams sent from distant places with an unsettling message in a time of crisis or perceived crisis had the potential to move politicians to probing inquiries and possibly bring to life the latent fears of a poorly informed public, thereby complicating the work of diplomats. Historian H. S. Ferns summed up the event succinctly: "Twenty years previously an incident of this kind would have been settled on the spot and become a matter of no account by the time news reached London."[22] Little came of this particular incident, but it was clear that the international telegraph system had the potential to magnify the importance of events and thereby cause discomfort to diplomats and politicians and, if the information appeared in the popular press, whip up the general public.

The temporary seizures of the *Plato* and the *Bessel* were only isolated events in the Argentine political crisis that continued to build until the onset of winter in the Southern Hemisphere in June 1880. Roca's opponents assembled an irregular army made up of gauchos that pushed its way into the city of Buenos Aires amid fighting that eventually left a total of 2,500 either killed or wounded. President Avellaneda relied on the army of the national government and ordered the navy to blockade the port of Buenos Aires. British minister Edwin Egerton dutifully reported on these events to the Foreign Office by telegraph as well as mail. British businesses—especially merchant houses—were threatened by the street fighting, and their owners put considerable pressure on Egerton to call for protection from the Royal Navy. Egerton, fully aware of the overall policy of the Foreign Office in South America, consistently refused to request such help. The Foreign Office and Egerton were in agreement that such intervention would be difficult to execute and would probably be counterproductive. Avellaneda's government put down the revolt in late June 1880, thereby clearing the way for Roca's inauguration as president in October.[23]

Egerton's telegrams to London documenting the crisis in Argentina provided more information about a diplomatic crisis in South America than the British Foreign Office was used to receiving. Telegraph facilities in Argentina were relatively modern, the government having completed the construction of a new headquarters for the nation's telegraphic services in 1878.[24] This up-to-date building coupled with the recently opened connection to London provided by Western and Brazilian and Brazilian Submarine made Egerton's telegraphic work quick and convenient. Perhaps these facilities were too convenient. The conscientious Egerton became involved in a dispute with the Foreign Office on another factor in the use of the new communications system—cost. Western and Brazilian and Brazilian Submarine charged large fees for the transmission of even relatively short messages between Buenos Aires and London. As the Foreign Office and Egerton discovered, messages sent in code (a practice of the Foreign Office) were even more expensive because the cable companies required that the message be transmitted twice in order to reduce the chance of error in sending the coded messages, which consisted largely of made-up words that typically did not appear in standard usage. The cable companies, therefore, charged double for each coded message. The Foreign Office rebuked Egerton for a lengthy and expensive £150 telegram during the Argentine crisis. Egerton responded with the suggestion of shortening the formal addresses of ministry officials to a single code word, as private businesses did in their coded corporate telegrams. The implementation of this practice would have meant the elimination of titles and other means of address that spelled out the status of the receiving party, which in turn would have meant more expense and time to get a given message to its recipient. Such wrangling made clear that the British Foreign Office was trying to find a proper balance between the rapid transmission of information and the cost of such transmissions.[25]

Barings Bank made more frequent use of the submarine cable than the Foreign Office, apparently with no squabbling about costs. Nicolas Bouwer, Barings' agent in Buenos Aires, customarily reported by telegraph on financial conditions in Argentina relevant to the bank's substantial investments there.[26] The content of Bouwer's telegraphic reports expanded to include politics in early 1880. The management of Barings in London assumed that the governor of Buenos Aires Province, Carlos Tejedor, would win the presidential election. Roca's victory and the ensuing tensions concerning Tejedor's decision to use force to challenge the new administration elicited several telegrams (supported by confirmatory letters) from Bouwer in Buenos Aires during the months between the election and the inauguration. On February 2 an optimistic Bouwer telegraphed that "political situation threatening but solution will be peaceful," but by mid-April he detected a definite deterioration.[27] In early June, Bouwer reported that President Avellaneda had withdrawn the national government offices from the city of Buenos Aries to the nearby community of Belgrano in anticipation of civil strife. Nevertheless, Bouwer concluded, "Hope still peaceful solution."[28]

A peaceful solution proved elusive. The outbreak of fighting in mid-June brought messages of distress from Barings' man in Buenos Aires. With an eye toward Argentine government deficits, Bouwer noted that civil strife would do harm to that nation's ability to repay its loans, many of which were held by Barings. A coded telegram dated June 21 informed the London office of "battle raging near town."[29] Interrupted cable services delayed further communications, but in July and August Bouwer noted in terse telegraphic language that the ups and downs of politics continued even in the wake of Roca's victory. Bouwer's record of telegraphic reports constituted a selective and condensed but up-to-date version of events in Argentina throughout most of 1880.[30]

The London *Times* also used the cable in an attempt to provide its readers with the breaking news on the crises in Argentina. Although these reports fell under the category of politics, the *Times* often placed them in the financial section of the newspaper, an indication that the editors believed that the readers most interested in these stories would be business people with investments in the Río de la Plata region. Brief telegraphic reports in June announced the closing of the port of Buenos Aires.[31] Cable services supplied by River Plate Telegraph, the only connection between Buenos Aires and Montevideo, ceased sometime around June 20 and did not resume until July 5. There is no evidence that either side in Argentina's civil strife damaged the cable. Reports from Western and Brazilian indicated that the River Plate Telegraph resumed operation without any major repairs.[32] No matter what the exact cause of the interruption, however, the London *Times* and, indeed, eastern South America and Europe lost electric communication with Buenos Aires for about two weeks. In that interval, the *Times* printed a report from Lloyd's of London insurance company datelined June 16 from

Buenos Aires. Trenchant accounts of the fighting and the announcement of the armistice came from Rio de Janeiro from an unidentified source.[33]

Immediately upon the reopening of cable services between Buenos Aires and London, the Avellaneda administration sent the following message by telegram to its representative in the British capital: "Rebellion ended with submission to the national authority and complete pacification. Great rejoicing." This message appeared in the London *Times* of July 6, one day after it reached the Argentine legation, although it was dated July 2.[34] Apparently River Plate Telegraph transmitted the telegram through its lines on July 5, the day on which cable service resumed.[35]

Throughout this period the *Times* exerted considerable effort to obtain news from Argentina from a variety of sources. In so doing, however, it employed a practice that provided confusing and seemingly contradictory versions of events, unless the reader noted the dateline of each article. For example, the July 5 issue contained a half-column account of the siege of Buenos Aires that mentioned the building of trenches and barricades in the city streets. This article was dated June 9 and originated in Lisbon (presumably carried by ship from the Río de la Plata). Below this article was a single paragraph on the political maneuvering in the national and provincial governments datelined July 3 from Rio de Janeiro, which was about two weeks *after* the rebel forces surrendered.[36] Only a careful reading of the datelines would prevent confusion for readers in London about the unsettled conditions in Buenos Aires.

The performance of the international cable network in the Argentine crises of 1880 was commendable for the increased speed of communication but often unsatisfactory in terms of the content of the messages. Egerton and Bouwer kept their headquarters informed, but the impressions conveyed in their terse missives were not always accurate. Although London complained about the cost of Egerton's cables, it was as concerned with the brevity and consequent lack of depth in the comments by Egerton; Barings voiced similar consternation about Bouwer's messages.

The London *Times* experienced problems in obtaining information even during times in which cable service was available. The record on the pages of the *Times* indicates that the newspaper did not have a correspondent stationed in Buenos Aires. The editors attempted to compensate by printing accounts from Rio de Janeiro and Lisbon and also by turning to Lloyds of London, which had agents in South America. The practice of drawing from an array of dispersed sources for information indicated a commendable effort, but this practice also created a confused version of events if read without a careful notice of datelines. In short, the press, as represented by the London *Times*, lagged behind the British foreign ministry and Barings Bank in the effective utilization of the cable. Overall, the London *Times*, Barings, and the Foreign Office had to struggle to understand the course of events in Argentina in spite of the increased speed of the new communications system.

THE "WIRE TREATY"

A third crisis furnished another opportunity for submarine-cable telegraphy to play a role in the resolution of an international problem. Argentina and Chile shared a 2,500-mile border, major segments of which were the subject of dispute. The uncertainties about these border areas arose from a lack of clarity on the maps of the Spanish colonial government in the eighteenth century and the inability of the governments in Santiago and Buenos Aires to resolve these differences after independence. As both the Chilean and Argentine governments drove the Native American populations further south and exploratory expeditions looked for coal and other resources in the region, pressure began to build in both countries to resolve the boundary disputes. Argentina's recently elected president, Julio Roca, had gained a national reputation for his prosecution of a large-scale military campaign in 1878–1879 against the Araucanians and other Native American groups, driving them southward into the colder, less hospitable region of Patagonia. During these campaigns, the Chilean government became suspicious of Argentina's ultimate goal. There was speculation in Santiago concerning the possibility of war between these two Andean nations. Linked to this boundary dispute was the uncertain status of the Cape Horn region and the strategically vital passage through the Straits of Magellan.[37]

Another factor in the tensions between Buenos Aires and Santiago had to do with Chile's recent victories in the War of the Pacific. Pitted against a coalition of Peru and Bolivia, Chile suffered defeats in the early months of the war, but by 1880 the course of the conflict had changed. Chilean forces conducted amphibious operations up the Pacific coast and in January 1881 captured the Peruvian capital of Lima. Although Chile faced the difficult task of pacifying portions of Peru and holding on to valuable nitrate regions in the postwar diplomatic negotiations, which could be expected to require its concentrated attention, the Argentines were nonetheless concerned about the intentions of its neighbor's victorious army and navy in the disputed territory to the south.[38]

This boundary issue assumed crisis proportions in 1881, less than a decade after Argentina and Chile had shared the convivial observance of the opening of telegraphic communications between Buenos Aires and Santiago, the presidents of the two countries proclaiming that the mountains were no longer a barrier to communications between their two countries.[39] It was a double irony, therefore, when nine years later Argentine and Chilean negotiators, aided by their diplomatic colleagues from the United States, used these same telegraph lines to locate a formal boundary between their two countries (most of which ran along the watershed of the Andes) and bring to an end this source of international tension.[40]

United States diplomats were concerned about the growth of diplomatic uncertainty in southern South America and the possibilities of another war. To make matters more difficult, neither Chile nor Argentina had formal representatives in the other's capital. At this point U.S. representatives in Santiago and Buenos Aires took the initiative. By coincidence both of these men were named Thomas Osborne (they were not related), and they both shared a determination to use the facilities of Transandino and the Argentine government to facilitate negotiations. Both U.S. ministers provided their "good offices," and soon Argentine foreign minister Bernardo de Irigoyen and his Chilean counterpart, Melquíades Valderrama, agreed to use the new communications device to proceed with the negotiations. Irigoyen and Valderrama consulted with each other by telegraph throughout the period of critical discussion from May 9 to June 25, 1881.[41]

During the six weeks of diplomacy by telegraph, negotiations in two particular periods stood out. On May 9, 10, and 11, the two foreign ministers agreed that the natural watershed (*divortie aquarium*) should set the boundary from the southern border of Bolivia in the north to 52 degrees south latitude. Secondly, the strategically important Straits of Magellan were to be neutralized. That these arrangements met approval in both capitals in a three-day period was the result in no small part of the frequent and rapid movement of messages along telegraph lines.[42]

The second crucial spurt of negotiations occurred on June 3 and 4. The Chilean Valderrama devised a complex six-part proposal to resolve the remaining areas of disagreement. One of the most important of these was the division of the island of Tierra del Fuego on terms that were acceptable to both governments. Again, both foreign ministers with the aid of both Thomas Osborns resolved virtually all of these points quickly by the diplomatic expectations of the day. Chile and Argentina found a middle ground via telegraph on the remaining key points in two days, negotiations that would have required weeks if not months had they been conducted by mail. Once the essential diplomatic issues had been resolved, the standard formalities moved smoothly. By the end of June, both governments had approved the comprehensive compromises and on July 23, 1881, the so-called "Wire Treaty" was official.[43]

CONCLUSIONS AND CONNECTIONS WITH LATER EVENTS

Judging from the outcomes of these crises in the Western Hemisphere, apparently the proverbial new day had dawned in international relations. The peaceful settlements in the *Virginius* affair and the "Wire Treaty" coupled with the life-saving arrival of the *Niobe* in Santiago harbor seemed to support the optimism associated with the new technology of long-distance telegraphy that permeated the celebratory inaugurals of the 1870s. London kept watch on the turmoil in Argentina in 1880 and decided intervention was unnecessary. The United States coaxed the Chileans and Argentines to use the telegraph to settle their long-standing boundary disagreement. The quick movement of words over wires, in spite of some transmission problems and the brevity of the messages, could contribute to the resolution of international disputes and save lives.

Along the margins and beneath the surface of these events, however, some rather ominous signs appeared—not obvious to contemporary observers but important enough to alert the student of history to track their presence over the following decades. In the *Virginius* affair, U.S. reporters cooperated with Secretary of State Fish, and the *New York Times* advocated negotiations between Washington and Madrid. On the other hand, the *New York Herald* explored the possibilities of jingoism and found, at least in the short run, an interested and pliant public. The *Times* criticized the *Herald*, but the latter's bellicose rhetoric found reinforcement in street meetings and political speeches. And Secretary Fish privately worried about the possibility of erroneous messages under similar tense circumstances. By the 1890s telegraphed accounts of crises in Latin America would find their way into the wire services of the United States and onto the front pages of big-city newspapers, where they would claim a large readership. These bellicose images and sensationalistic reports put pressure on diplomats, politicians, and policy makers, which would eventually have an impact on the public perception of international relations if not the relationships themselves.[44]

Another troublesome aspect of the use of the telegraph in international relations emerged from the "Wire Treaty." The close involvement—one might say intervention—of the United States in this event made for a peaceful solution in 1881, but Washington's activist diplomacy via long-distance telegraphy soon came to focus largely on Chile with significantly disruptive consequences. The frequent use of the telegraph in the delicate negotiations of 1881 set a precedent for rapidly composed messages and prompt responses that turned against diplomatic solutions in the crises that followed. In the aftermath of the War of the Pacific (1879–1883), U.S. diplomats and politicians assumed an anti-Chilean stance and favored Peru in its vain efforts to hold on to its rich nitrate territories of Tarapacá, Tacna, and Arica. The *Baltimore* affair of 1891 and the subsequent confrontations by cables in 1891–1892 accompanied by press commentary contributed to a rising trajectory of hostilities between the two nations that climaxed in a war scare. The use of the telegraph loomed large in all of these difficulties.[45]

On the other side of the Andes, Argentina drew closer to Great Britain, as economic and cultural ties between the two countries grew stronger after the resolution of the 1880 crises, but there was a cost to be paid for this international integration. On the surface, relations between London and Buenos Aires flourished. British investments poured into Argentine railroads and land. British investors' confidence in Argentina increased in the 1880s, as banking houses, led by Barings, attempted to keep close watch on that nation. This intimate relationship, enhanced through the expanding use of the cable system by diplomats, bankers, investors, and journalists, emerged as an important element in the financial disaster known as the Baring crisis, which caused a recession through the early 1890s. The lack of depth in the telegraphic messages of 1880 seemed to be forgotten later in the decade. Barings Bank placed too much confidence in its overly optimistic confidential cable reports from Buenos Aires. Barings, the London *Times*, and the British Foreign Office were caught by surprise, as business and political turmoil swept through Argentina. In 1890 the London *Times* decided to station a full-time journalist in Buenos Aires, and in the following year the respected Argentine newspaper, *La Prensa*, placed a correspondent in London. In the crucial period of 1890–1891 *La Prensa* carried frequent reports by cable and mail from London in an attempt to assess the attitude of investors and politicians and also the mood of public opinion regarding Argentina. In the same period, the London *Times* published numerous reports—many via cable—on the political and business environment in Argentina.[46] But these changes came after the onset of the crisis. By the early 1890s it was painfully obvious that the improved communications technology of the 1870s and 1880s did not necessarily mean an improvement in the reliability of the information transmitted. The international financial crisis was caused in part by misplaced confidence in the veracity of information transmitted over telegraph cables.[47]

The new communications technology brought positive results in the short run, but the leading historical protagonists faced a struggle to make good use of the international cable under the difficult circumstances that arose in the 1890s. The image of the heroic young telegrapher in the West India and Panama's Santiago cable office tapping out his life-saving message to the outside world was simply that—an image created in the collective imagination of the local public. The international cable network added to the frequency and ease of communication by technological advances that led to lower costs and greater access. At the same time, James Scrymser built cable connections from Galveston to Santiago, thereby closing the gap in the cable loop in the Americas.[48] As the flow of information increased, the people who attempted to use this system for their own political, national, or corporate purposes faced challenges that pushed governments to the brink of war and businesses to the edge of collapse.

CHAPTER FOUR

War, Diplomacy, and Propaganda

Chilean-U.S. Relations, 1866–1880s

CHILE FACED A DIFFICULT SITUATION. Valparaíso, its major port, was blockaded by warships of the Spanish Navy. The blockading force, inspired by its earlier success against Peru, demanded a financial indemnity and threatened the economic well-being of the entire country. In need of weapons and diplomatic support in what had become an open war with Spain, Chile turned to the United States. A young diplomat and writer of Irish-Chilean descent, Benjamin Vicuña Mackenna escaped the blockade and made his way to New York City in hopes of obtaining help in the homeland of James Monroe, the author of the famous doctrine of the same name. Vicuña Mackenna believed that the government in Washington, recently victorious in its own civil war, would seek to enforce the Monroe Doctrine by sending notice to Spain that its aggressive actions in South America violated the presidential pronouncement of 1823. Vicuña Mackenna had initial success in his venture with the purchase of four ships and began to recruit volunteers for the Chilean armed forces.[1]

More importantly for this study, Vicuña Mackenna initiated a publicity campaign to elicit broad support for Chile in the United States. He launched this campaign in New York City, the burgeoning metropolis that was the headquarters for the telegraphic giant Western Union and several newspapers of national prominence—especially the *New York Times* and the *New York Tribune*. Vicuña Mackenna was perhaps the first representative of a Latin American nation who made use of the mass press in the United

States to further his nation's cause. He also established a Spanish-language newspaper, *La voz de América*, and published a book in English called *Chili, the United States, and Spain* under the pseudonym Daniel J. Hunter.[2] The high point of his campaign was a public meeting at the Cooper Union in lower Manhattan. The chairman of the event was Ephraim G. Squire, noted scholar and diplomat. Speakers included Horace Greeley, the editor of the *New York Tribune* and an influential Republican, and poet and political activist William Cullen Bryant. Both of these individuals endorsed the Monroe Doctrine and urged U.S. support of Chile against Spanish aggression. Chairman Squire read letters of support from seventeen prominent citizens, including Civil War general Dan Sickles, former general and future president James Garfield, Speaker of the House Schuyler Colfax, and social reformer Robert Dale Owen. Their letters lauded Chile and enthusiastically endorsed of the Monroe Doctrine. Lawyer Daniel Dickinson, a prominent federal attorney in New York state, issued a resounding statement: "There is, perhaps, no question of national policy, either foreign or domestic, upon which the American people of all sections and parties are so firmly united and resolutely determined, as upon that of resistance to the encroachments of monarchy (meaning Spain) upon this continent (South America)."[3]

The climax of the evening was Vicuña Mackenna's speech. The Chilean praised the Monroe Doctrine as a bulwark against the retrogressive Europeans. He reminded his audience of Chile's current struggle against the Spanish fleet. His remarks were interrupted by "great cheering," and at the conclusion of his address Vicuña Mackenna enjoyed "a perfect storm of applause."[4]

Vicuña Mackenna's efforts to arouse support for Chile caused a stir in the United States, but not the type of response that the Chilean publicist wanted. In spite of his triumphs at the Cooper Union and in the pages of the *New York Times*, U.S. secretary of state William Seward was apparently more concerned about James Scrymser's quest for permission from Madrid to land his submarine cable in the Spanish colony of Cuba and chose to ignore Vicuña Mackenna. Even worse, federal officials in New York charged him with violation of the U.S. neutrality laws for trying to recruit a torpedo expert for the Chilean Navy. Ignored by Seward and thrust briefly into jail, Vicuña Mackenna found the U.S. government, unlike his Cooper Union audience, unconcerned about the Chilean cause.[5]

An even greater insult to Chile occurred on March 31, 1866, when Spanish warships bombarded Valparaíso. Eight vessels of the U.S. Navy were present off Valparaíso at the time, but their commander would not intervene without the joint action of the Royal Navy's contingent in the area. As a result, defenseless Valparaíso suffered a heavy bombardment that resulted in considerable property damage and the loss of two lives. The Chilean estimate of the value of the United States as a guarantor of its own Monroe Doctrine declined substantially.[6]

Washington's indifference to Vicuña Mackenna's publicity campaign and its passive response to Spain's bombardment of Valparaíso reinforced Chile's lingering animosity toward the United States. By 1866 political leaders in both countries were already caught up in a relationship that was characterized more by distrust and misunderstanding than cordiality and harmony. Even before Vicuña Mackenna's problems in New York, Chileans had been rudely expelled from the gold fields of California in 1849 and the early 1850s and returned home with tales of discriminatory treatment and even some incidents of brutality.[7] Given the troubles between the two countries, long-distance telegraphy came to play a vital, even decisive, role in Chilean-U.S. relations in the last decades of the nineteenth century. Vicuña Mackenna and his generation of Chilean leaders had learned that the mixture of politics, press, and public relations that swirled around the government in Washington was an important factor in the formulation of foreign policy. Although Vicuña Mackenna encountered rejection and failure, his countrymen would soon master the U.S. information system to the benefit of their national interest. This interface of communications, information flow, and diplomacy is the main theme of this chapter.

A DISTANT WAR

Chile's difficult relationship with the United States again became an issue of importance during the War of the Pacific (1879–1883). This conflict pitted Chile against a Peruvian-Bolivian alliance. Fought in the waters of the Pacific Ocean adjacent to South America and on the continent's rugged western coastline dominated by mountains and deserts, the War of the Pacific was a severe test for the combatants and for the newspapers that attempted to cover the fighting for their readers in Europe and the United States. The unavailability of submarine telegraph cables and the disruption of oceangoing trade routes in the vicinity of the fighting resulted in limited press coverage. In spite of its isolation from commercial centers such as London and New York, the west coast of South America was important for investors, farmers, manufacturers, and government officials in the nations of the North Atlantic. Offshore islands were covered with deposits of bird droppings known as guano—a valuable source of fertilizer. The guano supply diminished in the 1860s and was replaced by another export: nitrates. A dominant geographic feature of this region, the Atacama Desert, had rich deposits of nitrates that were used in the production of fertilizer and explosives. The Swedish chemist Alfred Nobel invented dynamite in 1867 and in the next decade continued to develop explosives with military applications. Chilean nitrates were the primary ingredient in these new weapons and became a valuable commodity on the international market. Business magnates and military strategists saw the potential in these arid coastal provinces, and so the war between the Peruvian-Bolivian alliance and Chile held the attention of commercial and political interests in the North Atlantic.

British journalists were experienced in the coverage of warfare in distant lands, often on the front lines or close to them. William Howard Russell, ace correspondent of the London *Times*, was preeminent in this group with his coverage of the Crimean War in 1854 and 1855 and subsequent coverage the U.S. Civil War and the Franco-Prussian War of 1870–1871. In the latter conflict, Archibald Forbes outdid Russell and other reporters in the use of telegraph wires, achieving several "scoops," or the first published account of a story.[8] Paralleling the extension of the London *Times* in its overseas coverage was the expansion of another type of news organization: Reuters. Julius Reuter pioneered the development of a news-gathering service based on the use of telegraph lines in the 1860s and, by 1870, had devised an arrangement with the French news agency Havas and the Wolff organization of Germany to coordinate wire-service journalism throughout much of the world. Havas and Reuters had a special relationship in South America; they had a joint-purse agreement whereby they shared the costs and profits from newsgathering and distribution, and they even operated joint offices in Brazil, Argentina, and Uruguay.[9]

This chapter compares the coverage of the War of the Pacific in the London *Times* and the *New York Times* in order to explore the flow of information on a foreign war in the international communications system. The press in the United States was inexperienced in contrast with the British press. The London *Times* and Reuters had covered wars in foreign countries by the 1870s, but with the exception the *New York Tribune*, for whom George Smalley provided accounts of the Franco-Prussian War, U.S newspapers had not. The New York Associated Press was mainly concerned with the gathering of domestic news, and the *New York Times* likewise had a national more than an international focus. Both the wire service and the newspaper tended to rely on foreign newspapers for international news. American journalists had earned battlefield experience during the U.S.-Mexican War of 1846–1848 and the U.S. Civil War of 1861–1865, but in those conflicts they had mainly been concerned with reporting on policies formulated in the United States and implemented by U.S. soldiers and diplomats.[10] In war reporting from foreign lands, the British had an advantage.

The naval and land battles on the west coast of South America, however, presented special problems even for experienced British journalists. The distance of the three belligerents—Peru, Bolivia, and Chile—from the North Atlantic centers of communications, trade, and finance coupled with the fact that most of the fighting at sea and on land took place in the vicinity of the Atacama Desert, a rugged, thinly populated six-hundred-mile stretch of coastline five hundred miles south of Lima and about the same distance north of Santiago, caused problems in communications. Nevertheless, newspaper editors in London and New York knew that their respective business communities had financial commitments in guano and especially nitrates. The overall economic development of Peru and Chile had enticed £36.2 million and £10 million in investments respectively from Great Britain.[11] The United States, after a slow start, also had a presence in Peru in the operations of W. R. Grace and Company and the railroad construction projects of Henry Meiggs in Chile and Peru.[12] In short, merchant houses, investors, and construction firms in London and New York had significant interests in the region.

At this point, it is helpful first to present a general sketch of the war. The conflict originated in the rivalry for control of the nitrate fields of the Atacama region. Bolivia was not landlocked at this point in its history, possessing a corridor that reached the Pacific Ocean at the port of Antofagasta. Peru joined Bolivia to form a secret alliance against Chile. For their part, Chilean entrepreneurs were aggressively promoting the extraction of nitrates through private corporations, many of which were supported by British investors. In 1879 Bolivia imposed a tax on the export of nitrates, which the Chileans claimed violated an agreement between the two countries. Diplomacy collapsed, and Chile faced the Peruvian-Bolivian alliance. When the fighting broke out, Peru and Bolivia held the disputed territory, and Chile faced the daunting task of gaining control of the ocean route to the Atacama before it could hope to mount an invasion of the disputed land. Chile's navy first attempted a blockade of Peru's nitrate port of Iquique, but this strategy went awry when in May 1879 Captain Miguel Grau of Peru deployed his two oceangoing ironclads against two of Chile's wooden-hulled ships. Peru won the battle, but when Grau's *Huascar* rammed Chile's *Esmeralda* under the command of Captain Arturo Prat, the young Chilean officer charged aboard the *Huascar* in a valiant but fatal effort to defend his ship. Prat's courageous example inspired a wave of Chilean nationalism that soon had a large impact on the war. In the short run, Peru's victory opened the way for Grau's *Huascar* to strike against Chile's coastal shipping. Peru's advantage disappeared in October 1879, when Chile's two modern ironclad ships avenged this defeat by cornering and capturing the *Huascar*. Now with command of the sea, Chile prepared for an invasion.

Chile launched an invasion of Peru's southernmost province of Tarapacá. Landing at the small coastal city of Pisagua, Chilean troops established a beachhead, scaled heavily defended bluffs, and secured control of a crucial portion of the coast. The Chilean army then pushed into the desert interior, but Peru offered effective opposition with its defense of the city of Tarapacá, to the point that the government in Lima claimed a victory. The combatants presented conflicting versions of events at Tarapacá in the British and U.S. print media. A little more than a year later (January 1881), Chile invaded the Peruvian capital, which, after fierce fighting, fell to a twenty-six-thousand man army. Chile had thoroughly defeated both of its opponents and eventually exacted large territorial concessions from its hapless foes.[13]

The London *Times* and the *New York Times* attempted to provide the basic facts concerning the war, but their coverage was slow, irregular, and at first, inaccurate. There were several problems in reporting this war, not the least of which was the lack of reliable telegraphic communication with the battle sites and nearby cities. West Coast of America had laid cable between Valparaíso and Lima in 1876, but the war broke out only a few months after the company opened for business. The Chilean armed forces either took control of the cable or severed it according to its military and naval interests.[14] To the north, there was no telegraphic link between Lima and Panama. These large gaps in the communications system offer some explanation as to why none of these news-gathering institutions in the United States and England sent a reporter to the war zone. The absence of direct communication with sources on the scene led newspapers to turn to other means to track events from the battlefronts. The result was a mix of brief telegraphic dispatches a few days old (and sent from cities in the region such as Santiago, Rio de Janeiro, or Buenos Aires with stations on the submarine-cable system) and lengthier accounts carried by oceangoing steamships to London or New York that were composed three or four weeks before they appeared in print. In reading such coverage, the dateline, or date and location of the original report, became crucial because the longer pieces delivered by mail often described battles that had been decided weeks earlier. For example, it was not uncommon to read accounts of the Peruvian ironclad *Huascar's* raids along the Chilean coast a few weeks after telegraphic dispatches briefly described that vessel's capture by the Chilean Navy.[15]

Given the absence of their own correspondents on the scene, both newspapers culled information from available sources. Government representatives of the belligerent nations stationed in London, New York, Washington, and Paris served as news sources close at hand, but the British insurance company, Lloyds of London, and several merchant houses with ties to the region were useful because they received reports by telegraph and steamship mail on a regular basis. Reuters news agency provided some stories for the London press, but the *New York Times* also found a fairly consistent source in the *Panama Star and Herald*, an English language organ with steamship mail links to a correspondent who was located in Lima for much of the war. The *New York Times* published extensive accounts from this uncredited correspondent, which were generally slanted in favor of the Peruvian point of view.

One example of inept coverage can be seen in the reporting of the sinking of Prat's *Esmeralda* by Grau's *Huascar*. The actual event took place on March 21, 1879, but first mention of it in both the London *Times* and *New York Times* appeared in their May 30 editions. One notice was in both papers, datelined "Santiago May 28 (via Rio de Janeiro)," and contained the cryptic, inconclusive, largely incorrect message: "The Chilean fleet has been attacked by the Peruvian fleet off Iquique. The latter was repulsed."[16] The second notice in both papers came from Lloyd's in London, which had received from an unidentified source another fundamentally incorrect version of the battle off Iquique "between Chilian wooden vessels *Esmeralda* and *Covadonga* and the Peruvian armour-plated frigate *Independencia*. . . . All three sank."[17] Neither telegraphic message named the *Huascar* as the victorious vessel. The first message, perhaps originating with the Santiago government, falsely asserted that the Chileans beat off the attack. The second message claimed three vessels sank when actually only one—the *Esmeralda*—went to the bottom. The other Peruvian ironclad, the *Independencia*, did not sink but ran aground, and Chile's *Covadonga* escaped. The *New York Times* did exceed its London counterpart by informing its readers in a small insert that "the

Huascar took part in the battle, and came out safely"—information it gleaned from "Chilean sources" in London.[18] Not until June 17 did the London *Times* provide a reasonably accurate version based on an account via New York and Panama from an unnamed source in Lima.[19]

The London *Times* seemed to adjust to the complexities of long-distance journalism from the west coast of South America with the story of the defeat and capture of the *Huascar.* The event transpired off the Cape Angamos on October 8, and four days later the *Times* ran a succinct account from a Lloyd's telegram: "An engagement has been fought between the Chilian and Peruvian fleets. The *Huascar* has been captured by the Chilians." There followed a three-line dispatch from Reuters datelined "Valparaiso, Oct. 8," which verified Lloyd's version.[20] The *New York Times* lagged behind its British counterpart in that it apparently had no coverage via telegraph of the event, but on November 4 it did publish a lengthy, somewhat sensationalized article that originally appeared in the *Panama Star and Herald.* Under the one-column headline "Trafalgar of the Pacific," the *Star and Herald's* correspondent in Lima wrote an account of the battle from Peruvian sources that reflected that nation's point of view.[21]

THE INSERTION OF PROPAGANDA IN THE FLOW OF INFORMATION

The beginning of the land war also marked a new phase in press coverage in Great Britain and the United States, as the War of the Pacific acquired a propaganda dimension that it had lacked in its early months. Countervailing tendencies resulted in a mix of fact and fancy in the London and New York newspapers. By late 1879 the London *Times,* Reuters, and the *New York Times* had improved the speed and accuracy of their reportage on the fighting, but both the Peruvian and Chilean governments began to insert their versions of events into the information flow in order to gain the maximum positive image for their respective causes in the centers of financial and political power in London, New York, and Washington. In addition, rumors and "wild stories" from unidentified sources entered the information flow, adding pieces of hyperbole to the press coverage. The Chilean army's establishment of an invasion foothold at the port of Pisagua received somewhat better coverage in both newspapers. News of this November 2 invasion appeared in both the London *Times* and the *New York Times* on November 11. The London *Times* story was based on telegrams datelined November 10 from Rio de Janeiro and Paris. The next day a telegram received by Antony Gibbs, a merchant house in London, confirmed that the Chileans had attacked Pisagua.[22]

The *New York Times* contained a similar brief account but from different sources—two telegrams from Valparaíso to London, one of which carried the date of November 7.[23]

A major problem surfaced, however, in the coverage of the fighting for Tarapacá in early December—particularly in London. Control of Tarapacá, Peru's southernmost territory, was essential in Chile's plan to move northward and invade Lima. The initial account seemed simple enough: "The town of Tarapacá had been captured by the Chilean troops. The losses were heavy on both sides."[24] Several weeks later the Peruvian minister to Great Britain announced an entirely different version of this battle by placing in the London *Times* a 117-word telegram from the Peruvian legation in New York that claimed that Peruvian forces had inflicted a staggering defeat on Chile. The invaders from the south had suffered 1,600 casualties in a force of 3,500, and the Peruvians had also taken a large number of Chilean prisoners. The Peruvians, according to this account, seized much of the Chilean weaponry.[25] Chilean diplomats denied this report by asserting that their forces had taken control of the province of Tarapacá. A careful reading of the details in the telegraphic reports that accompanied this battle of words in the pages of the London *Times* would have clarified the situation somewhat. The Peruvians held the high ground around the city of Tarapacá and did win an impressive victory on November 26, but they could not hold the city or the province for long. As a Chilean diplomat put it in late December, the "entire province of Tarapacá is in our power."[26]

Other metropolitan newspapers in the United States followed the example of the *New York Times*. By late December 1879 several of these newspapers began to report on the war. Their coverage was brief and sporadic at best. However, the *New York Tribune*, the *Boston Globe*, the *Washington Post*, the *Chicago Tribune*, the *San Francisco Examiner*, and the *San Francisco Chronicle* did establish that news from South America could spread across the nation via telegraph lines and that the Chilean role in the war and its aftermath was of some national interest. These dispatches that confirmed Chilean domination in the struggle were short (four to eight lines), datelined in Buenos Aires or some other South American city distant from the scene of the conflict, and about two or three weeks old. The piecemeal effort of the *New York Times* was better, but not by much. The New York newspaper carried a brief account on December 14 supplied by the Peruvian chargé d'affaires through a telegram from Lima, and a more detailed discussion arrived from Panama by mail steamer for the December 24 edition of the newspaper in which the limited nature of Peru's victory was placed in the context of Chile's ultimate control of the province. Another telling point emerged in a description of Lima's hurried preparations for a Chilean invasion, an invasion that would have been highly unlikely if Peru had retained control of Tarapacá.[27]

As the contrasting versions of the battle for Tarapacá indicate, both Peru and Chile took advantage of opportunities to put into print material favorable to their respective causes. Both of these countries and Bolivia circulated justifications for their actions in the United States, Great Britain, France, and other European nations in the months after the outbreak of the war, and this propaganda campaign soon expanded to include newspapers.[28] As early as July 1879 Peruvian diplomat José Antonio de Lavalle insisted in an interview that appeared in the *New York Times* that Chile had no chance of victory because of its relatively small population in comparison to the combined populations of Peru and Bolivia and because of its weakened economic condition brought on by declining prices for wheat and copper.[29] One day after de Lavalle's interview, the *Peruvian Mail*, "a new journal published in English in Lima," printed for general public consumption letters of protest from six consular representatives in Peru (including U.S. and British) concerning the Chilean Navy's bombardment of Iquique. While the letters were genuine, they were probably not written for publication.[30] And in a bizarre and boastful message, the Peruvian chargé in the United States issued what was supposed to be a point of clarification that may have added to the confusion: "No credit should be given to the intimation that the country is recruiting its forces by the employment of cannibal savages, to be armed with poisoned arrows. . . . Furthermore, Peru is not recruiting at all, the Government at Lima having issued a circular to the Prefects of the various departments that no more troops are needed."[31]

One of the most articulate spokesmen for Peru in Great Britain was Clements Markham, a student of Andean history who authored an extensive and slanted book on the War of the Pacific that appeared in 1882 after most of the fighting was over but while diplomatic negotiations were still under way. Openly stating a preference for Peru, Markham charged that Chile started the war on the basis of an unjustifiable claim to the Atacama and then employed brutal methods in the fighting. The Peruvian officers, from Markham's perspective, were courageous descendants of the Spanish Empire and their enlisted men were the offspring of the admirable Inca and Aymara peoples, while, by contrast, the Chilean's efforts indicated "rapid moral deterioration all for the sake of procuring some manure that belongs to her neighbors."[32]

THE CHILEAN PROPAGANDA EFFORT IN THE UNITED STATES

The focus on press coverage of the war and its aftermath now shifts from a comparison of the *New York Times* and the London *Times* to a study of the metropolitan newspapers in the United States. In efforts to counter negative propaganda, Chile had the advantage of access to the telegraph connection to London via Transandino, Western and Brazilian, and Brazilian Submarine, which it made good use of in 1879 and 1880.[33] Chile's communications with the United States were more difficult, however, and Santiago's diplomats were concerned that Peru, even though headed toward defeat on the battlefield, was gaining the advantage in press, political, and business circles in Washington and New York. Chilean affairs had become a political issue in the United States by the early 1880s. Leaders in Santiago feared that the United States would intervene in the postwar settlement to push for conditions favorable to Peru. In short, Chile was concerned that in spite of its victories on the battlefield, it might lose at the peace table. The openly pro-Peruvian stances of the *New York Tribune* and the *New York World* pointed in that direction. Chile's chief diplomat in Washington, Marciál Martínez, was a close observer of the interplay of the press, business, and politics. In letters to the government in Santiago in October and November of 1881, Martínez noted with concern the expanding influence of the Peruvian Company, a New York corporation with French connections that held claims to nitrate and guano properties in Peru's southern province of Tarapacá (which Chile had taken in 1879). A promoter for the Peruvian Company, Jacob Shipherd, had the audacity to send, according to historian Fredrick Pike, "an incredibly arrogant letter to Martínez" claiming that even though Chile had prevailed on the battlefield, its hopes of holding Peru's southern provinces were doomed. Martínez concluded that Shipherd had close ties to U.S. secretary of state James G. Blaine. The flamboyant Blaine was a prominent figure in the national press during the corruption-ridden "Gilded Age." He had been Speaker of the House, a senator, and a Republican candidate for the presidency (he ran and lost in 1884), but critics harped on his ties to railroad companies and his role in unethical if not illegal lobbying practices. The Chilean minister was aware of Blaine's reputation for behind-the-scenes dealings with corporate interests and expressed his concerns about both Blaine and his assistant secretary, Robert Hitt, to Chilean officials in Santiago: "The Peruvian Company claims may lead the United States to deny Chile's land acquisition rights. Every time I raise the matter with Hitt and Blaine, they are evasive."[34]

Throughout the fall of 1881 the Chilean diplomat found U.S. policy toward Chile to be shrouded by rumors and deceptions that overlay what appeared to be a poorly disguised drive for domination of the resources of the Atacama through the pretense of protecting the territorial integrity of Peru.[35] In a private letter he wrote a couple of years later, Martínez explained how he'd been afraid to address Blaine's policies in major newspapers across the United States. In addition, information forwarded to the press by Blaine's ally, Shipherd, gained extensive coverage in the press, while, at the same time, neither the U.S. public, nor politicians, nor reporters for its major newspapers understood the War of the Pacific from the Chilean point of view. In order to present his government's position to the political leaders and the public in general in the United States, Martínez assembled a team of writers and translators. These Chilean agents, directed by Martínez, issued a series of publications and press releases designed to explain their perspective on the war and the ensuing negotiations. Martínez and his staff produced a pamphlet containing a collection of these statements. In his preface, Martínez stated its purpose: "to enlighten the public opinion of this great nation, so friendly to Chile, to correct errors and misconceptions, which are so freely circulated owing to the very great liberty of the press."[36]

Operating in a media context in which Peru had more favorable press commentary, the Chileans focused on cultural and political symbols of importance to U.S. leaders and the general public.[37] For example, the pamphlet cited the nation's "geographical isolation" from international conflicts—an advantage symbolized by the revered Monroe Doctrine. It singled out the *New York Times* as a "conscientious organ" that counseled against intervention in the affairs of other nations. The *New York Times* also pleased Martínez with its expose of "a gigantic lobby scheme" that pushed Shipherd's claims for an alleged $1 billion to be guaranteed by the U.S. government. The *Times* also editorialized about the "firm but dignified language" employed by Chilean foreign minister José Manuel Balmaceda in asserting Chile's reasonable expectation of holding on to the nitrate-rich province of Tarapacá to secure reparations for the war it had successfully prosecuted against Peru.[38] In addition to the *Times*, other New York newspapers featured coverage that favored Chile and criticized the free-wheeling U.S. representative in Lima,

Stephen Hurlbut, who took a decidedly pro-Peruvian position. The *Sun*, the *Herald*, and the *Tribune* joined with the *Times* to support the Chilean position. Outside New York, the Chilean perspective gained widespread attention in the metropolitan dailies. Over a five-day period (November 25–29) the *Boston Globe*, the *New Orleans Picayune*, the *Atlanta Constitution*, the *San Francisco Examiner*, and the *Chicago Tribune* carried similar front-page stories based on telegraphic reports that a "gigantic lobby" favoring Peru was at work in Washington.[39]

The national exposure of these activities discredited Blaine's case for Peru and signaled a propaganda victory for Chile. Of course Martínez's propaganda was not the only factor in this political debate. The discussion of the fate of the nitrate and guano territories had become a partisan issue in the U.S. Congress by this time, and news stories in a given newspaper tended to reflect the paper's political orientation to some extent. The Democratic Party's criticism of the Republican secretary of state was echoed in the usually Democratic *Atlanta Constitution*, the *New Orleans Picayune*, and the *San Francisco Examiner*, but the *New York Times*, the *New York Tribune*, and the *Chicago Tribune* abandoned their generally Republican preferences in this case.[40] There is no claim here that the Martínez campaign by itself manipulated U.S. newspapers, but it is clear that he furnished them with selected information and well-crafted arguments that referred reverentially to the Monroe Doctrine (as Vicuña Mackenna had done in 1866) and that thereby fit comfortably with material that often appeared on the editorial pages across the nation and across partisan barriers. Several newspapers adopted the same positions advocated by Martínez in his pamphlet. Martínez made an explicit argument against the claims of the Peruvian Company, which, in this view, was operating in the interest of the defeated Peruvians. Furthermore, the Monroe Doctrine was being distorted by Peru and its supporters in the United States: "The Monroe Doctrine is being made a plaything for the amusement of triflers and fools. The tactics of the Peruvians and Bolivians in their attempts to interest the American public by arousing their greed, sordid ambition, and all the lowest human passions, is in the highest degree offensive to the nation and should be spurned with contempt and disgust." Martínez felt a sense of accomplishment in his propaganda work and reported that he received about fifty letters of congratulations from newspaper editors and politicians in the United States.[41]

Ultimately Chile overcame the machinations of Blaine-Shipherd and gained an extensive territorial settlement at the expense of Peru and Bolivia and over the protest of several political and business interest groups in the United States. Under the Treaty of Ancón, signed by representatives of Chile, Peru, and Bolivia on October 10, 1883, and ratified by the signatories on March 10, 1884, Chile took possession of the mineral-rich coastal provinces of Antofagasta from Bolivia and Tarapacá from Peru. Two additional Peruvian provinces to the north of Tarapacá—Tacna and Arica—were to be held by Chile for ten years, after which a plebiscite would determine their disposition.[42]

Chile's insertion of its perspective into the U.S. press coverage of the War of the Pacific exposed some of the limits of and gaps in the telegraphic network that underlay the new information system in the Americas. The newspapers' use of information that moved by submarine telegraph and steamship in the War of the Pacific was neither efficient nor effective, particularly when compared with the work of journalists and editors in the Franco-Prussian War of 1871.[43] It should be noted, nevertheless, that both the London *Times* and the *New York Times* improved from the bumbling coverage of the battle of Iquique to the more coherent reports on later battles. Both newspapers relied on information supplied by other sources—Lloyd's insurance company, merchant houses, other newspapers, and the governments of the belligerent nations—rather than taking on the responsibility and expense of sending correspondents to the scene. Indeed, expense may have been the crucial factor; this, at least, is what the brevity of the telegrams from the west coast of South America seems to indicate. The comparison of the London *Times* coverage with that of the *New York Times* helps to document the initial defects in the information flow from international sources into the U.S. communications system. Other big-city newspapers in the United States followed the example of the *New York Times* in its sparse, tentative, and sometimes flawed coverage of the War of the Pacific.

This collection of information and opinion from a variety of sources opened the way for the belligerents to place their messages alongside reports from journalists and other parties not directly involved in the fighting. The impact of this propaganda is difficult to gauge, but the new means of communication clearly provided opportunities for less powerful nations to insert their points of view into the print media of the United States and Great Britain. Astute diplomats such as Chile's Marciál Martínez grasped the importance of public debate in the United States and used the print media to attempt to influence this debate. The impact of Chilean (or Peruvian) propaganda in the United States is difficult to ascertain, but it is clear that Martínez, following in the footsteps of Vicuña Mackenna, explored the context of U.S. foreign policy formulation and deployed arguments that resonated well in U.S. politics and culture. In the era of the new communications system, propaganda was to play a large role in international conflicts.

DIPLOMATIC HUMILIATION VIA THE CABLES

An important dimension in the relationship between telegraphy and diplomacy allowed political considerations to be inserted into negotiations in unexpected ways with sometimes surprising consequences. The case to be considered here was a unique and important episode in the tense negotiations between Chile and the United States in the postwar period. Government officials who had access to the most recent and most accurate information held a decisive advantage. The use of the telegraph under these conditions did not automatically give the upper hand to the most powerful nation. Secretary of State Blaine had taken a firm stand against Chile in the short-lived administration of President James Garfield. The Chileans resented U.S. involvement in the postwar settlement and took advantage of an opportunity to discredit the two representatives sent to South America by Blaine. The assassination of Garfield brought Chester Arthur to the presidency and Frederick Frelinghuysen to the head of the State Department. Unfortunately for Blaine's two handpicked emissaries, William Trescot, an experienced diplomat, and Walker Blaine, the secretary's son, Frelinghuysen replaced the elder Blaine as secretary of state on December 19, 1881, after the pair had embarked on their mission. Frelinghuysen, a cautious strategist, opposed Blaine's assertive stance against Chile. The result was a balanced approach that sought compromise between Chile and Peru.[44]

Trescot and the younger Blaine did not know of this change in policy at the time they reached Chile in January 1882, but Chile's foreign minister, José Manuel Balmaceda, was acquainted with dynamics of policy and politics in Washington. A telegram from Chile's observant minister in Washington, Marcial Martínez, conveyed an accurate account of the alterations in U.S. policy that reached Balmaceda before the U.S. State Department's notification finally caught up with Trescot and Blaine by oceangoing mail steamer. It was because Martínez had read the publication detailing Frelinghuysen's new instructions in Washington (imposed by the U.S. Congress largely because of bickering within the Republican Party) that he and Balmaceda were able to engineer this coup. The use of submarine telegraph lines was crucial. Martínez sent his message by way of Paris, Lisbon, the

Brazilian Submarine–Western and Brazilian connection and then overland by the Argentine government system and Transandino. The expensive long-distance communication was much quicker than oceangoing transport and enabled Balmaceda to confront his visitors with their new instructions based on a new policy, both of which were unknown to the two norteamericanos. Furthermore, Balmaceda informed his two visitors that a translation of the new instructions from Washington would appear in the Chilean press before the slow-moving mail steamer would reach Valparaíso. Caught by surprise, both Trescot and Blaine were humiliated. The wily Chileans held the diplomatic high ground. The primary aim of the Trescot-Blaine mission, the preservation of Peruvian territorial integrity, became a lost cause. Balmaceda insisted that Tarapacá be ceded by Peru to Chile. Trescot and Blaine complied. Without U.S. backing, Peru had to give way to the Chilean demand. Chile also gained the right to occupy the former Peruvian provinces Tacna and Arica for ten years, giving them the chance to exploit their resources. Chilean officials made quick and clever use of the international communications system, which contributed to their diplomatic victories over the United States and Peru.[45]

CONCLUSION: CHILE TRIUMPHANT IN THE 1880S

Improvements in communications not only contributed to a larger flow of telegrams between Chile and the United States but also enhanced the movement of information between two nations with a record of unfriendly to hostile relations. Chileans were very much aware of the reach of U.S. power and also of the vagaries and volatility of politics, the press, and decision making in their neighbor to the north. The U.S. press often lagged behind the London *Times*. Benjamin Vicuña Mackenna had scored something of a victory in his publicity endeavor in 1866 even though Secretary of State Seward decided to ignore the public demonstration of support of Chile at the Cooper Union meeting. Vicuña Mackenna's efforts to recruit technicians for the Chilean Navy met with hostile rejection. In spite of these discouraging results, Marcial Martínez revived the public-relations approach in 1881 by circulating the Chilean government's position on the territorial settlement following the War of the Pacific among key U.S. politicians and newspapers. While there is no indication that Martínez manipulated the decision making in Washington, it is clear that the cause he advocated benefited from his work and ultimately triumphed. Martínez joined Foreign Minister Balmaceda in 1882 to outwit U.S. envoys by the use of information derived from public documents in the United States. Vicuña Mackenna, Martínez, and Balmaceda located vulnerabilities in the mix of telegraphic communications, politics, press, and publicity in the United States and exploited them to the benefit of Chile. This same system of telegraphic communications and the mass-circulation newspapers in the United States would continue to play an important role in Chilean-U.S. relations in the last decade of the century, but it would bring results less favorable to Chile. In both cases, better communications did not lead immediately to better relations.

CHAPTER FIVE

European Intrusions, Domestic Disorder, and U.S. Armed Intervention

Central America in the 1880s

EMPEROR PEDRO II OF BRAZIL expressed his enthusiasm for technological advancement in a telegraphed message to President Chester Arthur of the United States on September 26, 1883. The Brazilian emperor hailed the opening of the submarine cables laid by James Scrymser's Central and South American that tied southern South America to the United States by way of a system of connections that ran from Chile via Colombia's strategic province of Panama and along the shores of Central America. This new electronic line enhanced the speed of communication between Rio de Janeiro and Washington. In the words of Pedro II, "The new channel of direct communication between the two countries via Valparaíso and Galveston will contribute to the maintenance of the happy relations existing between them, and to their mutual benefit."[1] Less than two years later Haitian-born rebel leader Pedro Prestán was executed by orders of an international tribunal in Panama for inciting an uprising that resulted in the burning of the city of Colón (also called Aspinwall). The blaze destroyed most of the structures in the city, took the lives of eighteen residents, and left much of the surviving population homeless. The Panamanian uprising threatened the communications network that Pedro II had praised. Prestán, unlike Pedro II, had no cause to celebrate submarine telegraphy, for it was the flurry of cables that tied together Colón, Panama City, Washington, New York, and Boston that brought to bear a show of force from the U.S. Navy. In little more than three weeks, at least nine warships and approximately two thousand marines and armed sailors drove Prestán from the isthmus and captured, killed, or dispersed his followers. The Colombian military finally seized Prestán in Barranquilla and returned him to Panama for trial. His public execution was by hanging.[2]

Emperor Pedro II's warm welcome and Prestán's gruesome demise were indicative of the wide range of outcomes possible from the use of the international telegraphy. The emperor, diplomats, other government officials, and commercial and agricultural interests were eager to see the benefits that improved communications would bring to Brazil. The political and business elites in other countries shared similar views. Argentina, Uruguay, Chile, and Peru were also prominent in their use of the new system. For Prestán and others from the lower end of the socioeconomic scale, the international communications system also held out many possibilities—from the propitious to the ominous. Places of strategic importance, such as the Panama transit route that connected the Atlantic and Pacific Oceans, were now tied much more closely to the centers of decision making in Washington, New York, London, Berlin, and Paris, where powerful government and corporate leaders used the system to venture into new fields and thereby extend technology and employment opportunities into regions that had previously been regarded as commercial backwaters in the global economy. On the other hand, these same governments and corporations could act promptly to safeguard their interests, which meant protecting their property from both peaceful protests as well as violent demonstrations. Up-to-date information on diplomatic discord and political upheaval presented new opportunities for the peaceful resolution of disputes but also gave spontaneous local movements that challenged these structures of power a more circumscribed and closely observed arena in which to pursue their agendas. In short, the new technology of information flow would bring changes, changes that would lead to a number of different developments with unforeseen consequences for the diverse peoples involved.

The new information system expanded into Central America at a time when regional rivalries and foreign intrusions—especially the French canal project in Panama—created a high level of political uncertainty for the United States. French, German, and Guatemalan initiatives made the limitations of U.S. influence in the strategically important isthmus obvious to leaders in Washington and New York. Under these circumstances the United States adopted a more aggressive attitude, while a nervous U.S. consul in Panama warned officials in Washington that conditions were moving toward potential disaster: "We sleep on the brink of a volcano."[3] The recently built cable connections played an essential role in the most extensive U.S. military intervention in the region in the nineteenth century.

SCRYMSER ENTERS CENTRAL AMERICA

James Scrymser returned to the cable business in 1881 with a large concept and sound financial backing. Nine years after Western Union purchased Scrymser's first cable company, International Ocean (connecting Cuba with Florida), the persistent New York entrepreneur announced plans to run a submarine line from Galveston, Texas, to Veracruz, Mexico, under the corporate entity known as Mexican Telegraph and then, in the second and larger phase, extend this cable line across the isthmus of Tehuantepec and along the Pacific coasts of Central and South America as far south as Peru. Scrymser formed a second company for this purpose: Central and South American. Scrymser would eventually challenge British domination of the international cable business in the middle and southern sections of the Western Hemisphere. Elements of impatience and brashness surfaced in Scrymser's methods of operation, but he was also careful to obtain financial support and political backing from leading figures in the business and governing establishments in New York and Washington. He secured investment advice and financial commitments from New York banker J. P. Morgan. He obtained endorsements from New York Mayor W. R. Grace, who also headed a well-known merchant house in Peru, and U.S. senator Roscoe Conkling, a powerful figure in both New York and Washington.[4] Probably the most noteworthy approval of his project came from the illustrious French businessman and promoter Ferdinand de Lesseps, who had only recently begun work digging an interoceanic canal though the jungles and swamps of Panama. De Lesseps wrote to Scrymser: "Central America is on the eve of witnessing great events, notably, the cutting of the Isthmus of Panama, which will give those regions an incalculable importance and will cause your system of submarine lines in the Pacific to become indispensable."[5]

The ambitions of Scrymser and de Lesseps produced an intersection of communications and transportation technologies that had large implications not only for the isthmus but also North and South America, Europe, and the Pacific. Scrymser's cables moved north to south, linking the telegraphs networks of North America and Europe with Central and South America. The overall movement of de Lessep's work was east to west by joining the Atlantic and Pacific Oceans. The French phase in the history of the

Panama Canal has received extensive coverage by a number of historians, including David McCullough, Gerstle Mack, Maron Simon, and Thomas Schoonover.[6] The point of departure for this chapter is the intersection of the cable with the canal. The opening of Scrymser's new line coincided with the early digging on the waterway. As the attention of business and political leaders focused on the economic potential of the canal and the enhanced strategic importance of Panama and the five republics of Central America, the electronic cables that reached this region acquired a large significance as well.

The placement of the cable from the west coast of Mexico along the Pacific side of Central America moved forward in late 1881. In October of that year, a confident Scrymser reported to Central and South American stockholders that India Rubber, Gutta Percha, and Telegraph Works of London would soon begin laying over three thousand nautical miles of cable at the rate of twenty-two miles a day. Central and South American already held concessions from the governments of El Salvador, Nicaragua, Colombia (for its province of Panama), Ecuador, and Peru granting it permission to lay submarine cables on their shores.[7] Scrymser was concerned about possible complications from the War of the Pacific. Although the naval engagements in this conflict had come to an end with the Chilean invasion of Lima in January 1881, Scrymser took the cautionary step of securing a commitment from the U.S. Navy to observe the work on the Central and South American venture and to provide "assistance within [the] means" available to the ships of the U.S. Navy stationed in Peruvian waters. On October 10, 1882, Central and South American announced that the telegraph circuit between Galveston and Lima was open.[8]

Scrymser had placed his cables in strategic positions that would enable him eventually to compete with the British lines on the east coast of South America (under the control of John Pender) for the communications traffic that ran between and among Lima, Santiago, Buenos Aires, Rio de Janeiro, Washington, and New York and that would also include the political and financial capitals of Europe. The arrival of Central and South American in Panama City on the Pacific side of the Colombian province of Panama had an immediate impact on another British cable business, West India and Panama, which had an office in Colón on the Caribbean. Established in 1869, the company connected with Jamaica, other British colonies in the West Indies, and Cuba. Its tie with its sister British company, Cuba Submarine, established links with International Ocean (Scrymser's first company that was absorbed by Western Union in 1872) and thereby opened the way to connections with the U.S. telegraph system and eventually to that of Great Britain and Europe. Because of technical and environmental problems and the downturn in the West Indian economy, West India and Panama struggled through its first decade. By 1880, however, the company had begun to show a profit, only to have its future clouded by the arrival of Scrymser's Central and South American. Much to the dismay of West India and Panama managers, they had to reduce its cable rates in order to compete with Central and South American on messages to and from the United States and Europe. Indeed, from 1881 to 1885, West India and Panama cable rates between Colón and London dropped by 50 percent. With the Central and South American cable station in Panama City, that tropical metropolis not only became a major port for the movement of people and goods from California to the east coast of the United States and the Pacific opening for the future Panama Canal but, by the 1880s, an essential link for the cable lines of Scrymser's company.[9]

Scrymser had the advantage in this competition. His plans for his company were based on charging lower cable rates, and in 1884 he reported to his stockholders an impressive record of profitability. At the company's annual meeting in New York City on June 3, 1884, the company reported that revenue from January 1, 1883, to June 1, 1883, was $66,170, while revenue for the same period in 1884 was $149,556. Although figures for the expenses of 1883 were not included in the report, they were available for early 1884 and amounted to $48,000, leaving a profit of $101,556 for the first five months of that year. The company's board of directors—which included financial and political luminaries J. P. Morgan, W. R. Grace, and William Hamilton—was pleased with these results.[10]

THE BARRIOS UPRISING AND THE UNCERTAINTIES OF CABLE COMMUNICATIONS

The arrival of Central and South American in Central America and Panama coincided with the eruption of destabilizing movements in the region. President Justo Rufino Barrios of Guatemala proposed the revival of a unified government for Central America, which earlier in the century had included Honduras, El Salvador, Nicaragua, and Costa Rica. Factional and regional disputes tore the nation apart in the 1830s, but Barrios concluded that the time was ripe to reunite the five republics under one flag. A vigorous modernizer, Barrios had overseen the growth of railroads, highways, telegraphs, schools, government buildings, and hospitals in Guatemala.[11] He became one of the first leaders in Latin America to use the new cable network to attempt to gain support in the United States and Europe.

The Guatemalan caudillo employed dictatorial practices at home and adopted an aggressive foreign policy within the region. With a base of operations in Guatemala, Barrios initiated his campaign to rebuild a Central American government. One important motive for this movement can be found in the implications of de Lesseps canal project in nearby Panama. This project stirred interest in the United States for a canal through Nicaragua. Barrios, with encouragement in 1881 from U.S. secretary of state James G. Blaine, began his campaign for regional unification on the assumption that a unified Central America would provide greater security for a Nicaragua canal. Barrios also saw the advantages in sovereign authority over the canal. He launched a program of persuasion and diplomacy in 1882 and gained the support of nearby Honduras and also El Salvador under President Rafael Zaldívar. This peaceful campaign, however, soon collapsed because of the opposition of Nicaragua and Costa Rica.[12]

Barrios then turned to more aggressive means. With weapons purchased in the United States and an elite corps of army officers trained in the United States, Spain, France, and Germany, he attempted unification by force. This offensive began with a dramatic flair. Barrios sent one of his officers to the national theater in Guatemala City on February 28, 1885, to interrupt a performance of *Boccaccio* with the announcement of the campaign. Citing the recent military-led unifications of Italy and Germany, Barrios proclaimed that Central America should follow these examples. The audience in the national theater, presumably made up of some of Guatemala's wealthiest citizens, greeted this pronouncement with resounding applause.[13]

Barrios also sent his proclamation to the four other Central American nations and attempted to communicate his goals to interested powers outside the region. He was successful with the first set of messages, which reached all four Central American governments. Again, Honduras joined with Guatemala, but Nicaragua and Costa Rica defied Barrios and refused to comply. Most importantly, Barrios's ostensible ally, President Zaldívar of El Salvador, hesitated and then unexpectedly appealed to Mexico and Costa Rica for advice. If Barrios entertained hopes for victory based on intimidation, these hopes were dashed. At this crucial point, international telegraphy played an important if unexpected part in this campaign. Zaldívar turned against the use of military force. He telegraphed Mexico's president, Porfirio Díaz, for help. Mexico had a long-term interest in its Central American neighbors, and Díaz was determined to block Barrios. Díaz, the head of the largest nation in the immediate region who had a much larger army than Barrios, ordered Mexican troops to the border to force Barrios to face the prospect of war on two fronts. A thankful Zaldívar appreciated this maneuver.[14]

Fully aware of the larger implications of his campaign, Barrios turned to international cables. He formulated an announcement to be communicated to the United States, Mexico, and the powers in Europe via the facilities of Central and South American.[15] The United States, Great Britain, France, and Germany were concerned about any movement that might affect work on the canal as well as trade in the area. A crucial factor in Barrios's efforts to communicate with these nations was the fact that the nearest international cable station in Central America was located in El Salvador's Pacific coast port of La Libertad (or La Unión). Presumably unaware of the El Salvador's abandonment of his cause, Barrios sent these messages on telegraph lines that passed through San Salvador and the hands of his former ally, President Zaldívar. At this crucial moment, the Salvadorean leader apparently used the communications system to his own advantage in the transmission of the message. Only a garbled, unconvincing statement went out on the Central and South American lines to Galveston, then into the Western Union system, and onto the front page of the *New York Times*. Barrios had hoped to arouse support in the United States and Europe, but this flawed missive did only harm. An apparently complete version of Barrios's statement appeared in the *New York Times*, but nearly a month after the original pronouncement and only one day before his death in battle. Even then the *Times*'s heading discredited the statement as a "Peculiar Decree Issued by President Barrios." A lengthy supportive article written by an unidentified "special correspondent" appeared in the *San Francisco Chronicle*, but its publication came too late. Datelined from Guatemala City on March 16, the article did not appear until April 5, a few days after Barrios's death.[16]

Barrios's propaganda efforts in the United States failed. The United States joined with Mexico, Nicaragua, and Costa Rica and expressed opposition to the Barrios movement. Of course, this opposition came from causes more extensive and profound than the failure of Barrios's messages to convey his motives and intentions. For example, officials in the government of President Grover Cleveland were concerned that this outbreak of hostilities would further weaken the chances for stability in Central America. Cleveland and Thomas Bayard, his secretary of state, did not agree with Blaine's intrusive diplomacy in Central America. Bayard used Central and South American lines to inform Henry Hall, the U.S. representative in Guatemala, that Washington did not approve of a "display of force on the part of any one or more States to coerce the others." Bayard did share Blaine's preference for Central American unity, but he rejected the use of force.[17] Hall telegraphed intelligence to Bayard that indicated neither the government nor the public in El Salvador favored Barrios, revealing that the movement for unity was limited largely to Guatemala and Honduras and, in terms of resources, mainly to the former. Hall used his influence in the region to lessen the chances for war, in effect, to block Barrios.[18]

Scrymser's Central and South American was a participant in this crisis not only as a conduit for messages—both clear and garbled—but also as a corporation with important assets close to the zones of possible combat. On March 17, the company telegraphed Secretary of State Bayard that it had been informed a Guatemalan vessel was en route to cut the company's cable that lay along the Pacific coast of Central America. Scrymser, who had carefully cultivated friendly relations with politicians and government officials Washington, appealed to both Bayard and William Whitney, secretary of the navy, in the recently formed cabinet of President Grover Cleveland. Both officials responded promptly. On the same day, Whitney telegraphed Commander Alfred Thayer Mahan, telling him to proceed immediately from Panama City to the waters off El Salvador in order to "protest against any attempt to cut the cables or interfere with their use." Also on March 17, Bayard sent Scrymser a reassuring message:

> A telegram is sent today to the United States Legation in Guatemala holding that republic responsible for injuries by its authority or with its connivance to cables of interests of the United States citizens in Central America. The *Wachusett*, now enrooted to La Union, will be duly instructed.[19]

Further reassurances came from government officials in Mexico, El Salvador, and Nicaragua. Central and South American secretary James Beard reported to the *New York Times* that he had received information that "the authorities" in those countries had "posted guards at the cable landings" of the company to protect them from any attempt at violent action.[20]

The Barrios episode came to an abrupt end. The Guatemalan president assembled his armed forces, invaded El Salvador, and enjoyed a victory at

El Coco on Salvadorean soil on March 31, 1885. His drive seemed destined to triumph. Two days later his army reached Chalahuapa, but Barrios, who took a position close to the front of his forces, fell, mortally wounded. The Guatemalan campaign to unify Central America collapsed with the death of its leader. Relatively brief telegraphed accounts of his death appeared in newspapers from New York to San Francisco. Barrios hoped to rally support for his cause in the United States via the international cable system but was unable to do so. Washington used the cable system to help isolate the Guatemalan leader. Meanwhile, Scrymser's company reported no damage to its facilities during this brief war.[21]

EUROPEAN INCURSIONS: THE EISENSTÜCK AFFAIR AND THE FRENCH CANAL PROJECT

Central and South American survived the abortive Barrios offensive unscathed, but this outcome did not alter the fact that Scrymser's company was one of the few U.S. corporations active in Central America and Panama, a region dominated by the trading inroads that had been made by Great Britain, France, and Germany. The United States not only trailed the British in trade in the region but also ranked behind Germany and France. The aggressive policies of the Germans and the French offer a context that helps to explain the mounting concerns among officials in Washington. The activist policies of the European governments and the energetic practices of merchants and other immigrant communities in the region had managed to undermine the influence of the United States, despite its geographic proximity to Central America and Panama.[22]

German merchants, diplomats, and immigrants were more active in the region than most U.S. observers realized. U.S. representatives in Central America, however, were aware of the rising German presence. George Williamson, who served as minister to Central America in the 1870s, and his successor Cornelius Logan in the early 1880s both saw Germany as the leading threat to U.S. interests.[23]

The extent of the German presence in Central America was dramatized in the "Eisenstück affair," an unusual incident that historian Thomas Schoonover has examined in greater depth than we can cover here but that had ramifications that proved important to later events. The Eisenstücks, German businessmen, became entangled in a personal dispute with a prominent Nicaraguan family. The dispute escalated to the point that Berlin sent six naval warships to Nicaragua. In a preemptive strike, German armed forces landed in the port of Corinto and confiscated weapons that could have been used against them. The Nicaraguan government paid a large indemnity and rendered a formal salute to the German flag. The intervention lasted only about three weeks and drew little press attention in the United States. With its status recognized, Germany withdrew its naval forces in early April.[24]

George Williamson initially agreed with the Germans, regarding the alleged offenses against the Eisenstücks in a context in which similar problems of mistreatment by local officials might affect U.S. businessmen. But the intensity of the German protest and the arrival of the German Navy changed Williamson's attitude.[25] Williamson admitted that he had erred in helping the Germans. Siding with the Germans in their complaints against Nicaragua coupled with German naval intervention "tended to create the belief . . . that we have abandoned the Monroe Doctrine."[26] The recent growth of Germany's commercial stake in Central America viewed alongside the surge of European imperial expansion around the globe created an ominous situation for the United States in its own hemispheric backyard.

If Germany was the most serious threat to the United States in the Central America region in 1877–1878, France leapt to the top of the list of foreign interlopers in 1879 with its grand designs for a transisthmian canal. The French under Emperor Louis Napoleon had intervened in Mexico from 1862 to 1867, but this venture met failure and defeat and claimed the life of would-be Mexican emperor Maximilian von Habsburg. Louis Napoleon's government was soon brought down in the humiliation of defeat by Bismarck's Germany in the 1870–1871 Franco-Prussian War. The new French government took no initiatives in Central America over the next few years, but private citizens and an incorporated business took on one of the world's great engineering challenges with the announcement of plans to build an interoceanic canal across Panama. The protagonist in this enterprise was the most famous canal builder in the world, Ferdinand de Lesseps, the proud father of the Suez Canal, which had opened in 1869. De Lesseps also benefited from the international reputation of his homeland. French merchants and diplomats preceded de Lesseps in the region, and French culture claimed a high level of respect. French merchants enjoyed success in Central American and Colombia, but French language schools, cultural institutions, and military missions also prospered.[27]

The French also aspired to rival the cable enterprises of Scrymser and Pender in the Caribbean area but moved only in spurts for about two decades. A French company began to plan for a transatlantic cable to reach the Caribbean in 1869, but it never got beyond the planning stage. A decade later, in the same year that de Lesseps initiated his canal project, a second company, the Société du télégraph entre la France et les Etats Unis d'Amerique, included connections with the West Indies and South America in its prospectus. This company also languished, but in 1888 Télégraphs sous-marines opened lines that connected France with the Americas.[28]

De Lesseps eventually discovered that the canal was not only a monumental undertaking but also an overwhelming challenge. Supported by a large amount of capital invested by the French people and a public-relations campaign that inspired investors' confidence, de Lesseps seemed to be embarked on another successful canal-building venture. But his corporate managers and engineers found that their experience in the desert sands of Suez was of no help in coping with the jungles, swamps, and mountains of Panama. In spite of inadequate preparation and unrealistic expectations, the Compagnie universal de canal transoceanique moved ahead. On February 1, 1881, de Lesseps in Paris received a telegram from Panama: "Work begun." This brief message marked the start of what became a monumental disaster. Mudslides, tropical rains, and flooding rivers were devastating for the project. Even more calamitous were the lingering illnesses and untimely deaths of thousands of workers and dozens of engineers and supervisors from malaria and yellow fever. The company struggled on and did make some progress in moving large quantities of earth and succeeded in opening a slice in the Panamanian topography. One of the managers' great errors in judgment, however, was to keep the true enormity of the task from the French public in order to protect the reputation of the enterprise and retain the confidence of the investors. The opening of the transoceanic telegraph made the transmission of forthright information over cables that connected Panama with France possible. Instead, the company maintained a steady stream of favorable propaganda in its bimonthly publication, *Bulletin du canal interocéanique,* and in press releases.[29]

Unlike the Eisenstück affair and the Barrios uprising, the French canal project penetrated the information flow in the United States, but the reaction was muddled and the policy implications were mixed. The widespread assumption was that the project would succeed, but the discussants did not see it as a problem between the U.S. and French governments. De Lesseps organized the project as a venture in private enterprise financed by stock investments. The French promoter portrayed his undertaking as the proper object for national pride, but neither the French Foreign Ministry nor the French Navy had direct involvement. The discussion in the United States, therefore, was mainly concerned with the implications of a French-owned private corporation operating the canal. The issue was muddled by the fact that some U.S. businessmen invested in the project and several figures prominent in international commerce endorsed de Lesseps's plans. James Scrymser, W. R. Grace, and Charles Flint saw the canal as an opportunity to expand U.S. trade in Latin America, even if much of it had to move through a French canal.[30]

However, the administration of President Rutherford Hayes (1877–1881) and a large cohort in the business community took the opposite view. Hayes insisted that the U.S. government was obligated to protect the canal from damage that could result from armed conflict, whether it originated with a great power or from local instability. To stand aside under such conditions and to allow France to control the defense of the canal would invite a violation of the Monroe Doctrine and the security of the United States. Hayes was blunt: "It is the right and duty of the United States to assert and maintain such supervision and authority over any interoceanic canal across the isthmus that connects North and South America as will protect our national interests."[31] John Kasson, a former Republican congressman from Iowa and U.S. minister to Austria-Hungary at the time, took an even more belligerent stance: "To suggest control there by a naval power of Europe, is to suggest war with the United States. The connecting water-line between our Atlantic and Pacific coasts can, no more than the Pacific Railroad, be allowed to come under European control." Hayes, Kasson, and other public figures such as James G. Blaine sparked a public debate that came to focus not on overt opposition to de Lesseps but on the construction on an alternate route across the isthmus by the United States, perhaps a canal in Nicaragua or a combination of rails and ships across Mexico's Tehuantepec.[32]

Hayes and Kasson used strident rhetoric, but there was an underlying formal obligation for the United States to protect a canal across Panama. In 1846 the United States and Colombia had signed the Bidlack-Mallarino Treaty by which the government in Washington agreed to protect any transisthmian form of transportation. Manuel María Mallarino, Colombia's secretary of foreign relations, had entered the negotiations with the support of the Bogotá government not out of a sense of subservience to the United States but with a determination to use this special relationship to help fend off British and French diplomats and merchants. Panama had a record of social instability dating to the 1850s. The California gold rush made the isthmus a busy transit zone, and the intermingling of gold seekers and Panamanians created tensions that resulted in the Watermelon Riot of 1856, which took the lives of about twenty people and destroyed much of Panama City. In the aftermath, the United States carried out a brief armed intervention to help restore order.[33]

The de Lesseps canal project, along with the memories of the 1856 riot in Panama, the 1879 German intervention in Nicaragua, the Barrios uprising in early 1885, and the concerns expressed by Hayes and Kasson coupled with the obligations imposed on the United States by the Bidlack-Mallarino Treaty meant that the 1885 eruption of unrest throughout Colombia and within Panama created a crisis for the new Cleveland administration in Washington. Colombia had an unstable political history, and in 1885 Liberal Party president Rafael Núñez initiated a push for political centralization in a nation accustomed to federalism. Certain groups objected to Núñez's reforms, and in some cases violently opposed them. Panama especially was in a rebellious mood. Isolated from the rest of Colombia by the rugged northwestern corner of the Andes Mountains and the swampy lowlands of Darien, Panama inclined toward separatism. This centrifugal inertia was aggravated in the 1880s by the French canal company. The company's work force numbered 19,243 in October of 1884, and uncounted drifters hovered around the constructions sites in hopes of finding employment. Living conditions even for the employed canal workers were bad. The company did not provide commissaries or company stores, and local merchants tended to charge exorbitant prices for necessities. The small Panamanian elite, largely white with ties to European and U.S. businesses, was concentrated in the ports of Panama City on the Pacific and Colón on the Caribbean.[34] The large gap between the wealthy and the poor and the influx of West Indians seeking jobs created an unstable social environment exacerbated by Colombia's political turmoil. U.S. consul Richard Anderson's warning about a volcanic eruption of social unrest was correct. He predicted that the disturbances could provide an opening for "vile characters attracted here by the works of the Canal Co."[35]

REVOLUTIONARY UNREST AND U.S. ARMED INTERVENTION IN PANAMA

The month of April 1885 witnessed one of the most dramatic naval and military utilizations of submarine telegraphy in the relatively short history of the system. The small city of Colón, located on the north coast of Panama, was one of the busiest ports of the Caribbean. In recent decades, Colón and the Panamanian isthmus had become one of the world's most strategically valuable locations because of the expansion of the movement of goods and people between the Atlantic Ocean and the gold-rich communities of California and also because of French effort to build an interoceanic canal. Toward the middle of the day on April 1, 1885, the city of Colón burned to the ground; eighteen lives were lost and perhaps eight thousand people were left homeless.[36]

The conflagration was the outgrowth of civil disorder in Colombia (Panama was a province of this country until 1903). The twists and turns of late nineteenth-century Colombian politics are too intricate for full explanation here, but it is necessary to point out that in 1885 the bloody civil war in that country had an impact in Panama. Colombian government forces were stretched too thinly along the isthmus. Two insurgent leaders seized their respective opportunities. On the Pacific coast in mid-March, General Rafael Aizpuru staged an uprising and took control of Panama City. When the government's already beleaguered forces moved across the isthmus to Panama City, a small revolutionary band led by Pedro Prestán, an itinerant Haitian who had somehow worked his way into Colombia's political struggles, took control of Colón. For about two weeks Prestán's force grew, intimidating the entire community, and on March 29 the rebels took six U.S. citizens hostage in an effort to seize a cargo ship carrying weapons and ammunition. Foiled in this attempt, Prestán faced government troops that on the morning of April 1 defeated him on the outskirts of Colón. In the frustration of defeat, Prestán and his followers set the fire that devastated the city. In the confusion, the rebel leader escaped.[37] This outburst of disorder in Panama placed in jeopardy one of the transportation and communication centers in the Western Hemisphere.

Early warnings of the troubled times were very much in evidence in the Central American region (although a political province of Colombia, Panama was geographically a part of Central America) in the form of telegraphed messages spelled out to U.S. naval and consular officials. As early as January 18, 1885, Commander Lewis Clark of the USS *Alliance* reported deteriorating political conditions in Panama and a potential threat to U.S. property there.[38] Two months later at the urging of Scrymser of Central and South American, the U.S. Navy ordered Commander Alfred Thayer Mahan to take his ship, the *Wachusett*, north from its post off Panama to protect the cables of the telegraph company that were threatened by the Guatemalan government's planned invasion of El Salvador. Mahan's expedition apparently had a calming effect, but the disintegration of political order in Panama soon brought Mahan back to his original position.[39] In the meantime, Commander Theodore Kane's gunboat *Galena*, stationed off Colón, gained intelligence concerning conditions on the isthmus. Both Kane and Consul R. K. Wright in Colón reported on the collapsing situation to Washington.[40]

In spite of these early warnings, the magnitude of the crisis caught officials in Washington by surprise. On March 31 Pacific Mail Steamship requested that the U.S. Navy rescue its ship, the *Colón*, from the grasp of Prestán and his followers. William C. Whitney, secretary of the navy, responded immediately, issuing telegraphed orders to several U.S. warships including the *Galena* under Commander Kane in the port of Colón (not to be confused with the steamship of the same name). The U.S. Navy was already beginning to mobilize when late in the day on April 1 Consul Wright cabled the news of the burning of the city Colón to Washington. Overnight, Whitney telegraphed the Brooklyn Navy Yard, ordering it to bring "all available Marines . . . in readiness for immediate departure." The *New York Times*, the *New York Tribune*, the *Philadelphia Inquirer*, and the *Boston Globe* received telegraphed accounts of the conflagration of April 1 and placed them on their front pages the following day.[41] Aware of the growing public concern, Whitney telegraphed Pacific Mail Steamship to say that U.S. forces would have the transit from Panama City to Colón open in the "shortest possible time" and, in the same telegram, requested space on one of the company's steamships for U.S. troops.[42]

Over the next few days a flurry of Navy Department telegrams on Western Union wires to New York, Philadelphia, Boston, Providence, and Portsmouth (New Hampshire) gave orders for the navy's last large contingent of old-style wooden-hulled gunboats mixed with a few new iron and steel vessels to transport of force of more than two thousand marines and armed sailors to Panama City and Colón. Commander Bowman McCalla of the Marine Corps assembled his troops in New York, relying on the telegraph to stay in touch with officials in Washington and also to notify his men of their new mission. The poorly equipped U.S. Navy did not have transport ships, so it turned to private shipping companies. Pacific Mail Steamship did its part by carrying two hundred men on its *City of Pará* and six hundred on the *Acapulco*. These and the other ships did not arrive as a modern task force but rather in an irregular stream of war vessels and transports, reaching Panamanian waters between April 7 and April 15. Probably the most significant points in this movement of men and equipment occurred when the gunboat *Shenandoah* arrived in Panama City's harbor on April 7 and when the *Alliance* reached Colón the next day. On April 11 U.S. forces reopened rail service between Colón and Panama City.[43] The United States achieved a show of force of the isthmus within ten days, thanks to the interconnected international telegraph systems run by Scrymser and Pender as well the domestic lines of Western Union.

Once ashore, the U.S. forces established control of Colón but also discovered that they had not secured the situation in Panama City. Commander McCalla's troops, most of whom arrived on the *Acapulco*, faced the possibility of nasty fighting in the streets of Panama City. Aizpuru's forces began to erect street barricades that would have blocked McCalla's access to the Central and South American telegraph office. Faced with possibility the losing the ability to quickly communicate with Washington, McCalla decided to establish military occupation of the city. He informed the overall field commander Rear Admiral Jouett (who was on the other side of the isthmus in Colón) by telegram of his decision to move his troops into Panama City and received his commander's approval, also by telegram. In addition, McCalla used the local telegraph lines to communicate his requests for additional troops and equipment. The marines and blue jackets completed the operation on the afternoon of April 24 without incident.[44]

Officials in Washington were frustrated at times by the limitations of telegraphic communications with their troops on the isthmus. However, a subsequent dispute between McCalla in Panama City and Whitney in Washington indicated that there were some periods when centralized command was in place. Promptly informing his superiors of his action, McCalla received (through the lines that he had prevented Aizpuru from taking control of) a rebuke from Secretary Whitney, who deplored the commander's "interference" in local affairs. Officials in Washington were concerned about maintaining good relations with Bogotá. Questions regarding the extent of military actions in Panama continued to be the subject of debate in Washington and New York for several days. McCalla, whose efforts encountered criticism from several directions, sent coded telegrams to Washington to refute allegations from the U.S. consulate in Panama concerning the operation. On the other hand, Central and South American made clear its approval of the presence of U.S. troops by offering to send personal messages of marines and sailors to families back home free of charge.[45]

Scrymser's expression of corporate generosity quickly turned to cries of anguish when, on the night of April 25, McCalla, under pressure from Washington, withdrew his occupation force from Panama City. With a crucial station on his cable line to South America unprotected, Scrymser sounded an alarm in telegrams to Bayard and Whitney:

> The town is in the hands of the rebels. . . . The consequences cannot be foreseen. I have telegraphed as follows to the superintendent of the company in Panama: "Demand in writing from the American Consul or commandant of the United States forces protection to our property and men. At present the United States have by treaty all the rights and obligations that Colombia has, and must be held responsible for damage done to American property or for the failure to keep open communications."[46]

In the meantime, McCalla and Aizpuru began negotiations, and, when Colombian government troops arrived on the isthmus on April 28, the crisis came to a close, much to the relief of officials in Washington and corporate leaders—particularly Scrymser—in New York.[47]

Left: Commodore John Grimes Walker of the U.S. Navy in 1885 was one of the first to grasp the importance of both international cable communications and the U.S. public's reaction to the press coverage of the intervention (Stanley Waterloo, ed., *Our Living Leaders*. [Chicago: Monarch Book Company, 1896], 457). Right: Commander Bowman McCalla led the U.S. Marines in the 1885 intervention in Panama (James Rankin Young, *Thrilling Stories of the War and Returning Heroes* [Chicago: Providence Publishing Company, 1898], 128).

The execution of Pedro Prestán; he incited the ill-fated revolt and was hanged from a rope suspended between two telegraph poles, symbolic of the role of telegraphic communications in the intervention (National Anthropological Archives, Smithsonian Institution, Washington, D.C.).

MILITARY INTERVENTION, THE PRESS, AND THE U.S. PUBLIC

The Panamanian intervention brought attention to international telegraph cables as essential tools for naval and military operations, but this same episode also revealed another potential role for modern communications. Much like Scrymser, Commodore John G. Walker, head of the Navy's Bureau of Navigation, sensed the importance of the attitudes in the press and public about the intervention. Walker, who hoped to establish a larger role for the navy in similar situations in the future, was explicit in his confidential note to McCalla:

> It is of considerable importance in sending off such an expedition as that which you are to command to the isthmus that we should keep the country with us in the matter and for that reason you, with such officers to assist you as you think most judicious, should keep the Department informed of what occurs, how things are progressing and how they look to you, explain the political situation, go into the whole subject of the isthmus troubles as far as you can, and do not spare the telegraph in

> sending us information. We want to have everything of any importance in the way of information that it may be given out to the press, and the people kept in accord with the Department. Of course we must also avoid any appearance of overdoing the thing or trying to make capital.[48]

Walker's concerns revealed an awareness of what would later be called "public relations." The Navy Department wanted the support of the public without "overdoing the thing." Walker sought a fine balance: favorable coverage in the press but no heavy-handed propaganda. His memorandum to McCalla also indicated an awareness of the possibility that newspapers in the United States could present information critical of the intervention. With only two lines of telegraphic communication open between Washington and Panama (Central and South American's line via Mexican Telegraph along the western end of the Gulf of Mexico and West India and Panama and Western Union's linkage from Cuba to Florida), a few hostile commentaries from journalists at the scene of action might have a telling effect in the United States.[49] But, as it turned out, the press coverage was supportive and often patriotic. The *Chicago Tribune* and the *San Francisco Examiner* carried identical accounts of McCalla's occupation of Panama City (both datelined April 24 from that city): "At half past 2 o'clock this afternoon the United States forces appeared on the scene as if by magic, three taps on the drum being the signal by which they started. Three columns entered the city, and had full possession in about ten minutes, knocking down the barricades as they passed through the streets to the call of the bugle. The Columbian [*sic*] guards withdrew into the cuartels. This splendid coup de main has undoubtedly prevented another disaster similar to that of Colon."[50] The *Boston Globe* and the *Philadelphia Inquirer* provided supportive coverage. The *New Orleans Picayune* also praised the operation and editorialized against the withdrawal of McCalla's forces.[51]

One reporter, Irving King of the *New York Tribune*, accompanied U.S. forces and colored his coverage in favorable, even jingoistic terms. However, King also noticed signs of hostility toward the marines and sailors, and he found continued support for Prestán: "Many of the natives here are sympathizers with the rebel Prestán, or of the remnants of his gang, and are not particularly well disposed toward the 'Gringoes' as they call the Americans. It is not safe to go about the city at night unarmed, as all the natives carry revolvers, and many of them would count it a holy and patriotic thing to kill one of the Americanos."[52]

The execution of Pedro Prestán on August 18, 1885, was the culmination of the Panamanian drama. He escaped Panama before the arrival of the U.S. Navy in force and continued his rebellious exploits until captured in Barranquilla by the Colombian military in late July. He was transferred to Colón for military trial, and his conviction by the court for setting the city ablaze led to the most severe sentence. His death by hanging was witnessed by a large, restive crowd. Prestán's exploits made him a notorious brigand to some but a heroic character to others in Panama's diverse society—"a young mulatto with sharp features and a keen look."[53] Even his death was the subject of dispute and exaggeration. There is no doubt that his public execution was a spectacular event, but the accounts vary as to the methods involved. According to a young U.S. naval officer, Robert Coontz, Prestán was hanged by a noose attached to a telegraph cable that "was stretched between two telegraph poles across the railroad tracks." He stood atop a railroad car that, on orders of the executioner, moved out from under his feet and left him dangling from the noose attached to the telegraph cable.[54] This version of Prestán's demise contained a symbolic element. In a sense Prestán's fate was sealed by the strategic and economic importance of the rail lines that crossed the isthmus and the telegraph cables that brought U.S. forces so quickly to the scene. Similar local or regional rebellions in previously isolated areas were threatened by the arrival of modern transportation facilities and communications technology that incorporated them into the new international economy. Corporate leaders and government officials were prone to intervene to protect these installations from threats of disorder.

CONNECTIONS WITH LATER EVENTS

Unlike the Eisenstück episode in Nicaragua and the Barrios campaign to unify Central America, the U.S. intervention in Panama received front-page attention in the press in 1885. Nevertheless, it is often slighted by historians, who instead concentrate on subsequent events such as the 1895 Venezuelan boundary dispute with Great Britain, the Spanish-American War of 1898, and the U.S.-supported Panamanian revolt for independence from Colombia in 1903. Yet there were individuals and institutions connected with the 1885 intervention that played prominent roles in these later events and who applied lessons learned in Panama to these later events. The 1903 Panamanian revolt was organized and implemented by French engineer Felipe Bunau-Varilla and Panamanian doctor Manuel Amador Guerrero. These two men met soon after the young engineer arrived on the isthmus in early 1885. Bunau-Varilla took care to protect the doctor during the violent moments of Prestán's uprising. Both men witnessed the U.S. intervention, and its strategic and tactical implications could not have escaped them. Their relationship persisted over the years and, in 1903, they made extensive use of telegraph cables both to arrange and carry out the bloodless revolt that separated Panama from Colombia and to secure exceptionally rapid diplomatic recognition and military support from the United States.[55]

The significance of the intervention also captured the attention of W. L. Scruggs. Stationed in Bogotá as the U.S. minister in Colombia, Scruggs labored to maintain good relations with the administration of President Núñez. Scruggs was successful for the most part. Bogotá agreed to the U.S. intervention under the Bidlack-Mallarino Treaty, but this acceptance of naval action was formally communicated between Washington and Bogotá *after* the intervention was under way. An obvious problem was the absence of a direct cable between Bogotá and Panama, a problem complicated further by the breakdown of the telegraphic connection between Bogotá and Washington for three months during this crucial period.[56] While U.S.-Colombian relations remained on good terms, Scruggs realized that the conduct of diplomacy by 1885 depended on the quickness and reliability of electronic communications. He learned this lesson well and added to it an understanding of the growing importance of the print media in international affairs. When he left the U.S. diplomatic service and became a publicity agent for Venezuela in the 1890s, Scruggs drew on the traditional art of pamphleteering and the new technology of the wire services in the United States—especially their connections with popular big-city newspapers. His publicity campaign to protect Venezuela against alleged British expansionism entered the mainstream of the news media and public discussion in the United States and is the subject of chapter 8.

The U.S. Navy was the arm of the government that carried out the intervention. In the 1880s it was undergoing changes, both in terms of technology with the decision to build a world-class, steel-hulled battle fleet and in terms of strategic thought and naval doctrine with the creation of the Naval War College. The intellectual and historical underpinnings of these changes were reflected in the lectures and publications of Alfred Thayer Mahan. Events on the isthmus in 1885 had a major impact on Mahan's thinking. He saw the Panama Canal as crucial to the security of the United States, whether it was in the hands of the French or another foreign power. Ideally, in Mahan's view, the canal would be under the control of the United States.

Mahan's main doctrinal interests were the sources and projection of national power, and a strong navy was paramount in his calculations. A large part of the navy's responsibility was the protection of the isthmian canal to safeguard the nation's shipping as well as ensuring the mobility of its warships. Mahan's command experience aboard the *Wachusett* in 1885 weighed heavily in his deliberations on national strategy and security. According Mahan scholar Robert Seager, this "experience riveted in his mind an appreciation for the strategic importance of the isthmus, the chaotic political conditions within the weak little nations situated thereupon, and the importance of the transit rights the United States held there. Indeed, the problem of the isthmus would play a large role in Mahan's early studies of sea power and in the curriculum of the infant Naval War College after he became the institution's president in mid-1886."[57]

Mahan's seminal study, *The Influence of Seapower upon History*, appeared in 1890 and quickly acquired an admiring audience that included Kaiser Wilhelm II of Germany and Theodore Roosevelt in the United States, but Mahan was not the only perceptive analyst of strategic issues involved in the events of 1885.[58] Commodore Walker, chief of the navy's Bureau of Navigation, played an essential part in coordinating the navy's show of force in Panama and, at least on one point, offered insights that went beyond Mahan's analysis. While the former captain of the *Wachusett* focused on issues of national power and strategy and mentioned communications only in passing, Walker was aware of the relationship between such an expensive, distant, and risky operation and the need for rapid and reliable telegraphic communications as well as the need for favorable political and public reaction. According to historian Daniel Wicks, Walker explained to field officer Commander Bowman McCalla the importance of the Navy's press image within the United States. Walker cautioned McCalla: "It is of considerable importance . . . that we should keep the country with us in this matter. . . . Of course we must also avoid any appearance of overdoing the thing or trying to make capital." In his later years, McCalla indicated that he too had developed a sensitivity to press coverage of events during the Panama intervention.[59] Journalist Irving King of the *New York Tribune* became caught up in the momentary excitement and fell prey to a jingoism that constituted the "appearance of overdoing the thing"—an unwitting violation of Walker's cautionary note to McCalla. After McCalla and his troops secured the transisthmian railroad, King wrote that the "eagle screams from Colón to Panama." He further expounded on the expansive role of the nation as a hemispheric power: "The United States has asserted itself with a vigor that has not been equaled since the Stars and Stripes flaunted the Halls of Montezuma." Wicks concludes that "Walker's plans for press coverage . . . showed that he did want the expedition to affect public opinion. He evidently wanted intervention in Panama to appear a necessary and wise measure." With the overly enthusiastic assistance of reporter King, Walker saw the intervention applauded in the press at home, although the applause was directed less at the "wisdom" of the action than the pride of military triumph.[60]

Naval telegraphy played an important role in the extension of U.S. power. The Navy Department had advised its officers to adopt a conservative policy regarding the use of telegraphy before 1885 under regulations that had seen few revisions since their penning in 1866. For example, in 1884 a relatively short general order number 324 emphasized brevity and frugality but provided little in the way of instructions.[61] In 1886, however, the year following the Panama intervention, the Navy Department began to issue more extensive instructions regarding the use of long-distance telegraphy. In early 1892, Walker (by then rear admiral), in command of a detachment of warships off the coast of South America, made use of the telegraph as a matter of routine but also was careful to keep messages short and to the point.[62] The Navy Department gave revised instructions in 1896 that outlined a range of options in the use of coded telegrams. Most of the navy's electric communications concerned typical operational matters, but the Panama experience heightened officers' awareness of the importance of telegraphy in crisis situations.[63]

The deployment of naval power by the United States in Panama creates the impression that international telegraphy simply reinforced existing power structures and that it was a tool mainly used by rival great powers. Prestán's dramatic execution seemed to symbolize that point. The unsuccessful effort of Guatemalan Justo Rufino Barrios to use the communications system to promote his campaign for Central American unification was a pioneering venture that revealed the importance of security of the movement of politically provocative messages. Beneath the surface, however, knowledgeable people such as reporter King, entrepreneur Scrymser, Commanders Walker and McCalla, and diplomat Scruggs began to entertain doubts about the ability of powerful nations to control the new information system. The perception began to take hold that the system could bring controversial, potentially disruptive movements and events to the attention of effusive public commentators, ambitious politicians, and the amorphous, little-understood body of public opinion in an arena that previously had been the domain of a handful of diplomats, heads of state, military commanders, and corporate magnates who worked in accord with decades-old traditions. In 1885 most newspapers had little or no access to correspondents with first-hand experience in covering crises from distant locations such as Panama. Over the next two decades, however, roving reporters such as Richard Harding Davis, Nellie Bly, and Sylvester Scovel would claim a popular readership for their reportage on events in Venezuela, Mexico, and Cuba. Walker's admonition about "keeping the country with us" was not so much the issue as keeping up with the jingoistic mood of the press and the public. Newspaper readers and politicians expected prompt reports. The cable network fed information to the nation's newspapers at a speed that often matched communiqués from admirals, generals, and diplomats directly engaged in the crisis.[64] An interventionist policy could, in spite of short-term success, have ominous long-term consequences. Even the patriotic Irving King of the *New York Tribune* sensed the unexpected consequences of the Panamanian intervention when he reported on the expressions of "public hatred" there directed against the United States.[65] The next decade would reveal that the new international communications system and its connections with the popular press had the potential to sway events in directions unforeseen and unwanted by political elites in the United States and Europe.

CHAPTER SIX

Confrontation via the Information System

Chile and the United States, 1889–1892

A QUICK PERUSAL OF a world map reveals that Chile is one of the world's most remote nations. Its main port, Valparaíso, is much closer to the South Pole than San Francisco, New York, or London. While its geography implies international isolation, its history speaks otherwise. In the last decades of the nineteenth century two major developments brought Chile into close and often tense relationships with the major powers of the North Atlantic: the United States, Great Britain, and, to a lesser extent, France. The first of these developments was Chile's victory over Peru and Bolivia in the War of the Pacific. Not only did naval and military experts take notice, but business leaders in North America and Europe became interested in the rich deposits of nitrates in the provinces now under Chilean control: Antofagasta (from defeated Bolivia), Tarapacá (from defeated Peru), and, temporarily, Tacna and Arica.

The second development—or more precisely set of developments—was improvements in the technologies of transportation and communication. The arrival of international submarine telegraph cables in southern South America opened Valparaíso, Santiago, and even Iquique to the sending and receiving of messages from London, New York, and Paris in a matter of a few hours. In the same period steel-hulled, propeller-driven merchant ships replaced smaller and slower wooden-hulled, wind-powered vessels. Nitrate mining companies, export-import merchants, and industrialists could contract for the delivery of Chilean nitrates and other commodities and rely on shipments that would arrive at the designated port according to a schedule confirmed by international telegraphy.[1]

IMPROVEMENTS IN THE CABLE SYSTEM

The technology of submarine cables benefited from innovations in the 1880s and soon had an impact on Chile. The original cables were developed mainly to function in the depths of the North Atlantic, but soon shallow waters along coastlines began to pose problems. In some cases the cables went "dumb" or failed altogether. By the early 1880s engineers identified the sources of these problems in the coral reefs and rocky shorelines that, coupled with the tidal currents, cut into the cables and eventually did enough damage to render them useless. The 1882 directors' report for Western and Brazilian (essential for Chile's connection with Europe) explained the breaks as the result of "friction between the cable and coral reefs and obstructions of that kind under water."[2] Western and Brazilian sent cable repair ships to replace the defective lines with a defensive innovation—sheathed cables. As the name implies, this cable was wrapped in a new covering usually made of iron wire that was capable of withstanding the abrasion of coral and rocks. Over the next seven years Western and Brazilian replaced 162 miles of coastal lines "with the most satisfactory results."[3] Interruptions to service declined significantly.[4] Other companies did the same. James Scrymser's new cable enterprises, Mexican Telegraph and Central and South American, deployed their lines in the 1880s and boasted of their up-to-date equipment that most probably included sheathed cables.[5]

These protected cables brought a marked improvement to international telegraphy throughout the Americas both north and south, but changes in sending and receiving messages were limited largely to the busy lines between Europe and the United States. The main innovation, the siphon recorder, came into use in the 1880s, but messages sent along the network to and from and between Mexico, the circum-Caribbean, Central America, and South America continued to rely on the delicate instrumentation of the mirror galvanometer. The contrast between the two receiving devices was substantial. Instead of the human mental alertness and visual acuity required by the "mirror," the siphon relied on the effect of electrical impulses on a tube of ink that left a physical record of the signal—a vacillating line—on paper. The undulations in the line translated into letters and could be subjected to review for verification. The mirror reader, of course, had only his perception, his memory, and the hurried notations of what the tiny circle of light had touched on the horizontal array of letters.[6]

The continued use of the mirror galvanometer created a demand both from U.S.-based as well as British companies for trained mirror readers, and these could only be found in Britain. Scrymser's companies hired British personnel to operate their stations. William Eleroy Curtis, a promoter of U.S. business expansion in Latin America, was distressed to encounter only British telegraphers on his 1890–1891 trip that carried him through Mexico, Central America, and South America. Curtis concluded that Scrymser's Central and South American was actually controlled by British interests. U.S. businessman Charles Flint, an associate of Scrymser, chastised Curtis for this misunderstanding and then explained why that New York–based company hired British operators:

> As to the charge that the Company has Englishmen in its employ, if you will find Americans who have experience in operating the recorder instrument, in use exclusively on sub-marine cables, I am satisfied that the Company will gladly employ them, but the operations of the recorder instrument is unlike the American instrument, belonging to a branch of telegraphy unknown to American operators.[7]

Even Flint's knowledge of the system was vague. He probably erred when he used the term "recorder instrument," because that would appear to refer to the siphon recorder rather than the mirror galvanometer.

While Scrymser had to rely on graduates of Pender's Porthcurno school for telegraphers, the aggressive New Yorker did have his way on one very important issue—the lowering of the fees charged for sending messages via submarine cables. Scrymser was determined to cut telegraph rates on the west coast of South America to gain competitive advantage against the Western and Brazilian–Brazilian Submarine branch of Pender's cable empire on the east coast. Scrymser made this intention clear from the time he entered into an agreement with West Coast of America to extend the reach of his Central and South American from Lima, Peru, to Valparaíso, Chile, and thereby gain access to Buenos Aires and Rio de Janeiro. The board of directors of the West Coast of America refused to reduce rates, however. Scrymser's Central and South American and the independent and financially vulnerable West Coast of America had an acrimonious relationship for several years.[8] By the late 1880s, however, Scrymser's push for lower rates triumphed.[9]

These reduced rates opened the way for a new phase in the history of international journalism in the Americas. Western and Brazilian, in competition with Central and South American, reduced its rates in 1884 and again in 1887. These reductions resulted in a greater volume of business and a 23 percent increase in revenue by 1887.[10] Three years later Western and Brazilian and Brazilian Submarine reduced their rates by another 50 percent, while Scrymser continued to boast of his less expensive charges for telegrams to the United States.[11] By 1890 the electronic dispatch of telegrams between Santiago and Washington was affordable for the wire services and many big-city newspapers in the United States and more reliable because of sheathed cables. The looming diplomatic crisis between Chile and the United States would play out through long-distance telegrams and long-distance reporting.

CHILE'S PREDICAMENT

Fortunately for Chile, several of its leading political figures proved to be adept in the use of the new means of communication and transportation. The installation of the international submarine-cable system along with the larger, more efficient transoceanic ships that now frequented its ports brought Chile into the mainstream of international history at a time when that South American nation was trying to make the most of its victory in the War of the Pacific. In order to hold the territories annexed from Peru and Bolivia, Chile had to play a difficult diplomatic game of maneuver and intimidation with the United States. Washington sought to block Chile's expansion into Tarapacá to protect the interests of U.S. corporations in what had been Peruvian nitrate lands. Chile's minister to the United States, Marcial Martínez, employed public-relations techniques to influence the U.S. press and used submarine cables to keep Santiago updated on the political machinations in the United States.[12] With its control of Tarapacá and Antofagasta secure, Chile extended its influence by sending its warship, the *Esmeralda,* to Panama during the 1885 rebellion and by proclaiming the annexation of Easter Island in 1888.[13]

These achievements made Chile a continental power in South America, but this elevated status also brought problems. Nearby countries resented Santiago's successes. Obviously, Peru and Bolivia resented their loss of prestige as well as valuable territory. Argentina, which shared a long border and a long-term boundary dispute with Chile, was critical of and sometimes hostile to its western neighbor. Argentina joined with Brazil to sponsor a resolution against acquisition of territory by war at the 1889–1890 Inter-American conference in Washington, D.C.—a diplomatic slap in the face to Chile. That Brazil joined Argentina in this resolution meant that Chile was isolated in southern South America.[14]

Even more ominous for Chile was the smoldering resentment of the United States. Outwitted in the aftermath of the War of the Pacific, the United States in the 1880s and early 1890s was beginning to find a role for itself not only in the Western Hemisphere but also in the larger context of world politics. Key factors in this new identity were naval and military strength. Chile's relatively powerful navy proved its worth in the War of the Pacific. At that time the U.S. Navy was old and outmoded. It relied on wooden-hulled ships and borrowed merchant vessels during the Panama intervention of 1885. The balance of naval power in the Western Hemisphere was about to change, however. As Chile well knew, the United States had embarked on a naval buildup—it was in the process of constructing modern, all-steel warships with tonnage and firepower that would soon exceed Chile's best.[15] The geographical separation of Chile's business and political center in the Santiago-Valparaíso area from the nitrate-rich provinces of the northern desert meant that naval strength was important in the protection of Chile's exposed lines of communication and transportation along the Pacific coast. At the apogee of its international reputation, Chile's future was uncertain.

BALMACEDA AND THE ENTREPRENEUR FROM THE NORTH

Chile experienced a series of crises during the eventful and tragic presidency of José Manuel Balmaceda (1886–1891). An experienced politician, Balmaceda had held the position of foreign minister in 1882 and collaborated with Marcial Martínez in the defeat of U.S. diplomats Trescot and Blaine. Once he became president, Balmaceda directed his political acumen to the promotion of an ambitious agenda. With the government's treasury full from the taxes on the nitrate business, Balmaceda initiated the expansion of public education, the construction of bridges and roads, and the extension of the internal telegraph system. Expressed in terms of material progress for the benefit of Chile's future, Balmaceda's plans had a nationalist ring. However, he resorted to executive activism in the face of congressional opposition. By the late 1880s, he had to rely on a narrow base of support in the Liberal Party. The Conservative Party and an array of smaller political groups questioned his leadership.[16]

By the end of 1889 Balmaceda was facing opposition from a cross section of the politically dominant Chilean oligarchy. The national press was hostile to his aggressive leadership and large expenditures of government funds. The determined but besieged president encountered a major roadblock to his program in early 1891 when the opposition in Chilean Congress refused to approve his proposed budget. Undaunted, Balmaceda attempted to surmount the roadblock by using executive authority to decree that the budget for 1890 would apply to 1891. A group of nineteen senators and seventy representatives announced their opposition to the Balmaceda administration and, with the support of the Chilean Navy, departed Santiago for Valparaíso and then north to Iquique to make their headquarters in that nitrate-prosperous coastal city. The political battle lines were drawn and civil strife followed.[17]

In this exceptionally difficult time in Chilean history, Scrymser, the cable entrepreneur from New York, made his way into that nation's affairs. His company ran its cable south from Veracruz, across Tehuantepec and along the Pacific coasts of Central and South America, reaching Lima, Peru, in late 1882—only a year after Chilean diplomat Marcial Martínez contributed to the defeat of the Peruvian Company and James G. Blaine's machinations in Washington and a few months after the undoing of the diplomatic mission of Trescot and the younger Blaine. Next Scrymser formed a temporary agreement with West Coast of America, a small British company. These two companies completed the connection between Chile and the United States. Central and South American and West Coast of America continued their working relationship for nearly a decade.[18]

Scrymser's public statements and private correspondence revealed his ultimate intention was to expand Central and South American to connect its cables directly to Chile and then establish ties across the Andes to link up with the telegraphic systems in Argentina, Uruguay, and Brazil.[19] Chile, therefore, was crucial to his plans for South America. Central and South American's relationship with its erstwhile partner, West Coast of America, was tense. Scrymser wanted to reduce telegraphic rates to compete with the large British cable companies on the east coast of South America, but West Coast of America balked at this idea. In 1889–1890 Scrymser carried out a flanking movement that proved to be a corporate coup de main. He extended Central and South American cables from Lima to Valparaíso and then, in a move that surprised West Coast of America officials, he acquired Transandino, the Chilean-operated link between Santiago and the Argentine network.[20] These two moves converted West Coast of America into a coastal conduit for domestic Chilean telegrams while giving Central and South American control of the line across the Andes to Argentina.[21]

Scrymser's triumphs gave him a powerful position in Chile's international communications. Whether the telegraphed message moved up the east coast of South America via the British companies or through Central and South American cables on the Pacific side of the continent, the initial electric signals moved through lines owned and operated by Scrymser. There is insufficient evidence to support the generalization that the Yankee entrepreneur was part of a coordinated program to boost U.S. business in Latin America, but there is considerable documentation indicating that Scrymser often sought and received aid and reassurance from the U.S. State Department, several senators, and the U.S. Navy. His 1884 letter to Secretary of State Frelinghuysen recalled the support given his cable enterprises by Secretary of State William Seward in 1865, and in the early 1880s he enjoyed the endorsements of Senator Roscoe Conkling of New York and Secretary of State James G. Blaine.[22] Scrymser obtained the protection of U.S. warships off Peru for the cable-laying project the company undertook in December 1881 because of the unsettled conditions in that country during the War of the Pacific.[23] The political turmoil and social unrest in Central America in 1885 discussed in chapter 5 also prompted Scrymser to request the protective presence of the U.S. Navy. Again, U.S. government officials, including diplomats and consuls as well as sailors and marines, responded to these requests.[24] Scrymser turned to Washington when he perceived a threat to the operations of his growing cable network, and various branches of the government supported Scrymser. The close connections between Scrymser's cables and Washington's diplomatic and naval representatives were well established by the mid-1880s.

Nevertheless, Balmaceda welcomed Scrymser and Central and South American to Chile, granting the company the appropriate concessions. Perhaps the Chilean head of state wanted to encourage a counterweight to the extensive British presence in his nation's economy. The expansive Central and South American and the faltering Balmaceda government had established friendly ties.[25]

Both Balmaceda and Scrymser were concerned about basic telegraphic communications. The congressional revolt of January 1891 threatened Scrymser's lines as well as Balmaceda's government. The rebels, ensconced in Iquique, controlled an essential link in Central and South American's north–south telegraphic traffic. With the loss of the Iquique office, Central and South American could not communicate with Peru, Central America, or, most important in terms of the volume of messages, the United States unless it resorted to a time-consuming roundabout via the British-owned lines from Argentina through Brazil, Europe, and the North Atlantic. With Balmaceda's approval, Scrymser acted. Utilizing a cable ship under the protection of the U.S. Navy's cruiser *Baltimore*, Central and South American cut the cable off Iquique and spliced it to a new submarine line that joined Valparaíso with Lima and the American nations to the north. Scrymser's profits recovered and Balmaceda had regained an important cable connection.[26]

THE *ITATA* AFFAIR

The congressionalist junta in Iquique and the Balmaceda government in Santiago fought a bloody civil war over the first eight months of 1891. Meanwhile, in the United States, President Benjamin Harrison also faced troubled times. Although in a different environment than the one that engulfed his contemporaries in violence in Chile, Harrison found himself in a difficult situation. His political party, the Republicans, had suffered a humiliating defeat in the congressional election of November 1890. Harrison, the nation's head of state and the leader of his party, faced the task of trying to revitalize his supporters in preparation for the presidential and congressional elections of November 1892. A taciturn, cold individual in personal relations, Harrison was an able public speaker. Throughout 1891 he took the initiative. In a series of speeches he projected the image of an assertive chief executive in command of his administration and capable of dealing with whatever crisis it might encounter. Unfortunately for U.S.-Chilean relations, several of these crises involved Chile's civil war and its aftermath.[27] Harrison and his advisors discovered that confrontations with Chile had some positive resonance with the press and public in the United States.[28]

From the perspective of the congressionalists in Chile, the posturing of the Harrison administration was both frustrating and threatening. Secretary of State James G. Blaine favored Balmaceda. The U.S. minister to Santiago, Patrick Egan, was also sympathetic to Balmaceda and, in the months after the war, extended special treatment to the survivors of the defeated regime.[29] Furthermore, the supportive presence of the U.S. cruiser *Baltimore* under the command of Captain Winfield Scott Schley during the splicing of Central and South American cable off Iquique rankled the congressionalist rebels in northern Chile. Add to this mix an aroused president in Washington who wielded "a combination of excessive chauvinism (and) ethnocentrism" enmeshed in "a romantic, gallant, almost sporting view of war." The explosive potential was very much in evidence.[30]

At a critical juncture in the Chilean civil war, an event occurred that captured the attention of the U.S. press. The break between Balmaceda and the congressional rebels received only limited coverage in U.S. newspapers. The coverage increased substantially, however, once the press got wind of the dramatic story of the transport ship *Itata*. This story held a column on the front page of the *New York Times*, many West Coast papers, and other big-city dailies for most of May into June. The *Itata* case is considered by historians to be a fairly minor episode in the larger saga of Chilean-U.S. relations of the late nineteenth century, but its crucial contribution to this era can be found in the press coverage itself. The *Itata* episode attracted much more attention than the War of the Pacific, a much larger event, a decade earlier. The establishment of Scrymser's cable connections along the west coast of South America in 1882 made it possible to track the *Itata* via telegraphic reports, an important factor in the news coverage in 1891. This continuous coverage of mysterious dealings in weapons and a chase involving warships on the high seas appealed to the reading public. The excesses of the emerging yellow press may have played a role here, but there was indeed a secret arms deal, and the U.S. Navy did pursue a Chilean steamer down the Pacific coast from San Diego into Chilean waters, the threat of confrontation and even combat very much looming in the air. In this sense, the facts of the story provided sufficient drama for an exciting story.

The outline of events can be given quickly. The congressionalists purchased much-needed arms and ammunition in New York, shipped them by rail across to San Diego, where they were loaded onto a coastal vessel, the *Robert and Minnie*, and then shifted to the larger, oceangoing *Itata* for delivery to Chile. The U.S. government decided that while the sale of weapons to the congressionalists was legal, the shipment from a U.S. port to Chile might result in damage claims by the Balmaceda government based on the precedent of U.S. claims against the British government regarding the actions of the Confederate warship *Alabama* during the U.S. Civil War. So the U.S. Navy sent the new cruiser *Charleston* to overtake and seize the *Itata*. On the way, the *Charleston* had a close encounter with a congressionalist warship, the *Esmeralda*, off the Mexican port of Acapulco. The end, however, was anticlimactic: the congressionalists surrendered the arms to the United States to avoid further difficulties (the congressionalists had secured weapons from another source). It was not the outcome but the initial secrecy of the arms deal and the excitement of the sea chase that captured the attention of the newspaper-reading public.

Other factors that held newspaper readers' interest were the language used in the articles and the colorful images contained in these words. Drawing on a telegraphic dispatch from San Diego, the *New York Times* described the *Itata* as "the largest passenger boat that ever entered the [San Diego] harbor."[31] On May 5 Ricardo Trumbull, the business agent of the congressionalists, described the arms deal to the U.S. press and provided an account of "the mysterious cruise of the schooner *Robert and Minnie*" to a rendezvous with the *Itata* off the California coast. Over the next few days, the *Chicago Tribune*, the *San Francisco Examiner*, the *San Francisco Chronicle*, the *New Orleans Picayune*, and the *New York Times* reported on the departure of the *Itata*, referencing its crew's willingness to fight to defend the cargo.[32] The *Itata*'s abrupt departure was an affront to the United States because federal officials had ordered the seizure of the vessel for violation of neutrality laws and had placed Deputy Marshall A. L. Spaulding on board to enforce the law. The Chilean steamer left San Diego flaunting the law and carrying the deputy marshal on board against his will.[33] The report from other U.S. marshals to the press added to the tension: "At least eighty Chileans were drawn up in line on the decks, showing that while the whole vessel was in port she was plentifully supplied with men, arms, and ammunition."[34] Meanwhile, the *New York Times* reported by way of London that Ricardo (aka Richard) Trumbull was under arrest in San Francisco for violating U.S. neutrality laws.[35]

The sense of the dramatic heightened as Benjamin Tracy, secretary of the navy, chose a colloquial hunting metaphor to express his confidence in U.S. power. A reporter questioned the secretary in his Washington office:

> "Can you take this vessel on the high seas without violation of the law?" a reporter asked Secretary Tracy.
>
> "Yes, Sir," replied the Secretary emphatically.
>
> "Will you do it?" asked the reporter.
>
> "I have no answer to make to that," replied the Secretary. "You can't skin a hare before you catch him."[36]

Historian Osgood Hardy observes in his pioneering study of the *Itata* affair that California newspapers were "full of the story of the escape of the *Itata*." Among the press sources he cites are the *Los Angeles Times*, the *Los Angeles Examiner*, the *Daily Alta Californian*, and the *San Francisco Evening*

Bulletin. These papers gave close attention to the cruiser *Charleston*'s pursuit of the Chilean steamer. A *San Francisco Examiner* headline exclaimed that the Charleston was "flying through the water after the runaway *Itata*." Along the coast of Mexico a new element entered into the situation. The Chilean warship *Esmeralda* was waiting in the port of Acapulco, apparently to serve as an armed escort for the *Itata*.[37] Chilean sailors on the *Esmeralda* provided an inflammatory statement that was relayed through Mexico City and then to U.S. newspapers. Chilean officers were "reticent" with the press, but a Chilean sailor indicated "that should the *Charleston* attempt to capture the *Itata*, a naval engagement will take place, and the United States cruiser will get the worst of it." The *San Francisco Chronicle* anticipated naval combat under the headline "Will They Fight?"[38]

Press reports reached a nationwide readership from May 10 to June 5 of 1891, marking a contrast to the fragmented coverage of the War of the Pacific and its aftermath little more than a decade earlier and the narrow, limited reportage on the intervention in Panama in 1885. The height of the excitement came on the weekend of May 15–17, when front pages from New York to San Francisco discussed the possibility of combat involving the *Charleston* against the *Itata* and the *Esmeralda*. On Friday, May 15, the *Washington Post* printed the headline "The Sinking of the *Itata*" above a story that cited rumors from Mexico that the *Charleston* had overtaken the transport ship and sent it to the bottom. Other newspapers dismissed this story or gave it minimal coverage but found much of interest in the dispatches from the Pacific. The *Times* and *Tribune* of New York, the *Boston Globe*, the *Philadelphia Inquirer*, the *Chicago Tribune*, the *New Orleans Picayune*, the *Atlanta Constitution*, the *Washington Post*, and the *Examiner* and *Chronicle* of San Francisco relied on telegraphed wire-service reports to present essentially the same stories on the same days. Secretary Tracy remained in his office on Sunday because "the situation is regarded as so serious."[39] Tracy's anxiety stemmed from reports on Saturday describing a verbal confrontation between the captains of the two ships. The *Boston Globe's* headline summarized the dilemma of the *Charleston's* Captain Remy: "Fight or What?"[40] The *San Francisco Chronicle* printed a telegraphed dispatch from Acapulco that included comments from an officer on the *Charleston* describing events from Thursday, May 14:

> Later in the day a formal interview took place between Capt. Remy of the Charleston and the Captain of the *Esmeralda*, the latter stating that

the *Charleston* should never take the *Itata* until the *Esmeralda* was sunk. Capt Remey replied: "I have orders to take the *Itata*. The fact that the *Esmeralda* is present will make no difference whatever."[41]

These verbal salvos represented the height of hostilities. After two more weeks of uncertainty about what was happening with the *Charleston*'s pursuit of the *Itata*, submarine cables carried a message northward from Iquique to Washington announcing that the congressionalists had decided to surrender the *Itata* with its cargo. The *San Francisco Examiner* reported the story, which was datelined Iquique, Chile, on June 4, the day before it appeared on the front page. The *New York Times*, the *New Orleans Picayune*, and the *Washington Post* and other metropolitan dailies carried the same story based on cables directly from Iquique or on a similar report that had been released by the Navy Department in Washington.[42] The secret arms deal, the clandestine rendezvous of gun smugglers off San Diego, and a sea chase involving a confrontation between the navies of the United States and Chile came to a quiet conclusion. With some embellishment but generally not in the exaggerated language of the yellow press, these newspaper accounts took an interesting and, at times, dramatic story and placed it at the forefront the national news. An aroused sense of patriotism in the U.S. press was a preliminary indicator of the potential impact of this phenomenon as a factor in international relations.

THE *BALTIMORE* IMBROGLIO

The congressional forces won the civil war, and a dejected President Balmaceda committed suicide in August 1891. The combat ended, but bitterness remained between the victorious new government and the defeated members of the Balmaceda administration. Much resentment focused on the United States, which had worked with Balmaceda during the conflict. The anti-U.S. feelings erupted in an incident in late 1891. After about two months on board the cruiser *Baltimore* without a break, U.S. sailors were eager to enjoy shore leave in the Chilean port of Valparaíso. Schley, the ship's captain, thought the public mood in Valparaíso by mid-October 1891 had calmed since the cessation of the fighting in Chile's civil war. Schley and other U.S. officials, from George Brown, the rear admiral on board the USS *San Francisco* off Valparaíso, to Tracy, knew that the victorious congressionalists and their allies in the Chilean Navy were resentful of the close relations between the United States and the recently defeated

Balmaceda administration. U.S. officials were caught by surprise, however, with the explosion of anti-U.S. sentiment that took place on the afternoon of October 16 and shocked by the violent death of Charles Riggin, a boatswain's mate, on the streets of Valparaíso.[43]

Accounts of Riggin's demise and the subsequent death of William Turnbull, a coal heaver, vary according to the sources consulted. It is clear, nevertheless, that their deaths and several injuries to U.S. sailors and Chileans (both sailors and civilians) occurred in a riot that began in the True Blue Saloon and spread into nearby streets in the harbor section of Valparaíso. The *New York Times* printed a telegraphed version of the incident datelined October 17: "The anti-American feeling, which is very pronounced, reached a climax last evening in the streets of Valparaíso." In this version, the U.S. sailors were confronted by Chilean mariners and a fight ensued that soon became a riot.[44] The Chilean press saw the fight differently. According to *El Mercurio* of Valparaíso, two U.S. sailors started an argument with a Chilean sailor, threw him down, and then attempted to flee. A crowd of Chileans stopped them and the fighting escalated until the local police arrived and took several participants—both U.S. and Chilean—to jail.[45]

The deaths of Riggin and Turnbull ended whatever peace had returned to Chilean-U.S. relations. Both the U.S. Navy and the Chilean government conducted investigations into the incident, and neither had confidence in the outcome of the rival inquiry. As acrimony intensified, other questions, dormant for a few weeks, reemerged as international issues. Scrymser found himself caught up in a controversy, drawing fire from London and Santiago. Owing much of his preeminent position to concessions granted by then President Balmaceda, Scrymser tended to favor the incumbent over the rebels in Iquique. In a letter published in the London *Times*, the author, identified only as a Chilean, criticized Scrymser's Central and South American for "cutting the submarine cables between Valparaíso and Callao, cutting it in two places—north and south of the port of Iquique—so as entirely to exclude the latter place from communications with Europe, and making it considerably more difficult for the Junta to communicate with the rest of the world."[46] This isolation had an impact on the flow of information to the U.S. press. The writer accused newspapers in the United States of favoring Balmaceda. Especially damaging were false reports of congressionalist defeats on the battlefield. These reports spread across the Atlantic and "would seriously affect the credit of Chile in Europe and lessen confidence as to the final pacification of the country." According to the writer, these false reports "created great resentment against the United States in Chile." Scrymser's handiwork carried a cost in ill will.[47]

Scrymser defended the cable splicing. In a *New York Times* interview he claimed his company "remained neutral as long as it was safe." Balmaceda had requested that cable service from Valparaíso to Callao and thence the United States be reopened. By mid-July, according to Scrymser, this request became a demand backed up by the threat of revocation of the cable concession: "The demand for [reopening] communication was so strong that we could not refuse."[48] The splicing operation had the protection of the USS *Baltimore,* a fact that, to many Chileans, conjoined the cruiser and Central and South American in defending the Balmaceda administration.[49]

The victorious congressionalists added another name to their list of Balmaceda's collaborators: Admiral George Brown of the USS *San Francisco.* Brown sought to keep the Navy Department in Washington abreast of events in Chile during the decisive month of August by maneuvering his vessel off Quinteros Bay to observe the congressionalist invasion. He then sent his signal officer, Lieutenant George Dyer, to transmit a coded account of the invasion to the Navy Department via submarine cable. President Balmaceda heard of Brown's report and made a public statement in Santiago that implied his administration had seen this report. In the aftermath of the civil war, the congressionalists accused Brown and his staff of spying for Balmaceda. *El Ferrocarril* and *El Heraldo,* major Chilean newspapers, published material that reinforced the sense that these charges were valid. Brown's denials did not allay the growing antipathy in Chile toward him, the U.S. Navy, and the United States in general.[50]

The eruption of anti-U.S. feeling in Chile soon met a countervailing explosion in bellicose pronouncements and militant public opinion in the United States. The Chileans, both conservative and liberal, had managed to deal with politicians and public opinion in the United States through the 1880s—this coping was especially apparent in the astute efforts of Marcial Martínez. Congressionalists Augusto Matte and Augustín Ross made good use of the European press for their cause in mid-1891. By late 1891, however, U.S. politicians and press bristled with anti-Chilean, prowar rhetoric unlike anything the Chileans had experienced before in the United States or Europe. The electric and mechanical system of telegraph lines and printing presses spewed out the messages of a pugnacious patriotism in the United States on a nationwide scale that matched the Chileans in its vehemence but dwarfed the newspapers in Santiago and Valparaíso in its transcontinental extent.

The roots of this patriotic outburst lay with the sensationalist press and also in the heated accusations of President Harrison and Secretary Tracy. U.S. diplomatic representative Patrick Egan and Chilean foreign minister Manuel Antonio Matta had exchanged twenty-five sharply worded notes in the post–civil war period, but these communications were usually kept within government channels away from public view.[51] It was not until December 1891 that the language employed by Matta met the equally vehement words of Harrison in the pages of newspapers in the United States and Chile. These exchanges sparked a public outpouring of inflammatory bombast that continued for over a month. Even before the jingoistic commentary in the United States reached its full crescendo, Chilean congressionalist deputy John Trumbull expressed his concerns about distortions and excess in the U.S. press regarding Chile in a private conversation with Secretary Blaine. The two men spoke cordially on December 4, 1891, and shared the hope that their nations could find a path to more harmonious relations. According to historian Joyce Goldberg, however, Trumbull "insisted that U.S. newspapers had created a false impression" about the arrests of Balmaceda partisans by the new government.[52]

Trumbull's complaints concerned serious matters, but they involved fragmentary press coverage that constituted only a few scattered rain clouds before the arrival of a tumultuous storm. President Harrison, still troubled by his Republican Party's defeats in the 1890 congressional elections and frustrated by the pace of the investigation of two sailors' deaths in the Chilean judicial system, vented his anger in the annual State of the Union message to the U.S. Congress on December 8, 1891. His choice of subject matter and his vocabulary were deliberately provocative. He gave a lengthy description of the incidents at the True Blue Saloon, and then trampled on a sensitive nerve by demanding prompt legal action by Chilean officials. He saw the deaths of Riggin and Turnbull not only as criminal acts but also as affronts to the dignity of the United States. If Chilean justice did not satisfy his expectations, Harrison warned in threatening terms that he would "by special message, bring this matter again to the attention of Congress for such action as may be necessary."[53] In this context "congressional action" could well have meant a declaration of war.[54]

Tracy joined with Harrison by concentrating in his annual report on allegations of Chilean misconduct. He refuted congressionalist charges about the *Itata* affair, about the submarine-cable operations off Iquique, and about Admiral Brown's actions at Quinteros Bay. Like Harrison, Tracy insisted that the deaths of the two sailors were insults to the U.S. Navy and the nation as a whole. Meanwhile, Tracy used the telegraph system in the United States to organize a large naval force to send to southern South America.[55]

Chilean foreign minister Manuel Antonio Matta responded to the remarks of Harrison and Tracy with an equally forceful defense of his nation and its conduct in the dispute. Matta's message was not sent directly to Harrison or anyone in his administration but rather appeared in a telegram to Chile's minister to Washington, Pedro Montt. In addition, this telegram was read to the Chilean Senate and appeared in the government's official publication, the *Diario oficial*, as well as in several Chilean newspapers. Matta's response, therefore, was a public document in Chile and quickly found its way into print in the United States and was also inserted into a telegraphed report to the State Department from Minister Egan.[56] In short, it appeared to be a quick and heated retort to the provocative statements of Harrison and Tracy. The electronic-communications system accelerated the flow of angry rhetoric between Washington and Santiago.

Harrison, Tracy, and Matta had, with the help of submarine telegraphy, brought this dispute into the domain of public attention, and another new component of the communications system, the mass-circulation press in the United States, only added to the public notice it received. Newspapers from New York to San Francisco published accounts of war preparations in both countries. Even the usually sedate *New York Times* carried stories with such headlines as "The Chilean Controversy Begins to Look Warlike" on December 23 and "Looks Like a Chilean War" on December 25. The *Chicago Tribune* and the *San Francisco Examiner* reported on the mood in Chile with cabled accounts from Santiago and Valparaíso that carried datelines from only a day or two before they appeared in print. The *Times* and *Tribune* of New York published similar reports that originated in Chile datelined London.[57] A few weeks later the *New York Tribune* reported rumors of Chilean agents in the Puget Sound mapping the coastline. In this same issue, a front-page article cited unnamed "private sources" from Chile who described the preparations of the Chilean fleet for combat. The same column contained a discussion of the "deep concern" in the U.S. Navy Department regarding the high quality of the Chilean Navy.[58]

Two days later the *San Francisco Examiner* reported that the U.S. fleet in the south Atlantic faced a very real threat from the assembled Chilean force. The front-page story highlighted "startling rumors regarding Chilean warships" that placed "Admiral Walker's fleet in danger." A cipher telegram from Captain Robly Evans contained intelligence specifying that the Chilean naval force of four cruisers accompanied by torpedo boats was likely to depart Valparaíso for the Strait of Magellan. A British naval officer stationed in Chile reported that a Chilean counterpart informed him that their fleet was "on a special mission, and you will be astonished before long."[59]

The threatening image of Chile's naval prowess in U.S. newspapers was accompanied by assertions of the majesty and might of the new U.S. Navy, recently strengthened by the addition of several steel-hulled cruisers like the *Baltimore* and its sister ships the *San Francisco* and the *Chicago*. Tracy and his staff arranged to charter colliers and other support ships from civilian contractors.[60] Politicians and newspapers across the United States rallied to the cause with expressions of pride in nation and weaponry. The *Philadelphia Inquirer* used colloquial language in its headline: "The Country Getting Ready to Whip Chile." The *Chicago Tribune* surveyed the response in fifteen states from Georgia to Colorado and found a dozen to be enthusiastically committed to an assertive national policy—even war if necessary. The reluctant three states preferred diplomacy; they didn't argue against war but rather stated that war would not be necessary. A state official in Wisconsin, the *Tribune* reported, expressed a widely held sentiment: "Under no circumstances should this country allow its honor and dignity to be insulted, even though it be by an insignificant nation like Chile."[61] The *Chicago Tribune* raised its editorial voice several decibels:

> There can be no doubt that the large majority, both of Congress and the people, will sustain the action of the President and Secretary Blaine in making this final demand upon the bumptious Chileans and putting an end to their tricks and evasions. If they continue in their present mood there can be but one issue to the complication. It must be war.[62]

Preparations for war became a national preoccupation, as reflected in the pages of newspapers from coast to coast. The *Washington Post* listed the U.S. warships and assessed their readiness for combat in generally positive terms.[63] The *San Francisco Examiner* observed that General Nelson Miles had received maps of Chile's coastal fortifications and also pointed out that an aide to Miles opined that the U.S. Navy's new warships had the firepower to destroy these coastal batteries in short order.[64] The *New Orleans Picayune* speculated on the means by which the Harrison administration could enlist a portion of the National Guard for service abroad. The *Picayune*, located in a former state of the Confederacy, leaned to the Democratic Party on most issues. The idea of some of that state's National Guard units fighting alongside the federal army under a Republican president indicated a rising sense of nationalism in the South.[65] Likewise, the *State* of Columbia, South Carolina, discussed the buildup of national naval and military forces in supportive, excited tones.[66] Meanwhile, the *San Francisco Examiner* covered the shipment by rail of a "monster cannon" to be placed on board the warship *Monterey*. The *Examiner* reported that enthusiastic crowds gathered at stations in Utah, Nevada, and northern California along the rail lines to marvel at its size and that they applauded it with "enthusiastic cheers."[67]

"War talk" in the United States combined racism with nationalistic excess. The *New York Tribune*'s unnamed reporter quoted an unidentified U.S. Navy officer on duty in the Washington Navy Yard who was "familiar with the temperament of the Chileans." The officer insisted that war was inevitable:

> The Chilean population is an amalgam of Spanish, Indian, and negro blood and combines the bad qualities of all three. They hate our people, and are encouraged by the English and the Germans, who are quick to interfere with the interests of Americans. Of the 3,000,000 population of Chile, I believe that fully one-half would be roused to attack the government should it propose to indemnify the families of the American sailors who were killed.

The officer boastfully concluded that the Chilean Navy will "soon be flying the American ensign or . . . resting peacefully in the bed of Davy Jones at the bottom of the sea."[68]

Although the political rivalry between Republicans and Democrats dominated most issues, President Harrison's posture in defense of the nation's honor brought support from the opposition party. The *Washington Post* reported "it is clear that no party lines are to be drawn, and that the President will have the united support of Congress." The *Boston Globe*, allied with the Democrats, praised the preparedness of the Harrison administration—especially his secretary of the navy: "Tracy Ready: Prepared to Meet Chili at Every Point."[69] The same newspaper commented on the support for Harrison: "So patriotic is the feeling among members of Congress that it is reasonable to suppose that suggestions made by the President will promptly be indorsed by Congress."[70] Secretary Tracy received a telegram in which a loyal Republican and a lifelong Democrat boasted of their agreement on the "Chilian question." Democrat Austin Corbin of New York City captured the country's patriotic belligerence: "If the government decides to go down there & give these fellows a good thrashing[,] . . . [make] it last long enough until they get it."[71]

Drawing from past experience, Chilean observers were sensitive to the public mood in the United States as expressed by newspapers and politicians. The negative comments regarding Chile in the U.S. press in early December 1891 that Trumbull complained about paled by comparison with the image of a belligerent, bellicose United States as depicted in many newspapers across the nation in January of the next year. Augustín Ross used his experience in Europe and his vantage point in London to assess the situation. He concluded that the jingoism in the press meant that war was a genuine possibility. In an urgent telegram to President Montt on January 18, 1892, Ross saw the crisis as "very serious" because "American general opinion is in great excitement."[72] This ominous assessment found support in intelligence from Chile's representative in Paris, Augusto Matte, and also Gonzalo Bulnes, the envoy in Berlin.[73]

President Harrison used the telegraph to take advantage of the growing tension and impose the outcome he desired on the crisis. The Harrison administration had complained of the slowness of Chile's investigation of the *Baltimore* affair. Drawing from the complaints, Harrison placed a time limit on Chile's response to his message that required communication by submarine cable. The surge of jingoism in the U.S. press and in the speeches of U.S. politicians set the stage for the president's aggressive demands. While Secretary of State Blaine urged conciliation, Harrison chose confrontation. He rewrote Blaine's moderate message, turning it into an ultimatum, and sent it by cable to the U.S. legation in Santiago on January 21, 1892. Chile either had to agree U.S. demands and apologize for its "offensive" statements or else Harrison would order a break in diplomatic relations apparently in preparation for a declaration of war. The deciphering and translation of this message by U.S. minister Egan's staff in Santiago was completed on the morning of Saturday, January 23, and delivered to the Chilean Foreign Ministry at 10:00 a.m.. Sunday was a day of rest in most nations, including Chile and the United States. Harrison eliminated the opportunity for negotiations on Monday morning when he presented to Congress an indictment of Chile for its conduct in the *Baltimore* affair as well as for its failure to respond to his demands. He then turned the issue over to Congress with the implication that war was a strong possibility. The *Philadelphia Inquirer* employed a succinct, forceful headline: "An Ultimatum Sent to Chile by Harrison."[74]

The Chileans had barely forty-eight hours to respond. The fact that the Chileans had access to the international telegraph made it possible for the Harrison administration to impose a short deadline, by which it hoped to entrap the Montt administration. During previous diplomatic crises and confrontations, by contrast, the U.S. State Department moved at a more deliberate pace. The telegraph played a role in *Virginius* dispute between Spain and the United States in 1873 that required nearly two months of negotiations to resolve. The Samoan controversy involving the United States, Great Britain, and Germany extended from 1885 to 1889. The dispute about the harvesting of seals in the Bering Sea also relied on submarine telegraph lines between London and Washington. In all of these cases, the United States allowed other governments weeks and even months to respond whether by cable or mail.[75]

Faced with the possibility of war against a nation with twenty-two times its population, a much larger and more diverse resource base, a growing industrial capacity, and a new and expanding navy, Chile accepted the conditions in Harrison's message. The formal capitulation reached the U.S. legation in Santiago on January 25 in a letter signed by new foreign minister Luis Pereira. Legation officials immediately telegraphed the letter in Spanish to Washington. Chile agreed to apologize for the incident and indemnify the families of the two dead sailors. An unidentified Associated Press reporter cabled the story of Chile's accession to Washington's demands on January 25, and newspapers across the United States carried this story, datelined Santiago, on the their front pages the following day.[76] Thus the crisis ended quietly and quickly receded into history. Fortunately, historians Joyce Goldberg, David Healy, Emilio Menesses Ciuffardi, José Miguel Barros Franco, and Frederick Pike have rescued this event from the obscure margins of history and with good reason. The rise of jingoism in the United States and the threat of war are important points and deserve attention, as does the matter of how war was avoided. The decisions of President Jorge Montt and his new minister of foreign affairs, Luis Pereira, offer interesting contrasts to the aggressive diplomacy and warmongering that flourished in the United States.[77]

CROSSING A THRESHOLD

The United States and Chile lurched perilously close to war in the early 1890s, and the new communications system played a large role in this crisis.

Three trends in international communications emerged that help to explain the heightened tension: (1) the increased speed of diplomatic and journalistic communications, (2) factual mistakes and deliberate misrepresentations of events in news reporting, and (3) the insertion of nationalistic propaganda in the information flow by governmental leaders.

The acceleration of diplomatic messages between intemperate officials of the two governments added fuel to the fire. Harrison's angry comments about Chile in his State of the Union Address of December 8, 1891, reached Matta only a day or two after their original pronouncement, and the Chilean's equally angry response reached Washington in a similarly short period. The new electronic cables of Central and South American were impressive in their speed and technical reliability, which contributed to the emotional impact of these exchanges. Harrison's ultimatum of January 21, 1892, precipitated another period of tension. The president's message placed before the U.S. Congress a difficult question: whether or not the nation would declare war on Chile. Secretary of State Blaine, an advocate of a diplomatic solution, stood by helplessly. Tense and frustrated, congressmen debated the issue while the telegraphic missives moved in the form of electric currents on the cables between Washington and Santiago. Journalistic reports had an impact on these political deliberations. Representative James Blount, chairman of the House Committee on Foreign Affairs, expressed concerns shared by many of his colleagues on the acceleration of events made possible by recent improvements in the cable system and the reduction in cable costs: "I beg this House to remember that the line of safety is the line of deliberation. You can trust your selves as well to-morrow or next week, after deliberation, as you can in the confusion provoked by instantaneous discussion."[78]

Inadvertent mistakes and deliberate inclusions of rumors and false reports were likewise problems in the Chile crisis. The speed of the telegraph intensified the competition for "scoops" and often contributed to premature publication of reports later proved to be incorrect. The rapidity and length of these communications contrasts with the slow-moving, piecemeal reports on the War of the Pacific and its aftermath. A tendency to exaggerate emerged very early in the *Baltimore* affair. By late October 1891, the U.S. press had distinguished itself by inaccuracy. The London-based correspondent of the *New York Times*, aware of British interest in southern South America, complained that U.S. news coverage of the Chilean crisis reeked with errors that provided "a mortifying example of the worst that the Atlantic cable can do in propagating and dissemination lies."[79] Sensationalist reporting went even further in the second week of November with the claim that "the United States man-of-war *Baltimore* had been sunk off the coast of Chile." The rumor apparently originated with an unnamed reporter who, in interviews in the Brooklyn Navy Yard on November 7, furnished the workers with the concocted story instead of obtaining information from them.[80] As this false story moved through the wires in Europe, the London *Times* severely criticized the U.S. press for its use of the new communications system to spread bogus information:

> It is almost humiliating to a serious correspondent to have to contradict such an absurdity; but as, for want of better sources of information, the European press was obliged to accept better stories invented by Americans which were a disgrace to journalism during the Civil War with Chile, it is advisable to state that the report of a torpedo explosion is absolutely false. This malevolent invention . . . is only equaled by the accounts of battles which had never been fought . . . and interviews and events which had never occurred. All of these frauds upon the public must not be overlooked, and attention should be drawn to them to prevent further mischief.[81]

These flawed reports appeared fully a month before the Harrison-Matta exchange and two months before the war scare of January 1892. The irresponsible, inept performance of the U.S. press in this period was subjected to analysis in March 1892 by E. L. Godkin, a veteran newspaperman and editor of the *Nation*, a journal of commentary on public affairs. Given the popular pressures on representative government, the press became a vital but unpredictable and unreliable element in policy making. Godkin viewed these events as a sorry chapter in the history of U.S. journalism. Editors had too often relied on immature, overeager reporters to gather the news. This combination of inexperienced reporters and poor judgment by editors opened the way for outbursts of jingoism in newspapers across the United States. Godkin concluded: "There is no doubt that the newspapers are to blame for a great deal of this. . . . [W]e have all along maintained that a large part of the world talk on these matters in which the extreme youth of the writers to whom foreign topics are assigned in some offices." Enthusiasm is a characteristic of young males in most countries, he went on to suggest, "but in no country but our own are these youths allowed to stir up bellicose passions in the public press, and propose to fight first and negotiate afterwards."[82]

As for propaganda, the Chileans had made successful use of the foreign press in the United States in the past (through the efforts of Martínez in 1881–1882), but the virulent outpourings of war fever in the United States in late 1891 and early 1892 caught them by surprise. The patriotic statements of President Harrison and Secretary Tracy found reinforcement in newspapers across the country. Godkin identified an extreme variant of patriotism he termed "warriorism," a potent mix provided by military and naval leaders, politicians, and businessmen who saw advantages for themselves in the promotion of international conflicts.[83] The Harrison administration had taken the initiative and was able to set the agenda in the U.S. press. In cities across the nation, newspapers focused on the same issues (the *Itata*, the deaths of Riggin and Turnbull, the Harrison-Matta exchange, and the war preparations of January 1892) and also, in large part, shared a similar perspective on these issues. Although some newspapers held out hope for a diplomatic solution, there emerged a widespread outpouring of patriotism that advocated unwavering support for the Harrison administration across party lines.[84] The electronic-communications system displayed this sense of national unity.

Increased speed of communications, flawed reporting, and the insertion of propaganda were not the only new elements in the flow of information that had an impact on international controversies. The U.S. press added another one: coverage datelined from the nation embroiled in the dispute with Washington. Newspapers in New York, Chicago, and San Francisco reported on the pursuit of the *Itata* in May and June 1891 with information cabled from Iquique and on the clamor for war in Chile in December 1891 and January 1892 with telegraphic points of origin in Santiago and Valparaíso. These datelined sources seemed to add credibility to the press coverage. The tensions ended abruptly on January 26, 1892, when the Associated Press reported that the Chilean government had accepted Harrison's demands. Although most of the information in the U.S. press came from domestic sources, the reports that bore a dateline from within the boundaries of the potential opponent gave the information on the cable system a new level of urgency. Congressman's Blount's confusion was understandable.

Godkin's critique and the commentary by the *New York Times* and the London *Times* revealed a growing awareness that the press and its new electronic nervous system had crossed a threshold. The 1891–1892 press attention on the *Itata* chase and the *Baltimore* affair was more extensive and emotionally intense than the coverage of either the War of the Pacific or the Panama intervention. The 1879–1881 Chilean-U.S. crisis did not include the threat of war, but it is comparable in that U.S. business and political interests were at stake. The 1885 Panamanian crisis involved a temporary military intervention, but it is also comparable because the international prestige of the United States was on the line. The experience of the 1891–1892 Chilean crisis made clear that politicians could reach a large and excitable audience, as their jingoistic statements entered the telegraph lines that, by the 1890s, moved information quickly to three continents: North America, South America, and Europe. Furthermore, the network of telegraph lines and newspapers formed a new type of broadcasting system in the United States that allowed the same event to be covered quickly and simultaneously across the country. Given this interface between electronic and print information, the direction of U.S. foreign policy became intertwined with the inept methods and sensationalist predilections of the press and the credulity of its readers.

Events in the early 1890s offer a telling contrast with those of the previous two decades. The optimistic hopes for peace and harmony raised by the diplomatic settlements of the *Virginius* affair in 1873 and the Argentine-Chilean boundary agreement of 1881 and the general enthusiasm for the international cable network as a means of resolving international conflicts in the 1870s and 1880s weakened in the context of the early 1890s. Apparently Harrison, Tracy, and their advisors had not deliberately adopted a strategy to incorporate the press into a patriotic publicity campaign, but the effect was the same. Aggressive diplomats and politicians used the cables to accelerate their angry exchanges, and newspapers exacerbated the impact of these exchanges by playing to the patriotic emotions and martial instincts of the general public. This interaction of politics and press suggests an interesting question: could organized propaganda harness the public opinion and draw newspapers into a patriotic campaign that would contribute to the intensification of tensions and, perhaps, the threat of war? We return to this question in our examination of the crises that entangled the United States, Venezuela, and Great Britain.

CHAPTER SEVEN

Popularization of the Imperial Mentality

From Border Crisis to Hemispheric Hegemony

PRESIDENT HARRISON'S AGGRESSIVE approach to the Chilean-U.S. crisis of 1891–1892 departed from Washington's post–Mexican War diplomacy. Supported by Benjamin Tracy, secretary of the navy, several senators and congressmen, and a cross section of the nation's urban newspapers, Harrison made public statements that implied that the use of naval and military force was an option. Repeated in many newspapers and several speaking forums, these pontifications about the possibility of force stirred an excited response in the public. This outpouring of patriotic sentiment for executive aggressiveness offered a contrast to U.S. involvement in international crises of the 1870s and 1880s. Secretary of State Hamilton Fish in 1873 used quiet diplomacy and international telegraphy to bring to a peaceful conclusion the U.S.-Spanish dispute regarding the *Virginius*. President Grover Cleveland, his secretary of state, William Bayard, and his secretary of the navy, William Whitney, insisted on a limited and temporary intervention in Panama in 1885 to the point of countermanding their officers in the field.

Harrison's assertiveness in 1891–1892 was paralleled by a change in public attitude that manifested itself not only in newspapers but also in the popular press in general. That change, however, is difficult to measure in quantitative terms. The absence of public opinion polls makes statistical analysis difficult, although the publication record of popular writers who focused on questions related to U.S. expansionism gives some indication of the receptivity of the reading public to the idea of the United States as an imperial power. These writers chose themes that appealed to their readers and expanded on these themes in magazines and books to generate income for themselves and to enhance their reputations beyond what they garnered through their work for newspapers.

Three writers emerged during this time as proponents of U.S. imperialism: Captain John G. Bourke, Richard Harding Davis, and William Eleroy Curtis. Bourke's experiences on the U.S.-Mexican border led him to write an essay for *Harper's* magazine that endorsed imperialism. Bourke concluded that Washington should assume control of this underdeveloped and unstable region much as the European powers were building colonial empires in Africa. Richard Harding Davis agreed with Bourke on this point, but the popular reporter and novelist carried the case for imperialism in a different direction. Rather than outright territorial control, more indirect means of domination were at the heart of Davis's formula for expansion. He advocated investment in mining and the expansion of trade along with armed intervention if required—as in Panama in 1885, for example—as the means by which to establish U.S. hegemony in the Americas. Davis found support in the voluminous writing of Curtis, a champion of U.S. domination of the Western Hemisphere through trade.

Davis was the best known of the three authors. He was a journalist who also wrote popular fiction, and his book sales, reportorial exploits, and photogenic good looks enabled him to reach the status of celebrity. His choice of journalistic subject matter ranged from the U.S.-Mexican border to Central America, Venezuela, and Cuba. His perspective included an approving account of the expanding influence of the United States in this region that also embraced military intervention and endorsed the vague idea of a U.S. "protectorate" in Central America. It was in Davis's novel *Soldiers of Fortune*, however, that he elaborated on his ideas regarding informal imperialism. The main character was a U.S. mining engineer who brought technical expertise and something like a stable government (by a mix of force and manipulation) to a region that, in Davis's view, needed both.

REVOLT ON THE BORDER: GARZA VS. DÍAZ

Davis decided to cover the dramatic events that were unfolding on the U.S.-Mexican border at about the same time the Chilean-U.S. confrontation was taking shape in the fall and winter of 1891–1892. The news-seeking reporter chose the lower Rio Grande Valley because of the revolutionary actions and public-relations acumen of a Mexican who was virtually unknown in the

United States prior to the border crisis. Mexican revolutionary Catarino Garza was another type of media personality. Like Davis, he was a journalist by profession with a flair for the dramatic. But Garza had barriers to overcome in order to gain exposure in the mainstream U.S. press, including his nationality and language and the fact that most of his activities were directed against the Mexican government on the other side of the border. Garza's success in gaining the attention of Davis and the U.S. press merit further examination. Like Chilean Marcial Martínez, Garza understood the connections between the press in the United States and politics and policy making in Washington.

Garza launched his revolt against the administration of President Porfirio Díaz on September 15, 1891, the day commemorated as the beginning of Mexico's struggle for independence in 1810, crossing the Rio Grande into Mexico from Texas. Garza chose this date deliberately to appeal to his compatriots' sense of nationalism. His revolutionary proclamation denounced Díaz as a dictator and warned his fellow Mexicans of the growing influence of the United States in the nation's business and political affairs. Garza was well known for his political activism and mercurial personality in the lower Rio Grande on both sides of the border, but his campaign did not immediately inspire concern in Mexico City and was virtually ignored in Washington and New York. Earliest reports in the *New York Times* mistakenly called him "Garcia" and then "Garzia" and included vague rumors and misperceptions about his movement.[1] In spite of this obscure beginning, in little more than three months Garza's name (correctly spelled) would appear on the front page of the *New York Times*. Based on telegraphed reports from the border, the New York editors assembled a picture of the revolution that gave it credibility as a threat to Díaz.[2]

Garza's revolt was the subject of concern in both Mexico and the United States for several months, even though the revolutionary army seldom numbered more than a few hundred men in the field and did not pose an immediate threat to the Díaz administration. Garza's greatest strength was to be found in his widespread public support in Mexico's northeastern states of Tamaulipas, Nuevo León, and Coahuila. Elliot Young in his study of this movement observes that "although many Mexican government officials believed the reports of Garza's revolution to be exaggerated or completely fabricated, they also considered him a public relations genius."[3] Garza's experience as a newspaper editor and journalist served him well, and grassroots support from Mexicans on both sides of the border sustained his movement in spite of the small number of men under arms. It was the apparent viability of the revolution in the U.S. press that concerned Díaz.

By 1891 Mexico under Díaz (1876–1880, 1884–1911) and short-term president Manuel González (1880–1884) had received large investments from the United States that contributed to the revival of mining operations, the purchase of landed estates, and the building of railroads.[4] James Scrymser's Mexican Telegraph had been one of the prominent U.S. corporations to promote Mexico as a field for New York investors.[5] Scrymser, J. P. Morgan, and James Stillman had much in common with Díaz, Matías Romero (Mexico's representative to the United States), and other Mexican political and business figures. Therefore, the Garza uprising worried this newly formed network of investors, corporate executives, and political leaders, who were concerned about the value of Mexican bonds on the New York market, land evaluations south of the border, and the reemergence of Mexican mining. Garza fed information into the international telegraph network that, in turn, supplied U.S. newspapers with information that could damage Mexico's fragile image as a safe harbor for business.

Garza understood the cultural and business values of the United States as well as his native Mexico. Born on a farm outside Matamoros, he moved across the border to Brownsville in 1877 at the age of twenty-two. He was a clerk in a dry-goods store before moving further north to St. Louis where he became involved in a dispute with the Díaz government about the position of the Mexican consul in that midwestern city. After his application for consul was met with rejection by Díaz, Garza turned against the government. During the late 1880s he built a political movement against Díaz in the lower Rio Grande Valley. He was instrumental in the founding of the Spanish language newspaper *El comercio mexicano* that, in addition to promoting business, also featured articles by Garza critical of Díaz. The revolution of September 15, 1891, was in many ways not a surprise to the residents of the region but was the culmination of Garza's efforts over several years.[6]

Early press accounts in the United States bungled Garza's name and often presented conflicting descriptions of him, but by late December 1891 newspapers north of the Rio Grande were taking him seriously. Relying on telegraphed accounts, the *New York Times* and the *Washington Post* reported on December 25 that Garza's men had dispersed after their initial engagement with Mexican army forces and regrouped on the U.S. side of the border. The rebels reportedly numbered between five hundred and six hundred and, according to the *Times*, constituted a threat to stability in the region.[7] Three days later, the *Times* again drew on telegraphed accounts to conclude that Garza's movement was "spreading rapidly" and that U.S. troops were unable to police their side of the border because telegraph lines connecting local army bases were down.[8]

Garza took full advantage of his media opportunities. In the interview that appeared on the front page of the *New York Times* on January 4, 1892, he stated that his experience as a "newspaper man" gave him respect for the role of the news media in public discussions of important issues and that he chose this interview as the appropriate forum to communicate the goals of his movement to the people of the United States. Garza directed his criticism at the dictatorial methods of the Díaz government—especially its persecution of political opposition. He scored the corrupt business practices of the regime that demanded bribes in exchange for monopolistic concessions. The *San Francisco Chronicle* also paid attention to him. Although that newspaper's editorial labeled him a mere troublemaker, the news coverage of January 3 presented another image: "All the people on both sides of the river are in sympathy with Garza, and it is impossible to get any assistance looking toward his capture."[9] Garza avoided direct criticisms of the U.S. interests that were making extensive land purchases in Mexico in spite of the fact that he was critical of similar extensions of U.S. power and influence elsewhere. Again, Garza proved himself to be adroit at public relations, tailoring his comments to fit the audience.[10]

Garza's uprising gained national press attention in the United States. By January 1892 newspapers in New York, Boston, Philadelphia, Washington, Atlanta, New Orleans, St. Louis, Chicago, Cleveland, Los Angeles, and San Francisco were covering his revolution. The *Chicago Tribune* featured a map of the border region, although the text of the article quoted the Díaz government's assessment of the movement as a mere "flash in the pan." The *New Orleans Picayune*, by contrast, reported that Garza's movement was gaining strength. The *Picayune* carried news of Garza on ten of the first fourteen days in January, the largest number of days out of the ten metropolitan newspapers in this study. On January 3 the New Orleans newspaper presented the portrait of a rising revolutionary movement: "All advices agree in reporting that the followers of the reckless rebel are rapidly increasing in numbers, and that he is abundantly provided with money and supplies."[11] Two days later the *Picayune* emphasized that the Mexican army could not capture the "daring rebel."[12] On January 11, the *Picayune* found that the absence of news from the border was no indication of a weakening of the revolt.[13] By contrast, the skeptical *New York Tribune* minimized the strength of the uprising in the headline "Not Yet Captured," which implied the inevitability of the movement's collapse.[14]

The *New York Tribune*, the *Boston Globe*, the *Philadelphia Inquirer*, the *Atlanta Constitution*, and the *San Francisco Examiner* gave Garza more scattered coverage than the *New York Times*, the *Washington Post*, the *Chicago Tribune*, the *San Francisco Chronicle*, and the *News Orleans Picayune*, but all of these newspapers saw him as an elusive rebel with popular support and therefore as a problem for the Díaz government.[15] Pro-Díaz journalist H. W. Allen in a retrospective essay in the *Review of Reviews* denied that the Porfirian inner circle took the border unrest seriously. Allen quoted a business publication, the *Mexican Financier*, that criticized the wire services in the United States—specifically the Associated Press—for spreading questionable accounts of Garza's revolution:

> American journals [are] deceived by Associated Press dispatches sent out from San Francisco, Cal., by some skulking enemy of Mexico and its Government. No name is ever given to substantiate the reports, and it is for that reason that we wonder why Mr. William Henry Smith, the head of the Associated Press in the United States does not put a stop to the circulation of injurious rumors. . . . Mr. Smith owes it to his position as chief of an important news agency to refuse to publish manifestly absurd reports regarding Mexico.[16]

Garza persisted in his efforts to address the public of the United States through an open letter to President Harrison. The Mexican rebel and journalist described himself as "an interpreter of the public sentiment of my country" and insisted that his critics were using the "subsidized press" in the United States to attempt to discredit him. The letter itself followed an interesting route. Dated January 31, 1892, it first appeared in print in the Spanish language newspaper *El Chinance* of San Antonio. The *New York Times* published a translation on February 9, 1892.[17]

Garza's publicity campaign enjoyed a degree of success that his military campaign did not. His forces never established a stronghold in northeastern Mexico, as he had hoped, and eventually the Porfirian army dispersed them. Garza left Mexico in February 1892 for Houston, then New Orleans, and then Central America and the Caribbean, where he assumed a roving existence.[18] In spite of the collapse of his movement, the mercurial Mexican rebel had had an impact on the print media of the United States.

THE PORFIRIAN PROPAGANDA RESPONSE

Garza's penchant for the use of the printed word met its match in the public-relations skills of Matías Romero, Mexico's representative to the United States and a longtime advisor to Díaz. Romero was Mexico's representative in Washington for many years. He was sensitive to his nation's image in U.S. and European press accounts and the effect of that image on foreign investments in Mexico. Learning from Romero, Díaz himself had become a persuasive booster, serving as the commissioner general of the Mexican delegation to the 1884 World's Fair in New Orleans. Díaz, Romero, and Mexico's political leadership worked to project a positive image abroad, especially at the 1889 World's Fair in Paris and the 1893 Columbian Exposition in Chicago.[19] Given the government's worries about public image, Garza's revolution was a matter of concern.

The close attention given by the administration in Mexico City to newspaper coverage in the United States was evident in an unusual public apology issued by President Díaz to U.S. Army general David Stanley, who served on the border. Díaz's letter also indicated the importance of telegraphy and the repercussions of errors in the enciphering, deciphering, and translation of messages that found their way into newspapers in the United States. Apparently a combination of faulty enciphering and erroneous translation gave the appearance that Díaz was critical of the work of Stanley's command. That Díaz had not meant to make such a statement was evident in his letter of apology to Stanley that appeared on the front page of the *New York Times* on December 29, 1891. Díaz wrote of "the pain that has been caused me by the alterations in my telegram sent in cipher to Mexican Minister Romero in Washington[,] which suffered in publication transmission and translations." The translation conveyed the opposite meaning of what the Mexican president intended. In an attempt to undo the damage caused by the mistranslation, Díaz praised the "friendly offices" and efficiency of Stanley's troops and expressed his "grateful thanks" for Stanley's help with the Garza revolution.[20]

A more clandestine phase of the Díaz government's efforts to manage the flow in information included the dissemination of false rumors and outright intimidation of the press. Governor Bernardo Reyes of Nuevo León, a close ally of Díaz, spread rumors that Garza's movement was bankrolled by speculators who wanted to profit from a precipitous decline in Mexican bonds on the international market. These and other rumors found their way onto the pages of the *Chicago Times* and the *Washington Evening Star*.[21] The Mexican government also fed its share of stories to the press through news conferences. And historian Elliot Young has uncovered evidence that Díaz and Reyes used flattery, cajolery, and monetary rewards to sway editor Justo Cárdenas of *El correo de Laredo* (a Spanish-language newspaper) to their cause.[22]

Executives in U.S. corporations with connections to the Díaz regime also supported the government in statements to the press. In reply to accounts of Garza's destruction of telegraphic communications in the border region, Mexican Central Railroad president S. W. Reynolds said his company had its own telegraph line to the border, where it joined with Western Union. The Mexican Central line was open and the company used its own cipher without fear of interruption. In conclusion, Reynolds claimed that the trouble on the border was in the exaggerated stories in the U.S. press.[23]

The chief defender of the Díaz administration was the articulate Matías Romero. His post in Washington gave him convenient access to newspaper editorial offices, and he was also a respected figure in the capital's political and diplomatic circles. During the six months in which the Garza revolution was at or near the crisis stage (September 1891–February 1892), Romero authored several articles and gave numerous speeches and interviews that appeared in the U.S. press. For example, on January 8 Romero took advantage of an interview in Washington to point out that in his view the strength of the revolt was overrated. The Mexican minister to the United States saw an advantage for Garza resulting from the rebel's exploitation of the border. The insurgent would cross the boundary and escape the pursuit of the one nation's army's by taking refuge on the opposite side of the border. Romero was confident "that capture of Garza might be effected without doubt if the troops of either country were allowed to follow him into contiguous foreign territory."[24] Although these interviews, articles, and speeches covered a variety of topics, Romero's close association with Díaz and his opposition to Garza made these pronouncements an effective tool by which to counteract the revolution.[25]

Romero's main contribution to the propaganda war was an essay that was both diatribe and analytical commentary. Romero included the expected verbal assault on Garza and movement, but he concentrated much of his attack on the U.S. press and its preference for slipshod journalism and sensationalism. He saw "improbable and absurd stories" in print in major national newspapers based on "false reports" that originally appeared in Texas newspapers. The pressure of deadlines and the convenience of the telegraph moved these unreliable reports onto the front pages of newspapers across the nation in a few days.[26]

Romero argued that these false reports tended to undermine the reputation of the Mexican government in the United States and to cast doubts on the viability of the Mexican economy. Both Romero and Díaz jealously guarded the recently won improvements in Mexico's standing with the U.S. public—especially among officials in Washington and investors on Wall Street.[27]

Mexico's representative in Washington unleashed his most severe criticism on the U.S. press, which, in his view, had made the revolution into something that in reality it was not: a serious threat to the Díaz administration. Romero charged that

> the desire of the journals in the United States to give sensational news, however absurd, carries them often to the extreme of accepting without reserve, and without judgment and due discretion, whatever is told them, no matter how foolish it may be, and this fact has greatly contributed abroad to detract from the reliability of the news published in the United States, and if this evil is not corrected it will end in throwing increased discredit upon the newspapers of the country.[28]

Romero's critique of the press received support in the pages of *Literary Digest*, a national journal that printed editorial statements from publications throughout the United States and some foreign countries. The editorial faulted the U.S. press for its exaggeration of the Garza movement and its use of unreliable telegraphed information: "The way in which the American press magnifies the ridiculous revolutionary attempt of Catarino Garza is laughable. Day after day columns of [the] most absurd and improbable matter about him are telegraphed."[29] Editor Albert Shaw's *Review of Reviews* echoed this perspective. Contributor H. W. Allen belittled the U.S. press and wire services: "The absurdity of attaching any more national importance to that movement than could be given to the 'rustlers' war in Wyoming is now admitted by everybody." The so-called war correspondents sent to the border by their editors were on a pointless mission.[30]

BORDER INSTABILITY AND THE IMPERIAL MENTALITY

Romero's criticism of sensationalism in the press was an outgrowth of the Garza-Díaz public-relations conflict, but there was another perspective offered in mainline publications that placed the affair in a larger context. Some U.S. observers—notably army captain John G. Bourke and *Harper's* foreign correspondent Richard Harding Davis—saw this struggle as an opportunity for the United States to exert its growing power in the Rio Grande Valley and elsewhere, enabling it to join the front ranks of the world's imperial nations. They added another dimension to this border crisis by connecting it to the issue of imperialism and by attracting a broad readership in magazines and books.

Bourke was on active duty in pursuit of Garza during the border uprising, but the army captain was also a trained observer who served as an analyst and publicist. His reports on the lower Rio Grande Valley not only furnished the military with a record of events but also supplied the press with the captain's perspective. Bourke was convinced that this region was one of the least developed areas in North America—lacking in essential transportation such as railroads, subject to frequently interrupted communications owing to faulty telegraph lines, and virtually bereft of modern industry or schooling for its residents. In an 1894 article for *Scribner's Magazine* Bourke made his case for this region as an "American Congo," by which he meant an isolated, backward area in need of the firm guiding hand of a "civilized" power. Bourke's reports for the army (written in 1891–1892) and his article in *Scribner's* argued for imperial intervention not only by launching military operations but also by the building of railroads and telegraph lines to promote economic development and by the establishment of schools to eradicate the ignorance and superstition that Bourke saw as barriers to modern civilization.[31]

The Garza revolution also attracted the attention of Richard Harding Davis, the roving correspondent who often sought out dramatic stories in locales that contrasted with the settled farmlands and cities of the East Coast and Midwest that were familiar to his readers. The Rio Grande was obviously "out west," where, in the popular imagination, Native Americans, cowboys, and Texas Rangers roamed the countryside engaged in adventurous exploits. Davis believed that he had a line on a good story, but his travels in Texas proved him wrong. Garza eluded Davis. *Harper's* eminent correspondent sent back stories on the arid landscape and the inadequate rail system, but there was no account of the capture of the Mexican rebel. Davis's interpretation of events shifted from the particular to the general and followed the track laid out by Bourke, the lower Rio Grande emerging as "darkest Texas," "an unmistakable analogy to the popular nineteenth century phrase 'darkest Africa,'" according to Elliot Young.[32]

Davis and Bourke saw this episode more as a justification for the United States to assert itself as an imperial power than as a political and social uprising of interest in its own terms. Both placed emphasis on imperial opportunities for the United States in their subsequent publications. Bourke's title for his *Scribner's* essay—"American Congo"—was obviously a call for a replication in the Western Hemisphere of assertions of imperial power the Europeans had made in Africa. Davis went even further in his nonfiction book *Three Gringos in Venezuela and Central America.* Accompanied on his adventures by two young friends, the thirty-year-old Davis explored the jungles and mountains of Honduras, coastal Nicaragua, and the remains of the ill-fated French canal effort in Panama. His conclusions were consistent with his findings on the lower Rio Grande Valley. Central Americans, in his view, were a hopeless lot, from their affluent, corrupt leaders to the poverty-stricken peasantry. Their attitudes toward politics and government were indicative: "The value of stability in government is something they cannot be made to understand."[33]

In the judgment of Davis, what the Central American citizen "needs is to have a protectorate established over him either by the United States or by another power; it does not matter which, so long as it leaves the Nicaragua canal in our hands." Davis held to the biases typical of this era of imperial expansion: "There is no more interesting question of the present day than that of what is to be done with the world's land which is lying unimproved: whether it shall fall to the great power that is willing to turn it to account, or remain with its original owner who fails to understand its value. The Central Americans are like a gang of semibarbarians in a beautifully furnished home of which they can understand neither its possibilities of comfort nor its use. They are the dogs in the manger among nations."[34]

SOLDIERS FOR EMPIRE

Already well known for both journalism and fiction, Davis enjoyed widespread popularity and even critical acclaim. As early as 1891 the *Review of Reviews* commented that "no one has won public favor more rapidly or more completely than Mr. Richard Harding Davis."[35] Davis adjusted the subject matter of his writing to fit the current public interests. His *Three Gringos in Venezuela and Central America* rolled off the press of Harper and Brothers in 1896 at the time the Venezuelan boundary controversy pulled the United States and Great Britain close to war and placed Venezuela on the front pages of newspapers across the country. His novel *Soldiers of Fortune,* drawn from many of the same experiences detailed in his travel book, appeared the following year in magazine serial form and then as a book. According to biographer Arthur Lubow, it "was the most popular book of Davis's career."[36] In 1895–1896 Davis entered William Randolph Hearst's stable of correspondents on the *New York Journal* as the star reporter and soon began covering the Cuban insurrection that would lead to the Spanish-American War. His large public following, combined with his skill as a writer, gave him a place of singular importance in press history. In particular, his role in the "yellow press" competition in New York City and his penchant for a dramatic and at times sensationalist style gave his writing a sense of immediacy, and his established popularity gave him credibility with many of his readers. As literary scholar Karen Roggenkamp has elucidated, Davis used the imaginative techniques of fiction to embellish his reporting: "Journalism and fiction, reality and artificiality had become utterly entwined."[37] Furthermore, Davis's personal experience in Mexico, Central America, the Caribbean, and the northern part of South America gave a certain gravitas to his writing. Davis was capable of producing what Matías Romero described in 1892 as "sensational news" that was at once convincing and unreliable.[38]

It was in his novel *Soldiers of Fortune* rather than in his journalism that Davis produced his most sweeping prescription for U.S. domination of the circum-Caribbean. The novel is set in a fictional country—Olancho—in northern South America (probably modeled to some extent on eastern Cuba) and has a cast of largely stereotypical characters caught up in intrigue, revolution, and romance.[39] The central character, Robert Clay, is an engineer from the United States in charge of an iron-mining enterprise financed by the Langham family. Davis chose engineering as Clay's profession because it was common at the time for men with technological expertise in railroad construction and mining headed south from the United States to Latin America to open that area to the expanding industries of the United States and Europe. Construction engineer James L. Baldwin pioneered a railroad line across Panama in the 1850s. Henry Meiggs gained a reputation as a daring railroad builder in Peru's Andes Mountains in the 1860s and 1870s, and his nephew, Minor Cooper Keith, laid track from the tropical coast to the interior highlands of Costa Rica.[40] British mining engineer George Chalmers redesigned and reopened the Morro Velho gold mine in Brazil in the 1880s and 1890s. These modern frontiersmen, who applied technology to harvest the mineral resources of Latin America, were part of a growing profession that also included specialists in electrical and mechanical engineering.[41] *Engineering Magazine*, a publication intended for this new profession, touted the opportunities available in South America in the 1890s.[42]

Davis's fictional engineer, Robert Clay, exemplifies the audacious males who exploited these opportunities.

Like Baldwin, Meiggs, Keith, and Chalmers, Clay is not a university-trained engineer but the product of on-the-job training and field experience, which includes projects in Spain, Germany, Zanzibar, and Central and South America. Clay is introduced in *Soldiers of Fortune* as a seasoned professional with an international reputation for skill, tenacity, and honesty, and his stature as a leading man is reinforced when he impresses visitors to the iron mine for whom he is conducting a tour on horseback by "stooping to pick up a piece of ore from the ground in cowboy fashion."[43] His commanding presence is evident even after he dismounts and let his assistants carry on the presentation:

> He would stand for minutes at a time up to his boot-tops in the sliding waste with . . . his thumbs in his belt, listening to what his lieutenants were saying, and glancing quickly from them to Mr. Langham to see if he were following the technicalities of their speech.[44]

Clay's expertise is not limited to engineering and business management; he also understands the nuances of imperial politics. Early in the novel he confronts an abrasive officer in the Olancho army, warning, "Try to break that concession; try it. It was made by one Government to a body of honest, decent business men, with a Government of their own to back them, and if you interfere with our conceded rights. . . . I'll have a man-of-war down here with white paint on her hull, and she'll blow you and your little republic back up there into the mountains."[45] But subsequently, Clay reveals a grasp of the complexities of U.S. interventions in Latin America. Speaking with close associates, Clay shows calculated restraint. He cannot call for a U.S. warship without provocation from the local malevolent revolutionary: "I have to wait until he interferes with me, or my mines, or my workers."[46] In other words, Clay has to be faced with a manifest threat to the mines before calling in the marines.

Clay also demonstrates insights regarding the unbalanced social structure of Olancho. On the way to the mines, Clay, Langham, and their party pass by the "mud cabins thatched with palm leaves, and alive with naked, little brown-bodied children who laughed and cheered them as they passed." Apparently happy in their poverty, these peasants-cum-miners and their families contrast with the ne'er-do-well male offspring of the Olancho elite, "the dark-skinned dandies . . . in white duck suits and Panama hats with tortoise shell canes."[47]

Clay also has an understanding of local politics; he is especially acquainted with the uncertainties associated with the common practice of changing governments by revolt rather than election. The sinister revolutionary in the novel is called Mendoza and is not only opposed the aging constitutional president Alvarez but also announces his resentment of the Langham mining interests. He broadly implies that his new government might seize or in some way menace the mines and issues the more overt threat that his forces might set fire to the Langhams' home.[48]

As Clay predicts, Mendoza carries out his coup d'etat. The rebel kills President Alvarez and imprisons his kindly vice president. It falls on Clay to execute a countercoup to restore constitutional government (of a sort) and to protect the U.S. investments in the iron mines. Clay does all this without calling on U.S. naval or military forces. He organizes a small army of his own mine workers (immigrants and locals) and a discontented regiment of the Olancho national army. Drawing on his technical superiority over the South American opposition, he utilizes the local telegraph and rail lines first to take a shipment of arms intended for Mendoza's forces and then to move his army into place for the surprise attack against the usurpers in the nation's capital. Clay's plans work to near perfection, and his courage in the face of long odds and personal harm is undaunted. Vice President Rojas assumes the presidency, and order, prosperity, and progress are restored to Olancho.[49]

Clay's role as a protagonist of imperialism is reinforced in the novel's concluding scene, where the counterrevolutionary receives a heartfelt patriotic salute from a detachment of U.S. Marines who had arrived too late to participate in the countercoup but in time to help maintain order. The marine lieutenant orders his troops to salute Clay: "Even though you haven't worn our uniform, you're as good and better, than some that have."[50] Davis himself had pulled off a coup by asserting the informal imperial authority of the United States through the fictionalized, symbolic actions of a mining engineer who boosts the local economy and opens the way for constitutional government.

The combination of ethnocentrism and condescension typical of Davis was not unique. Many leading politicians, writers, and business leaders of the late nineteenth century held similar beliefs.[51] Probably the most innovative aspect of Davis's writing was the articulation of the surging imperial ambition of the United States in the popular press. Bourke joined Davis in the early expression of this mentality regarding the border region along the Rio Grande River in their coverage of the Garza revolt of 1891–1892, but it was Davis who carried the imperial ideal to a popular audience.

THE AWKWARD PROPAGANDIST

William Eleroy Curtis was a contemporary of Richard Harding Davis, but, even though the two men shared a firm belief in U.S. expansion, there were differences between the two. Davis was a prominent public celebrity and a popular writer. His novels sold well—especially *Soldiers of Fortune*—and William Randolph Hearst regarded him as the nation's premier reporter. Curtis was a competent correspondent for the *Chicago Inter-Ocean* who was publicly recognized, but only by business leaders and politicians. Davis was outgoing and stylish and won friends easily. Curtis, by contrast, was often difficult in personal relations and had a tendency to express controversial opinions that angered people crucial to his work. Davis wrote in a fluid, easy-to-read style. Curtis used a lively, sometimes overstated prose in his travel books, but his publications on commerce bristled with statistics reinforced by a repetitive, colorless text. In short, the two men wrote in contrasting styles tailored to fit their respective audiences.

Nevertheless, Curtis, like Davis, was persistent and productive. From 1886 to 1900 Curtis wrote and/or edited six books concerning Latin American–U.S. relations. He also compiled, edited, and wrote several government reports and was in charge of the publications of the Bureau of American Republics (later the Pan-American Union). Curtis was a promoter of trade patterns that would place the United States in a dominant position. U.S. investment dollars and manufactured goods would flow southward into Latin America, while the region's primary products—sugar, coffee, and copper—would support the industrial expansion of the emerging power north of the Rio Grande.[52] In spite of his erratic judgment and sometimes blundering social skills, Curtis became a leading authority on Latin America—especially its foreign trade and its commercial relations with United States. For a business executive or politician in the United States who wanted to learn about contemporary Latin America, William Eleroy Curtis was the author to consult.

Curtis emerged as the central member of the ill-fated Latin American Trade Commission dispatched by President Chester Arthur and the U.S. Congress in 1884 to promote increased trade between the United States and the nations to the south. Originally designated as secretary for the three-member commission, Curtis rose to full membership when two of the commissioners dropped out. Ostensibly a nonpartisan body representing the U.S. government, the Latin American Trade Commission nonetheless had partisan origins—the result of pressure applied and political deals made by Republican senator James G. Blaine. Curtis and the remaining original commissioner, Solon Thacher, followed a meandering course from Mexico to New Orleans and then on to Venezuela, Central America, Peru, Chile, and Argentina. Their visits to Mexico, Guatemala, and Venezuela were congenial, and Curtis gathered information and met with officials on a reasonable schedule. By the time they were concluding their stay in Chile in late May 1885, however, Curtis and Thacher received unexpected orders to return to New York by July 1, an abrupt change that gave them barely a month in Argentina and no time for Brazil. The credibility of the commission suffered in Buenos Aires, where Argentine government officials and press commentators ridiculed their fast-paced efforts at fact-finding. The *Buenos Aires Herald* parodied their work as a tour of "The Argentine Republic in 35 Hours," possibly jokingly alluding to Jules Verne's popular contemporary novel *Around the World in Eighty Days.* This near farcical end to the brief career of the Latin American Trade Commission called into question the entire venture, a verdict that was unfair to the earlier phases of the trip. This denouement also subjected Curtis himself to sharp criticism from Argentine officials, who saw the brevity of his stay as an insult to their nation.[53]

Curtis compiled a lengthy report on the efforts of the commission, but his most publicized work on this venture was *The Capitals of Spanish America* published by Harper and Brothers in 1886.[54] This volume, which was over seven hundred pages long and contained many illustrations, was intended to reach a popular audience as well as political leaders and the business community. The book created the impression that the traveling norteamericano thought the United States would and should be the major force in hemispheric trade. Curtis did not hesitate to express his opinions. He included commentary on social customs, religion, and political conditions, but his main interest was the potential for increased commercial relations with the United States, his favorites candidates in this respect being Mexico, Guatemala, El Salvador (if it could develop an accessible port), and Venezuela.[55] Like Richard Harding Davis, he found Nicaragua and Honduras to have limited potential.[56] Peru, in the aftermath of its defeat by Chile in the War of the Pacific, was in need of U.S. support but had little to offer in the immediate future.[57]

Curtis was in Chile during May 1885 when anti-U.S. sentiment was still strong. Curtis emphasized the "intense prejudice against the United States growing out of the attitude assumed by our government during the late war with Peru."[58] His relationship with James G. Blaine encouraged an assessment of Chilean character that was consistent with former Secretary of State Blaine's unfriendly policy toward that nation. In a revealing passage, Curtis began with a compliment but slanted his text toward insults: "The Chillano [*sic*] is the Yankee of South America—the most active, enterprising, ingenious, and thrifty of the Spanish American race—aggressive, audacious, and arrogant, quick to perceive, quick to resent, fierce in disposition, cold-blooded and cruel as a cannibal."[59]

Curtis reserved some of his most effusive praise for Argentina, but it was the Argentines who, more than other Latin Americans, including the Chileans, saw his visit as objectionable. He praised the government and politics of Argentina, finding many positive parallels between that nation and the United States. His comments on Argentine wheat, wool, and meat exports were also positive, and he saw a healthy business environment produced by the energetic inhabitants of Buenos Aires.[60] He found the Argentines to be interested in the United States as evinced by the sale on the streets of Buenos Aires of newspapers and books imported from the neighbor to the north: "The people are right when they assert that their country is the United States of South America, and there is nothing else that they are so proud of. They study and imitate our institutions and our methods and in some cases improve upon them."[61] This portion of *The Capitals of Spanish America* first appeared in *Harpers' Monthly Magazine* in 1886 and stirred a wave of interest. In a footnote to the second edition of his book, Curtis observed that he received requests for more information on Argentina "from several hundred persons" and that the Argentine consul in New York and diplomatic mission in Washington "have also been overwhelmed with similar inquiries."[62]

Curtis created favorable publicity for Argentina, but the representatives of the government in Buenos Aires were his most articulate and adamant critics during the Inter-American conference of 1889–1890, a meeting for which Curtis handled the planning and local arrangements. The brainchild of Secretary of State Blaine (returned to that office after an eight-year absence), under the newly elected president Benjamin Harrison, the conference was to showcase the industrial might, business acumen, and national power of the United States. The delegates of all the Latin American nations (except the Dominican Republic) assembled in Washington in

September and October of 1889 and soon departed on an extensive railroad tour that included the sprawling city of Chicago, the Menlo Park laboratory of Thomas Edison, the military academy in West Point, and the naval academy in Annapolis.[63] Curtis worked diligently to promote a favorable image for the United States as a partner in commerce, but he failed to win his own personal acceptance among the Latin Americans. The Argentines Roque Sáenz Pena and Manuel Quintana adroitly outmaneuvered the U.S. delegation at every turn. The meetings often became sparring matches from which no major resolutions emerged.[64]

Critics of Curtis had a field day. W. R. Grace, head of the South America merchant house and the mayor of New York City as a reform Democrat from 1880 to 1886, used the word "committee," but by implication he placed the blame on Curtis in an interview with a *New York Times* reporter:

> Well when they arrived here, they were put in charge of a committee which understood neither the people the delegates represented nor their ideas. You know these people have certain forms and etiquette which mean much to them. I need hardly tell you these were not sufficiently considered. They did not understand our brusque practicality.[65]

Mexico's delegate Matías Romero was not so harsh. Romero concluded that Curtis did win some favor among the Latin Americans who, after personal contact, were able to see beyond his blunt personal style. Nevertheless, Curtis remained persona non grata among much of the Latin American diplomatic corps.[66]

Curtis frequently bungled his opportunities to engage in personal diplomacy and in the building of bridges between the business communities of the Latin American nations and the United States, but in many of his publications his strong-willed bluntness expressed itself as a determined advocacy for the cause of increasing inter-American commerce, at the vanguard of which would be Wall Street and Pennsylvania Avenue. One of his encyclopedic books, *Trade and Transportation Between the United States and Spanish America*, was a qualitative and quantitative call to arms for more extensive commercial exchange within the Western Hemisphere.[67] As historian Stephen Topik has observed, Curtis became "the leading propagandist for Pan-Americanism."[68] In the run-up to the 1893 World's Fair in Chicago, Curtis provided a series of newspaper releases for the Associated Press that featured predictions by prominent figures in business and politics as to what the United States would look like in 1993. Curtis envisioned an informal imperialism in the Americas that would be supported by a railroad from Chicago to Buenos Aires and multiple voyages of merchant ships tying together hemispheric ports:

> The manifest destiny of the United States is to dominate the American hemisphere. This will be accomplished not by political intrigue, not by diplomatic negotiations, not by force of arms, not by annexation of territory, and not by the establishment of protectorates—but by the influence of example and by commercial relations. The tie that will bind the American republics and colonies will be the tie of trade. And in 1993 American commerce, to a very large degree, will be confined to American waters.

His prognostication of hemispheric self-sufficiency under U.S. domination took on a religious quality: "The Creator intended there should be an exchange of products between the American continents. He distributed their natural resources so that their population can live in prosperity and contentment without an ounce of European or Asiatic merchandise."[69]

THREE VOICES FOR EMPIRE

Curtis, Davis, and Bourke apparently never met, but they shared a vision of the United States as the dominant power in the Western Hemisphere. They were sensitive to public opinion. Davis cultivated a broad readership that purchased his novels and followed his reportage, making him one of the highest-paid writers of his day. Curtis concentrated on politics and business, while Bourke focused on his military experience. Curtis and Bourke had smaller readerships, but they included individuals involved in the policy-making processes in Washington and New York. All three wrote in anticipation of the rising public interest in U.S. expansion in the Western Hemispheric, but they had different perspectives. Bourke discussed the advantages of formal empire. Davis and Curtis advocated informal empire. Davis personalized the marginal risks and substantial rewards of direct investment in mining through the fictional character of Robert Clay in *Soldiers of Fortune*. Corporate control of resources in Latin America was one of the building blocks of informal imperialism. Curtis cited statistics and composed prose descriptions to build his case for increased trade, the second building block for a structure of imperial domination. Davis and Curtis projected business expansion, a direction their readers greeted with some enthusiasm.

This imperial mentality was no longer represented by a set of seldom discussed, vaguely formed policy ideas but instead had acquired a formidable presence in big-city newspapers and popular magazines. Matías Romero's critique of Garza's rise to prominence as a result of what he termed "the desire of the journals in the United States to [print] sensational news" has relevance here. Garza of course opposed U.S. interference in Mexican affairs and could be classified as an anti-imperialist. But the main point is that the story of the border conflict and its imperial implications (i.e., the alleged need for the United States to assert its authority to maintain order) caught the attention of readers across the United States. Davis achieved wide public recognition in developing this theme in later publications through the sensationalistic methods that Romero found both extreme and threatening. Romero apparently was not so disturbed by the plans and publications of Curtis, but this tolerance most probably arose from the close relations between the Díaz government's business clique and the Washington–New York axis.

The coalescence of the imperial mentality as espoused by Davis in his popular fashion and by Curtis and Bourke in less compelling prose not only found reinforcement on the front pages of Joseph Pulitzer's *New York World* and William Randolph Hearst's *New York Journal* but also contributed to the headlines of other prominent and more respectable newspapers. Davis and other foreign correspondents reported from exotic locales often via cable telegraph with the excitement of immediacy and, by implication, the weight of eyewitness authority with respect to crises that impacted the politics and policies of the U.S. government. Curtis made a less spontaneous but equally convincing case for greater U.S. involvement in Latin American affairs as a means of safeguarding the paths of commerce. In the next chapters, I explore how events in Nicaragua, Venezuela, and Cuba brought the U.S. press and public into a new international community in which the imperial mentality was a guiding force for national leaders as well as ordinary citizens.

CHAPTER EIGHT

Propaganda, Public Uproar, and the Threat of War

The United States, Great Britain, and the Venezuelan Boundary Controversy

PRESIDENT HARRISON MASTERED the use of international telegraphy in his dealings with Chile, but the new international information system had the potential to move messages and circulate news at a speed and intensity that frustrated and confused most government officials. This chapter explores the role of the information system in the Venezuelan boundary controversy of 1895–1896 that escalated to the level of a war scare. While government leaders seemed to operate in slow motion in their reliance on steamships to deliver their correspondence, newspapers exploited the speed of electric communications, which amplified the potential for hostility. And, in a bizarre twist, Secretary of State Olney broke with the sluggish methods of traditional diplomacy and employed two prominent journalists to use the cables and the front pages of newspapers to carry on an unorthodox form of public diplomacy. Yet, in a retrospective essay, Olney joined other officials and press commentators in admitting that for a tense period of several months, diplomacy had failed at the same time that the cables and newspapers were inspiring a dangerous belligerence across the United States.[1]

The abrasive assertion of the Monroe Doctrine by Olney and President Cleveland found reinforcement in the propaganda work of the Venezuelan government and its agent, William L. Scruggs. Venezuela's minister to the United States, José Andrade, collaborated with former diplomat and newspaperman Scruggs, who worked as a lobbyist and propagandist for Caracas. While neither Andrade, nor Scruggs, nor their supervisor in Caracas, Ezequiel Rojas, minister of foreign relations, mentioned the public- relations efforts of Chilean Marcial Martínez in the U.S. print media, they took similar advantage of the open access to the press and the telegraph news network with its ability to broadcast across the nation. Scruggs and Andrade played on familiar themes in U.S. politics and popular culture—especially the Monroe Doctrine, which had been a part of the nation's core political values since 1823.[2]

Caracas had placed publicity in the U.S. press on the boundary issue in the 1880s, and Rojas mounted a campaign directed at the State Department and other Western Hemisphere nations citing British violations of the Monroe Doctrine on the eastern border of Venezuela in early 1894.[3] Seen in this context, the propaganda of William Scruggs emerged as part of a well-orchestrated publicity effort directed by Andrade and Rojas based on earlier experiences of Venezuelan diplomats. Andrade arrived in Washington in 1893 at the age of forty-five after having served as speaker of the Venezuelan Congress. He was a quick learner. His ability to understand the nuances of politics and the national culture of the United States was in evidence over the next few months. His collaborator Scruggs was formerly the U.S. representative in Caracas. Their cooperation produced a pamphlet that had a resounding impact not only in the United States but also Great Britain and eventually throughout Latin America. Scruggs gave credit to Andrade in a letter to Rojas: "Before placing the ms. in the hands of the printer, I submitted it to Dr. Andrade who kindly made some suggestions which I adopted. I am indebted to him also for data which enabled me to trace the history of the controversy to date."[4] Published in October 1894, *British Aggressions in Venezuela; or, The Monroe Doctrine on Trial* circulated widely. Scruggs, with the financial support of the Venezuelan government, sent copies to President Cleveland, members of the Senate and House of Representatives, state governors, and state legislators. He also targeted the press, including the editors of leading newspapers and magazines. Copies also appeared at newsstands and in bookstores.[5]

By the fall of 1895 the Venezuela boundary question was thoroughly interwoven with the Monroe Doctrine. The truculent rhetoric of Scruggs and like-minded champions of Monroe's principle had entered the national arena first in this pamphlet and then gained resonance in a series of bombastic speeches by U.S. politicians reported by the press, including magazines as well as newspapers. Salisbury's five-month delay in responding to Olney's memorandum of July and the silence of the U.S. State Department and the British Foreign Office created a vacuum in the news coverage.

Olney's communiqué was confidential, but the U.S. press built on stories of an impending conflict between Britain and Venezuela with rumors of a strongly worded demand from Washington. In October and November 1895 rumors and partial reports featured discussions of U.S. involvement in hostilities. The print media from New York to San Francisco sounded the drumbeat of jingoism. The recently constructed wire services and their client newspapers abhorred the vacuum in the information flow on an issue that had risen spectacularly from the low status of a distant, obscure boundary dispute to an international crisis. Analogies, rumors, and half-truths along with some responsible reporting rushed in to fill the void.

While the British Foreign Office and the U.S. State Department seemed content to use traditional methods of communication (letters sent by ship), the U.S. press sent its messages by electricity, converting the wire services and newspapers of the United States into a nearly instant information system. The result magnified the contrast between the ponderous diplomacy of an older generation of leaders on both sides of the Atlantic and the dynamic, headline-generating coverage in daily newspapers from New York to San Francisco.

AT THE CENTER OF THE STORM

These outbursts of anti-British feeling stunned both British and U.S. officials, but there had been signs that the combination of submarine telegraphy and the popular press had explosive potential. Nearly a quarter of a century earlier, in 1872, an editorial in the London *Times* highlighted the dangers as opposed to the advantages in the new communications system. The event that led to the editorial was a dinner at Mansion House in the heart of London to honor Cyrus Field, the chief promoter of the first transatlantic telegraph cable. The prime minister, William Gladstone, was in attendance and joined in the congratulations lavished on Field. The *Times*, while respectful of the promoter and the prime minister, nevertheless expressed grave doubts about the consequences of rapid communications that, in light of the events of 1895–1896, had a prophetic quality. Gladstone opined that the common language of the two nations combined with the new means of communication constituted "potent safeguards of peace and goodwill" between the two countries, but the *Times* editorial disagreed:

> It is precisely the community of language between nations whose modes of thought and social organization are so different which gives meaning and point to every harsh word or piquant criticism of English

writers or speakers about Americans. It is precisely the extension of this electric telegraph across the Atlantic, largely due to Mr. Cyrus Field's efforts, which has facilitated the instant publication of all such words and criticisms, generally without their context and not infrequently with malicious additions, in every city of the United States. The mischief thus occasionally done can hardly be overstated. . . . The danger of rupture was possibly greater, on the whole, when months passed and seas rolled between the dispatch and the reception of a pacific overture; but telegraphic communication has its evils too, against which the American Press, which uses it so liberally, ought to be on guard.[6]

This prescient warning was forgotten in 1895. The Cleveland administration had been waiting for a response from the British since July 20, 1895, and it arrived too late to be included in the president's scheduled speech to Congress on December 2. Cleveland and Olney were angry, and their emotions came through in the president's special statement in defense of the Monroe Doctrine and Venezuela on December 17. According to historian R. A. Humphreys, the content of Cleveland's message "startled and shocked Great Britain" and electrified the United States.[7] The press and politicians in Britain were caught by surprise, but British diplomatic officials in the United States were aware of the ascending level of hostility. As early as March 1895, British ambassador Julian Pauncefote's secretary, Cecil Spring Rice, had commented in personal letters to his sisters about the "floods of abuse" being heaped on Great Britain.[8] Spring Rice was more explicit in a letter to a friend: "The South Americans are in with all the low press men and every sort of lie is propagated about British aggressions." In the same letter he expressed exasperation: "You might as well conciliate a jackal or, let us say, a tiger." The animosity against Great Britain is so "acute" that there is "no hope of ameliorating the situation."[9]

The first major outburst occurred two months before Cleveland's December statement. The *Washington Post*, the *New York Times*, and the *San Francisco Examiner* published stories on British and Venezuelan preparations for armed conflict. Both sides had purchased or arranged to purchase Maxim guns to be deployed in the disputed territory.[10] The *New York Times* also claimed that British troops were moving into Venezuelan territory. The *Times* wove together a compelling mixture of facts and rumor:

It is now believed by well-posted diplomats that this report is well-founded,

> that the force in question consists probably of a limited number of British police in British Guiana . . . and that it is now crossing [the disputed area] with the view of entering a part of Venezuela not occupied by the native troops. If it shall be shown that this is really the case, the purpose of the British Government in sending this force into Venezuela can only be conjectured, but it is not doubted that a definite purpose lies behind it.[11]

Stories of war preparations indicated that the press demand for information—whether rumor or fact—was especially intense from October 19 to 22. While details in the accounts varied and some of the information was accurate, most of the press coverage centered on the alleged British and Venezuelan decisions to prepare for war in Guiana and the possibility that the United States would enter the dispute. The similarity in the coverage of and speculations about the outbreak of hostilities indicated that newspapers across the country were drawing most of their information from the transatlantic cable. For example, on October 20 and 21 ten metropolitan newspapers—the *Washington Post*, the *New York Times*, the *New York Tribune*, the *Boston Globe*, the *Philadelphia Inquirer*, the *Chicago Tribune*, the *New Orleans Picayune*, the *Atlanta Constitution*, the *San Francisco Chronicle*, and the *San Francisco Examiner*—all used the word "ultimatum" to describe the British note to Venezuela. In all ten newspapers, this story carried a dateline of October 19 from London. The international communications system was feeding information to the AP and other wire services in the United States, conveying news from London across the Atlantic Ocean and North America in a day or two.[12]

Somewhat reserved in its jingoism in October, the *New York Times* abandoned moderation and devoted the entire front page to Cleveland's message on December 18 and included coverage of favorable responses to the president's statement datelined Richmond, Virginia, Columbus and Cleveland, Ohio, and Topeka, Kansas.[13] The editorial staff surveyed newspaper opinion and, on December 24, the *New York Times* reprinted excerpted editorials that reflected its opinion of forty-six newspapers from nineteen states. Connecticut and Georgia on the East Coast and Montana and Utah in the Rocky Mountains were included, and a large number of smaller cities, such as Binghamton and Utica (New York), Toledo (Ohio), Quincy (Illinois), South Bend (Indiana), Butte (Montana), and Little Rock (Arkansas), were represented.[14]

The typical pattern of news coverage and opinion pieces in the *Washington Post*, the *New York Tribune*, the *Boston Globe*, the *Chicago Tribune*, the *San Francisco Examiner*, and the *New Orleans Picayune* consisted of telegraphed stories from Washington and London combined with bellicose statements from local officials, often active or retired military personnel. The *Boston Globe* quoted Civil War veterans who demanded unbending support for the Monroe Doctrine.[15] The *San Francisco Examiner* was the most overt in its call for war, offering comparisons of the British and U.S. navies and armies that were accompanied by several drawings of warships. Logic was not a major concern at this juncture for the *Examiner,* which noted in its comparative listing of warships that "the United States makes a poor showing, but she has some magnificent warships." The *Examiner's* conclusion that the U.S. Navy was "decidedly weak" in comparison with that of the British did not shake its stand in favor of war.[16]

Cleveland's December message brought these anti-British feelings into a clear focus. The Colonial Office in British Guiana followed the responses of the press and the public in the United States. Joseph Chamberlain, head of the Colonial Office, was concerned but resolute; however at least one of his high-ranking assistants was shaken by the outpouring of anti-British feeling. Lord Selborne, Chamberlain's undersecretary, could not understand this reaction: "The hatred . . . of England by Americans is quite unaccountable. We expect the French to hate us and are quite prepared to reciprocate the compliment if necessary; but the Americans, No!"[17]

Probably the most extensive discussion of Cleveland's missive and its massive public support came from the pen of British ambassador Julian Pauncefote. A veteran diplomat at the age of sixty-seven, Pauncefote had served as the British minister in the United States since 1889 and became the first British diplomat to serve at the rank of ambassador to Washington in 1893. He had weathered the Bering Sea dispute in the early 1890s, but nothing in his dealings with the government in Washington prepared him for the stridency of 1895. Pauncefote was disturbed by the enthusiastic applause with which congressmen greeted the belligerent portions of Cleveland's message, but the British ambassador was even more shaken by the ease with which the general public could be aroused to a condition of fanatical patriotism. He concluded that the U.S. government's formulation of foreign policy was immature and downright dangerous.[18]

Pauncefote found himself in a vulnerable position with respect to the print-media storm. The British ambassador had a duty to review the newspapers of the capital and those of other large East Coast cities, and he also witnessed the methods and manners of the nation's growing press corps that produced the coverage. In a brief letter to the Foreign Office, Pauncefote made clear his avoidance of contact with the press of the United States.[19] Instead he concentrated on the end product of the journalistic endeavor—the newspapers themselves, in which he detected alternating cycles of media intensity. After the first outburst of mid- to late October 1895, Pauncefote reported that "the press was raging and lying outrageously about Venezuela," but in the same letter (dated October 25), he also noted that this particular surge was beginning to diminish.[20] On November 8 he commented on another wave of "fantastic speculation . . . indulged in by irresponsible chatterers . . . as to the time and character of the British reply to what they termed 'Olney's ultimatum.'"[21]

Cleveland's aggressive reply to Salisbury on December 17 spawned the third and most intense media storm of this chapter in U.S.-British relations. A member of Pauncefote's staff noted the close connection between the press and politics when he observed members of the House of Representatives who "eagerly bought up" late-afternoon copies of newspapers that contained Cleveland's message even before it was formally presented to that body. These newspapers were on sale in "the approaches" to the House of Representatives wing of the capitol building on the day of the president's pronouncement.[22] In the excitement, other issues seemed to disappear from view. Pauncefote observed that "nothing is heard but the voice of the jingo bellowing out defiance to England." He confided to the prime minister that "we must wait until the noise had subsided to judge the real attitude of the country."[23]

Defying Pauncefote's expectations, an unnamed official (or officials) in London attempted to explain the unpredictable Americans. An unsigned internal memorandum in the Salisbury Foreign Office, dated December 20, 1895, offered an analysis of the political motives behind Cleveland's address and the factors that connected politics, the press, and the general public uproar in the United States. According to this analysis, Cleveland was deeply troubled by the scathing criticism directed against his administration because of its inaction in the Nicaragua crisis earlier in the year. The arrival of Olney as head of the State Department and Cleveland's urge to earn public approval as a forceful leader led to the unexpectedly aggressive position on the Venezuela boundary issue. The result surprised even the Secretary of State with "popular enthusiasm exceeding Mr. Olney's expectations." But expanded press coverage and popular excitement did not bring about an improvement in public understanding of the issue. The agitated public simply did not read the basic documents and did not fully grasp the issues.[24]

As the Christmas holidays approached, the excitement did not diminish.

Cleveland's pronouncement was followed by a brief but sharp sell-off of stocks on the New York Stock Exchange. The combination of a belligerent foreign policy and a sudden financial downturn seemed to intensify the emotionalism. Pauncefote was critical of the absence of basic information and the skewed thought processes in the cultural-political environment in which the government made its foreign-policy decisions:

> This episode presents an illustration of the singular want of knowledge which prevails generally in the United States in regard to pending diplomatic questions, and of the facility with which the people can be roused to a state of feeling akin to political delirium by such an appeal to their patriotism as the President has resorted to in the present instance.

Later in the same letter Pauncefote implicated the communications media as the nervous system that elicited the patriotic excess that "had produced in Congress and among the public a condition of mind which can only be described as hysterical."[25]

THE BRITISH ANALYSIS OF ANGLOPHOBIA

In addition to Pauncefote's frequent notes from the center of bluster and hyperbole, Salisbury received reports from a trusted political ally who quietly traveled in the United States in March and April 1896. Sir Stafford Northcote, a Conservative member of Parliament, interviewed E. L. Godkin, editor of the *Nation*, a well-known journal of political and social commentary that opposed jingoism. The veteran editor emphasized that the people of the West—meaning the states and territories west of the Mississippi River—were generally ignorant of international affairs and seemed especially resentful of the British. Westerners disliked and distrusted easterners of the United States and, rightly or wrongly, associated them with the British. Godkin noted that school history textbooks treated Great Britain harshly, emphasizing the American Revolution. Northcote asked if anything could be done to improve the British image. Godkin responded in the negative. Only the passage of time and the arrival of a new, more enlightened generation would reduce the hostility.[26]

About a month after his interview with Godkin, Northcote summed up his impressions of the press, politics, and the general culture of the United States. He sought to understand why such strong anti-British feelings abounded. The tendency to defend the Monroe Doctrine was important, but, by late April, it seemed to be a less compelling issue than it was a few months earlier. Northcote noted, however, a deeply entrenched perception that resonated with some aspects of the Monroe Doctrine—a resentment against the British on the grounds that they were international bullies who took advantage of their financial and commercial supremacy to despoil the lands and resources of weaker nations. He repeated Godkin's point about anti-British history texts and westerners' resentment toward upper-crust New Yorkers and New Englanders, whom they regarded as protégés of the British.[27]

Several British commentators outside the Salisbury government also expressed concern about the outburst of hostility toward the British in the United States. Henry Morton Stanley, a veteran British journalist and renowned explorer with extensive experience in the United States, held that the jingo press represented something more widespread than the regional phenomenon Godkin described. Stanley arrived in New York City in mid-September 1895 to discover American journalists in that city "impatiently awaiting" Salisbury's response to Olney's memorandum. Stanley observed that the East Coast "press had already prejudged the question." As to why such a precipitous judgment had already taken place, Stanley found the answer in "the bitterly unfriendly tone" of much that appeared in the newspapers and in his discussions with fellow journalists. This "dislike of England was provoked by many things, and was of such long-standing that nothing short of war would satisfy the majority of Americans."[28] Stanley was as shocked as his readers on the eastern side of the North Atlantic and used the familiar mountains out of molehills metaphor to try to dissuade his American colleagues but to no avail. The Olney message and Cleveland's public statement were not simply electoral politics and transient extremism but "the expression of American sentiment and American conviction" that were ingrained in its political culture on both sides of the Mississippi River.[29]

British writers such as G. D. H. Gossip and W. T. Stead reached similar conclusions. Taking an historical approach, Gossip saw this hostility as the outgrowth of more than a century of U.S.-British relations.[30] He stated bluntly: "Anti-English sentiment among the American people, born with the American Revolution and fostered by the War of 1812, has never died out."[31] His observations of massive patriotic celebrations in the United States and the generally respectful if not reverential attitude toward the Monroe Doctrine led him to conclude that "there is no disguising the fact that the situation is alarming."[32] And the press aggravated the distress: "On the whole the tone of the American press is hostile to England."[33]

An editor, reformer, and practitioner of sensationalism in his *Pall Mall Gazette,* Stead made a distinction between the jingoism of Henry Cabot Lodge and the more reasonable assertions of Albert Shaw. The latter, according to Stead, was neither a politician nor a jingo but a serious thinker of mature judgment whom Stead had placed in the editorship of the *North American Review.* Shaw's calm arguments in support of the Monroe Doctrine represented an element in the political culture of the United States that would respect the British point of view and perhaps, in time, accept British efforts to educate the U.S. public as to their intentions in South America. Unfortunately, according to Stead, it was Lodge who furnished "a fair sample of the way in which the case against England is constantly urged in the American press."[34]

Writer Edward Dicey, a British academic, offered his insight into the political culture of the United States.[35] He observed that the president's defense of the Monroe Doctrine, no matter how crude, effectively touched on a widely held belief in the United States that European expansion in the Western Hemisphere should be precluded. In his admonition, he posited that schools and the press had a lot to do with this sentiment:

> It is only fair to remember that belief in the "manifest destiny" of the Great Republic to become the dominant power over the length and breadth of the American continent is implanted in every true American by the education he receives, by the traditions on which he is nurtured, by the very air he breathes.[36]

These essays, letters, and reports testified to the surge of hostility in the United States. Salisbury's Foreign Office even edited a volume of documents intended for the U.S. public, but it had little impact.[37] In the words of Harry Whates, an early biographer of Salisbury, the prime minister saw the Olney-Cleveland policy largely in terms of the upcoming presidential election: "Another move in the sordid game of American domestic politics . . . and a bid for the Irish-American vote."[38] Pauncefote evaluated the storms of patriotic excess in a larger framework. He saw successive waves of blustery, often fallacious criticism in the press followed by troughs of relative calm. Northcote, Stanley, Gossip, Stead, and Dicey took longer views at a distance from the center of the political storm and located cultural factors that they believed underlay the fluctuating outbursts of emotionalism that dominated the front pages of metropolitan newspapers. They saw resentment against perceived British wealth and power in the western states at a time when the populist movement was challenging the two-party system by responding to the antielitist attitudes among struggling farmers and small town folk who identified the financiers of Wall Street with their counterparts in England. They also saw anti-British feelings among reporters and editors of the East Coast who shared with their big-city Irish-American readers' animosity toward British expansionism in the Western Hemisphere. The populism of the West and anti-British sentiments in eastern cities reinforced veneration of the Monroe Doctrine in newspapers and magazines across the country.

CLEVELAND, OLNEY, AND THE PUBLIC

As the uproar in the press became a patriotic crescendo, two of the chief instigators commented on the disturbing force of public emotionality. President Cleveland and Secretary of State Olney "had foreseen nothing like the boisterous acclaim which their critics described as a popular outburst of jingoism."[39] Cleveland understood the need "to stem the tide of 'jingoism'" but did not know how to do so. He and Olney had hoped their defense of the Monroe Doctrine would deflate jingoism, but when that tactic backfired and produced a war scare, they saw the need for accommodation with the British.[40] Cleveland was surprised by the persistence of the aggressive mood as expressed in the inflammatory language of the print media. The business community—especially Wall Street—was shaken by the decline of stock prices and supported the president's turn toward a diplomatic solution. This shift among corporate leaders brought criticism from champions of patriotism. Theodore Roosevelt was disgusted by the advocates of peace among Wall Street businessmen.[41] A few months later Cleveland, in an interview with Stafford Northcote, confessed that he had felt pressure from the jingoes who accused him of lacking patriotic zeal. As a consequence, he had spoken in haste with unfortunate results.[42] Ambassador Thomas Bayard objected to the Cleveland-Olney hard-line diplomacy, and his revulsion for the U.S. press was unequivocal. He damned the "mischievous and reckless sensationalism of the Press in continuing to give currency to rumors" that could only worsen British-U.S. relations.[43]

Prominent figures outside the Cleveland administration were also shaken by the uproar. Harvard professor Charles Eliot Norton, like Bayard, condemned the press. His criticism was contained in a brief but far-reaching litany of complaints about what he saw as the decay of public civility and social order in the United States. Norton cited as examples the cowboy ruffians from the West and the ruffians of the football fields back East as well as corruption in politics and government. He blasted major metropolitan newspapers that in his view catered to the lesser instincts of their readers in order to boost circulation. An example of this sort of moral turpitude was jingoism—the appeal to "that barbaric lust of conquest and dominion." To Norton the possibility of war with Great Britain was absurd, an absurdity made more likely by Cleveland's "astounding message respecting the Venezuela boundary dispute" and its "popular reception."[44]

A longtime friend of Norton, E. L. Godkin had a similar reaction to the patriotic hyperbole in the press. In a letter to Eliot, Godkin attacked the anglophobia of the public and the information system that blasted it across the country:

> The present crisis is really a fight between the rational business men, and the politicians and the newspapers, and the rational business men are not getting the best of it. The press is the worst feature in the situation, and yet the press would not be what it is without the public demand for what it is.[45]

A few days later Godkin wrote to British commentator James Bryce on similar themes:

> You greatly, in England, overrate the popular intelligence here. A generation has come into the field which has since it left school been fed on a very vile and silly press. It is cut off from all other communities. It knows nothing of Europe, and knows nothing of any history before 1776. It does not read books at all. It is made perfectly drunk by the knowledge that it is 70,000,000 strong, and has a large revenue. Its morals are very low, and it has no religion to speak of, and no public men who act the part of teachers, or exact any moral authority.[46]

For Olney, the novice secretary of state, the uproar surrounding the Venezuela boundary controversy was a learning experience that revealed the power of the national press in the conduct of foreign policy. He decided in early 1896 to publish the U.S. government's perspective on the controversy, asserting that the "effect on the public mind both in this country and in England . . . would, I think, be of great value."[47] Two years later Olney wrote an essay for the *Atlantic Monthly* that emphasized the importance of well-informed public opinion and of well-vetted channels of communication, which furnished information and opinion to the public. From 1894 to 1896 this system of communication revealed its flaws: "Perhaps the importance of enlightened public opinion as well as the lamentable absence of it was never more strikingly demonstrated than by the circumstances attending what has come to be known as the Venezuela boundary incident." The public rallied to the Monroe Doctrine and the Cleveland administration while "only most dimly comprehending what the government had done or why it had done it."[48]

DIPLOMACY BY CABLE AND NEWSPAPER

As the Venezuelan boundary crisis reached a critical phase in the early months of 1896, Secretary of State Olney employed a type of diplomacy that made use of transatlantic telegraphy and newspapers. His use of journalists in unorthodox negotiations that led to sensitive information being placed on the front pages of newspapers seems almost absurd to the historian. While his decision to pursue this approach was perhaps motivated by his frustrations with Ambassador Bayard in London, the idea of diplomacy through cable and press certainly did not fit into the expectations of diplomats such as Salisbury and Pauncefote given their recent experience with the commotion in U.S. newspapers and magazines.[49]

Olney decided to work with Henry Norman, the special correspondent for the *Daily Chronicle* of London assigned to Washington.[50] Norman agreed to meet with Olney several times during the first two weeks of 1896.[51] The British journalist, coached by the secretary of state, presented Washington's view of the Venezuelan boundary issue to the readers of the *Daily Chronicle*. Norman pointed out that the general public of the United States took the Cleveland-Olney position seriously and that British policy based on the assumption that the American demands for arbitration could be ignored had little chance of success.[52]

The *New York Times* and Ambassador Pauncefote followed Norman's writing in the pages of the *Daily Chronicle*, although they took different views of his dispatches. Apparently neither the *Times* nor Pauncefote knew that Norman was working for Olney. The *New York Times* described him as "Great Britain's best teacher, . . . who tells the truth about the Venezuelan situation." The *Times* judged Norman's dispatches to be a much needed counterbalance to the unfavorable image of the Cleveland administration and termed them "a burst of sunshine through the fog" in Britain.[53] From his vantage point, Pauncefote found little to respect in Norman's publications. The ambassador wrote to Salisbury that the journalist had "cabled something foolish to

London yesterday which has been published under the Head Line of 'England Breaks Down.'"[54] Norman eventually lost favor with Olney as well, however, when he stressed the willingness of the government in Washington to reach a settlement in Venezuela in order to avoid conflict with Great Britain.[55]

The Olney-Norman collaboration, brief as it was, used public diplomacy, that is, attempted to sway government leaders by making a convincing case to that nation's newspaper-reading public. After this failed effort, Olney moved on to another venture in public diplomacy with another reporter, American-born anglophile George Smalley. As the London *Times* correspondent, Smalley had reported on the coverage of the Venezuelan crisis in the major newspapers of New York City in terms favorable to the Cleveland administration. A native of Worcester, Massachusetts, and a graduate of Yale, Smalley had pioneered the creation of a coordinating European news bureau in London for the *New York Tribune* during the Franco-Prussian War of 1871 (see chapter 1). He continued with European assignments before returning to the United States to serve as the correspondent for the prestigious London newspaper.[56] Smalley downplayed the jingoism in the New York press early in the crisis.[57] Cleveland's threatening reply to Salisbury on December 17, however, brought what Smalley called a "volcanic reaction" in the New York press.[58]

In spite of the difficult circumstances, Smalley worked with Olney for a settlement of the dispute.[59] However, the journalist's haste to reach a settlement and his reliance on the telegraph contributed to a misunderstanding that turned into a diplomatic blunder. On January 24, 1896, Smalley telegraphed to London *Times* editor G. E. Buckle details on the proposed arrangements for a U.S.-British arbitration commission, and the Salisbury government promptly agreed to these conditions. But, unfortunately for Smalley, his conversation with Olney on January 26 indicated that the Cleveland administration would not accept some of the provisions in the proposal that the journalist had originally sent to London two days earlier. The British acceptance of the first set of proposals and Smalley's amendments to these same proposals passed through the submarine cables on the same day in opposite directions. This "fiasco," as historian Joseph Matthews termed it, set negotiations back during a critical time.[60]

About three weeks after this embarrassment and for reasons that are difficult to ascertain, Smalley used a new and, from the point of view of Salisbury and Pauncefote, an even more irresponsible method. Smalley sent a long telegram to editor Buckle which contained an earlier proposal for a settlement along with some additions made by the *Times* reporter. Buckle published this proposal in the London *Times* along with his response, giving the appearance, in a way, of speaking for Salisbury. Over the next several days, British-U.S. negotiations seemed to be conducted in full public view on the pages of the London *Times*.[61]

The main themes of this book do not necessitate a review of the details of the Smalley proposals (they concerned mainly the nature of the arbitration agreement for the resolution of the dispute), but the assertive journalist used his sway with the London *Times* to try to bring pressure to bear on the British prime minister. Smalley warned that the "jingo press in America" and its stepchild, "popular agitation," could easily limit the options open to Cleveland and Olney and thereby place Salisbury in an awkward position. The multitude of proposed policy options, patriotic bombast, and homilies on the Monroe Doctrine had become common currency in the U.S. press, but it might die down if Salisbury would accept the general notion of arbitration on the boundary issue.[62] Smalley himself began to feel pressure, as he responded to accusations concerning the origins of the proposals he placed in print. According to his evasive responses, these proposals came from unnamed sources in the State Department but were of nebulous origin as indicated in the following quotation: "The plan which I sketched is the one which the government is ready to propose officially as soon as it has reason to believe that it would be accepted by the British government."[63] Smalley struggled to maintain what little credibility his work had in Washington and to convey that to the offices of the *Times* in London through the use of the transatlantic cable.[64]

The image of the peripatetic Smalley contrasted with that of the deliberate, sedate Salisbury who moved according to his own schedule.[65] In the wake of Smalley's forays into diplomacy Salisbury wrote to Joseph Chamberlain that "this attempt by Smalley to negotiate in print is absurd."[66] Chamberlain agreed: "Smalley is getting mischievous and has certainly ceased to be useful."[67] In general, Salisbury condemned Smalley and Norman as "amateur diplomatists." According to historian J. A. S. Grenville, the usually taciturn Salisbury rendered a "rather scathing judgment on the effects of the unofficial go-betweens."[68] The information system allowed reporters to take on the roles of diplomats, thereby complicating an already inflamed situation. In a lighter mood after tensions had diminished, E. L. Godkin ridiculed the diplomacy by newspaper and telegraph as performed by Smalley and Norman:

> There is only one way out of this imbroglio, and that is the communication to more newspapers—say twelve—of the real secret of the negotiations. Delays are proverbially dangerous. In the multitude of newspapers there is safety, and no plan which has only one newspaper behind it can command the confidence of a great people.[69]

CONCLUSIONS

Salisbury acceded to the will of his cabinet on January 11 and agreed to begin discussions that soon led to the creation of a five member joint British-U.S. arbitration committee that, over the next three years, worked out a settlement of the disputed territory. The detailed work of the arbitration committee and the routine of diplomatic correspondence replaced the inflammatory rhetoric of Scruggs, Lodge, Olney, and Cleveland and the intrusive forays of journalists/amateur diplomats Norman and Smalley. The final arbitration awarded most of the disputed territory in the interior to British Guiana, but the key port of Bairima went to Venezuela.[70]

An unexpected stimulus for Salisbury's flexibility with the United States was British entanglement in a second diplomatic crisis. The second crisis also involved submarine cables and jingoistic journalism. Long-simmering differences between British entrepreneurs and Dutch farmers (called Boers) in South Africa reached critical proportions in late December 1895 when the headstrong Leander Starr Jameson, a close confidant of mining magnate Cecil Rhodes, led a reckless, unsuccessful raid against the Boers. Their military defeat was bad enough for the British, but on January 3, 1896, Kaiser Wilhelm II of Germany sent a congratulatory telegram to Paul Kruger, the leader of the victorious Boers. This telegram, which was published in the British press, and the outpouring of newspaper stories in Europe seemed to suggest that the colony-seeking Germans might be well on the way to some sort of alliance with the Boers against the British. Furthermore, a public controversy arose when it was alleged that the head of the British Colonial Office, Joseph Chamberlain, conspired with the raider Jameson, a charge later partly verified by the revelation of the content of telegrams from Rhodes's agents in England to Jameson in South Africa. While the details of this controversy do not directly concern us here, it is important to note that the speed and unpredictability of events and the hyperbole and sensationalism employed by journalists were factors in the coverage of events in southern South Africa as well as northern South America. The impact the kaiser's telegram to Kruger had on U.S.-British relations is difficult to measure, but a reasonable assessment is that this cable brought on a spurt of energetic diplomacy from London that had been absent from July 1895 to January 1896.[71]

The importance of the Venezuelan government in the boundary controversy is often overlooked as a result of the massive outpouring of press and political commentary on the momentary confrontation between Great Britain and the United States. The government in Caracas began its efforts to win the support of Washington in defense of its territorial claims as early as 1889. The Crespo administration placed much responsibility for its publicity campaign in the hands of José Andrade, who proved to be a capable representative in Washington and an able collaborator with Scruggs in assembling information for the 1894 pamphlet. Andrade and Scruggs continued to push Venezuela's case with Cleveland's State Department, key U.S. congressmen and senators, and the nation's press corps. Viewed in this sense, the Andrade-Scruggs collaboration triggered an intensification of newspaper and magazine interest in the dispute, but they quickly lost control of the media coverage, displaced by the spontaneity and emotionality of jingoism. Unfortunately for Andrade, Scruggs, and Venezuela, the two major powers decided to resolve the dispute according to their own interests. Venezuela learned of the results of the arbitration much as the rest of the world did—through government announcements from Washington and London.

The actions of individuals such as Scruggs and Andrade, Olney and Cleveland, Salisbury and Pauncefote were important in this affair, but the role of the relatively new international communications system—telegraph, wire service, and press—should not be overlooked. The Scruggs-Andrade collaboration injected Venezuela's message into the U.S. media through a variety of forms: letters to newspaper and magazine editors, a printed pamphlet, and congressional lobbying. The telegraph lines of Western Union and the transatlantic cables relayed news to and from Washington and London and across the North American continent on a daily basis, creating a drama that held the attention of a large audience of newspaper readers on both sides of the Atlantic. Cleveland biographer Richard Welch's conclusion reinforces the nature of the president's role in the press:

> The significance of Cleveland's Venezuelan diplomacy was largely confined to its influence on the American public. Not only did it encourage

an initial spasm of militant nationalism; it also had the consequence of stimulating public acceptance of the Monroe Doctrine as an eternal principle of United States foreign policy.[72]

Equally important, none of these political figures, including the president of the United States or the prime minister of Great Britain, was able to master the situation, as the press and what was understood to be public opinion in the United States seemed to be running out of control. The flow of telegraphed stories from London, Washington, New York, and occasionally from Venezuela and British Guiana fed the anglophobia that erupted on the pages of newspapers across the United States. Pauncefote, Chamberlain, and Salisbury were shocked and dismayed, as were Olney and Cleveland, who watched in consternation while their stand against the British not only failed to disarm U.S. jingoism but actually strengthened it. Furthermore, journalists Norman and Smalley stepped across the boundary between reporting and diplomacy (with the collaboration of Olney) and attempted to conduct foreign policy via cable and newspaper. The balance in the information flow during the Venezuela boundary controversy had shifted toward the press to the distinct disadvantage of official diplomacy. The uncertain balance between journalism and diplomacy gave way to a situation in which the press not only dominated the flow of information but also had an input in diplomacy.

The flow of information over the telegraph lines reflected the values of two communities. One was the national community in the United States that encompassed the Midwest, the South, and the Pacific coast, along with the historical centers of national power, the mid-Atlantic and northeastern cores that included the cities of Washington and New York and the states of New England. The second was the transatlantic community of politicians, diplomats, businessmen, and university professors in London, Washington, New York, and Boston, making the mid-Atlantic and northeastern cores overlapping regions within both communities. This second community tended to reject patriotic extremism for negotiated solutions that would satisfy business leaders and diplomats. The Cleveland-Olney administration attempted to work within both of these communities and discovered, to its chagrin, that their respective political expectations and public attitudes were different and, on some issues, incompatible.[73] Meanwhile, the recently constructed telegraph lines and the newly organized wire services supplied newspapers in the Midwest, South, and West Coast of the United States with a steady stream of reports from Washington and London. The minor friction that arose in 1894 regarding a remote boundary controversy became, during the next year, a public obsession within the national community that celebrated the Monroe Doctrine as a source of security and a symbol of recently revived unity. The emotional forces unleashed in this shared sense of community collided headlong with the staid, slow-paced practices of diplomats and politicians, who, although they had used the international telegraph system, seemed to prefer the resolution of such differences through messages delivered by steamship and railroad. The combination of the telegraph, mass-circulation newspapers, and a news-hungry public made these practices obsolete and, in a crisis situation, perilous.

CHAPTER NINE

Information Flow and Revolution

Cuba, Spain, and the United States

THE VENEZUELA–BRITISH GUIANA boundary controversy shared the news columns and editorial pages with another crisis—that of the Cuban rebellion for independence from Spain. The two events unfolded simultaneously in 1895 and early 1896, so that coverage of the Venezuelan affair often appeared alongside stories about Cuba in many metropolitan newspapers in the United States. Although there is no hard evidence on this point, common sense leads to the assumption that there was a spillover effect, meaning that newspaper readers interested in the border dispute in northern South America also read about the unrest in Cuba. Once the Venezuelan crisis was resolved, newspaper editors and their readers turned to Cuba for exciting stories from foreign lands. The rapid movement of newspaper reports on the U.S.-British confrontation in the information system made clear that provocative, sensationalist journalism could leap ahead of the efforts of government officials following traditional methods of negotiation to avert a crisis. It could be expected that the rebellion in Cuba would ignite a similar outburst of jingoism because of the threat to U.S.-owned property on the island and Washington's preference for political and social stability in nearby lands. Spain's imperial dominion over Cuba predated the Monroe Doctrine, so the assertion of U.S. strategic hegemony in the Western Hemisphere that it insisted on did not apply in this circumstance. Nevertheless, Spain was a European power with imperial possessions that posed a challenge to U.S. power within a region that Cleveland and Olney insisted the United States was "practically sovereign" over. The Cuban revolt for independence, therefore, was an opportunity for the information system to supply newspapers with reportage that could emulate the jingoism of the Venezuelan boundary dispute.

The onset of the Cuban revolution in 1895 furnished the New York press and newspapers nationwide with a stream of stories that not only lent themselves to sensationalistic coverage but also resonated with a decades-old interest in Cuba as a close neighbor that often experienced political instability. The underdog Cuban rebels struck at plantations, sugar mills, and provincial cities, and the Spanish Army was unable to put a stop to their attacks. Joseph Pulitzer had pioneered sensationalist journalism after his purchase of the *New York World* in 1883, but after October 1895 he had a rival. William Randolph Hearst acquired the *New York Journal* and proceeded to copy Pulitzer's sensationalistic methods and also raided the staff of the *World* with offers of higher salaries. Pulitzer and Hearst built on the long-term U.S. preoccupation with Cuba. When the news was not sufficiently exciting, the two rivals and their staffs found ways to twist or falsify the information to create the desired results—an exciting front-page story that would boost circulation. The competition between these New York newspapers has been examined by three generations of historians: Joseph Wisan and Marcus Wilkerson in the 1930s, W. A. Swanberg and Charles Brown in the 1960s, and Joyce Milton, John D. Stevens, W. Joseph Campbell, and John Maxwell Hamilton more recently.[1] With these studies as reference points, this chapter considers the coverage of the civil strife in Cuba in the ten metropolitan dailies that form the basis for much of this study. The Hearst and Pulitzer versions of events figured prominently in this mix, but an important question to ask is how these ten big-city newspapers presented events in Cuba and how they represented U.S.-Spanish relations. Did their coverage differ from the coverage offered by the sensationalist press?

The parallel between the situations in Venezuela and Cuba was superficial. The wilderness area of eastern Venezuela and the northwest corner of British Guiana were lightly populated and distant from major U.S. business interests. Also, there was very little civil strife in the region. Cuba, on the other hand, was only ninety miles south of Key West, Florida, and in the 1880s and early 1890s it had become closely tied to the U.S. economy through Wall Street investments in railroads, sugar mills, sugar and tobacco plantations, and trade. For example, in 1890 Cuba exported 632,000 tons of sugar to the United States. By 1894 that figure exceeded a million tons. And, the most telling contrast, Cuba was the scene of a major social uprising. Foreign investments—largely from the United States—and the boom of the 1880s and early 1890s resulted in the consolidation and modernization of agricultural estates and sugar mills, depriving manual laborers of their means of earning a living. Poor peasants, many of whom were recently liberated from slavery or recently dispossessed small farmers, initiated an uprising in response to economic hardships and what they perceived as inept Spanish rule. The Venezuelan situation was resolved after much diplomatic tension and press attention by negotiations between Washington and London with occasional reference to Caracas, but the Cuban revolution was beyond the control of both Spain, which had putative administrative authority over the troubled island, and the United States, which had no direct political influence in spite of its sizeable economic presence.[2]

THE CABLE COMPANIES AND THE CUBAN CRISIS

The roles of the international cable companies in the Venezuela–British Guiana boundary dispute and the crisis in Cuba were likewise quite different. Most of the telegraphy in the boundary dispute involved communication between Washington and London, with a few messages sent south to Caracas. Both journalists and diplomats used the same cable lines. With Cuba, however, the center of attention became the island itself, as New York and other big-city newspapers sent young, aggressive male reporters to forage through the Cuban outback in search of dramatic tales of civil strife. In order to send their "scoops" north, these reporters had to gain access to a telegraph office with connections to the Western Union network. The Spanish often closed the Havana cable office and other stations in Cuba to foreign reporters, so that reporters not only had to cover the story itself but find a way to convey the "copy" to the most convenient open telegraph station, which came to include Key West, Tampa, and eventually the British stations in Kingston and Port Royal, Jamaica, and the French station at Môle St. Nicolas, Haiti.

To set the events of 1895–1898 in perspective, telegraphic communications between Cuba and the United States began with Scrymser's founding of International Ocean in 1867. In fact, Cuba had a functioning internal telegraph system in the 1850s, originally erected by Samuel Kennedy of New York. The system carried about fifty thousand messages annually. By the end of the 1850s, Havana and south Florida had a regular boat service that ferried mail between Cuba and the United States. After Scrymser's line opened, the volume of telegraphic messages continued to expand, and at the same time many U.S. investors began putting their money into Cuba's economy. Historian Louis Pérez has documented Cuba's intimate relationship with several cities in the eastern United States, particularly Tampa and New York.[3] International Ocean became a profitable operation, only to be swallowed up by Western Union in 1873. Telegraphic communications between the United States and Cuba became more reliable after the laying of a new cable in 1875. There were few interruptions in service; the main disruptions were those imposed by the Spanish colonial administration during the uprising.[4]

In addition to the International Ocean/Western Union offices in Havana, Key West, and Tampa, there were three British companies and one French firm with cable operations in Cuba and/or nearby Caribbean islands. Since reporters came to use these facilities to send their missives from Cuba, it is helpful to review briefly the activities of these companies in the 1890s. Of the two original British companies in the area, Cuba Submarine had a record of greater prosperity starting in the 1870s. The company benefited from the insurrection in Cuba because the Spanish government's landlines were frequently cut by the rebels and, as a consequence, either did not work or were, at best, unreliable. Cuba Submarine's relatively secure offshore cables carried a much larger number of messages in the 1895–1898 period than in previous years, and the company's income reflected this increase in business. The average annual profit for this three-year period was £40,852, which, given that Cuba Submarine was capitalized at £220,000, was a handsome figure. The company reported that its lines were in good working order and that it had in fact laid new cables in these years.[5] The larger West India and Panama was not so fortunate. The average annual profit in this period was £30,976—barely three-fourths of that of Cuba Submarine. West India and Panama's capital base was £1,355,530 (about six times that of Cuba Submarine), making its rate of return only a fraction of that of its sister company. The company, however, did earn profits; 1898, when it found itself with a margin of £44,902, was an especially a good year.[6]

Executives in both British companies greeted the arrival of competing submarine cables with dire predictions of lost business, falling revenues, and diminished profits. The Direct West Indian Cable Company, a British firm, and Télégraphs sous-marines, a French company with U.S. investors, opened their offices in 1897 and thereby provided alternative routes for the cabled reports of journalists in and around Cuba. The Direct West India line connected with existing West India and Panama lines, offering an all-British route to London via Bermuda and Nova Scotia, while the French company connected Haitian stations with French transatlantic cable in New York. These five cable companies became vital conduits in the information system. Journalists covering the Cuban revolution usually returned to Florida or New York to file their stories, since the Spanish authorities denied them access to their telegraph lines, but with the explosion of the *Maine* on February 15, 1898, and the onset of hostilities between Spain and the United States in April, they turned to the cable offices in Jamaica and Haiti. Journalists based in Cuba used them to telegraph their coded reports to editorial offices in the United States, where they were decoded and put in prose tailored for newspaper readers from New York to San Francisco. The sending of the messages was only one step, however, in what became an increasingly difficult news-reporting process in revolutionary Cuba.[7]

THE CHALLENGES OF REPORTING A REVOLUTION

More and better cable connections did not mean that the quality of reporting improved. To the contrary, reporting became a greater challenge and, given the public interest in Cuba, newspapers turned to marginal and, in many cases, unreliable sources. Faced with a difficult military struggle, the Spanish imposed censorship on the island, thereby closing off much of the direct information flow to the United States. The governor-general of Cuba, Arsenio Martínez de Campos, established press censorship in May 1895. Colonial officials controlled the telegraph offices and also prohibited reporters from moving into the war-torn interior.[8] The *Journal*, the *World*, and other New York newspapers turned to the Cuban junta (a rebel propaganda and lobbying group in New York) as a source of information on the rebellion. The combination of rebel propaganda embedded in the jingoist slant of influential papers disturbed Spanish officials, who saw the New York publications as indicators of U.S. press and public opinion.

Viewed as a necessity by Spanish officials, their censorship policy had enforcement gaps. A few audacious reporters from the United States found ways to circumvent government regulations and maintain a narrow stream of information flow into the U.S. press. In 1895 a trio of eager reporters from New York newspapers penetrated the Cuban interior. Metropolitan newspapers from New York to San Francisco usually covered the Cuban situation through interviews with diplomats, recently returned businessmen, and Cuban expatriates, and handouts from the Cuban junta of New York, but these three reporters actually reached the combat zones on the island. Richard Harding Davis, arguably the preeminent reporter in the United States, was not among them. One was Sylvester Scovel, an ambitious twenty-six-year-old and the son of an Ohio minister, who sought not only employment but also fame and fortune as a reporter for the *New York Herald*. Scovel obtained a letter of introduction from the Cuban junta in New York addressed to General Maximo Gómez, leader of the rebel army. Spanish colonial officials blocked Scovel's entry to Cuba through Havana, causing him literally to eat the letter of introduction to Gómez in order to avoid incrimination and possible arrest. The dogged Scovel, armed with new advice from Horatio Reubens (the Cuban junta's attorney in New York), entered Cuba through the south coast port of Cienfuegos, avoiding detection by Spanish officials and, at last, reaching the interior of Cuba in mid-October 1895.[9]

Scovel eventually met George Bronson Rea. Electrical engineering and not journalism was the original chosen profession of the twenty-seven-year-old Rea, who had been in Cuba for five years before the outbreak of the revolution. As the fighting spread from the eastern end of the island to the more developed west, Rea decided to take up an opportunity to send reports to James Gordon Bennett's *New York Herald*. Unlike Scovel, Rea did not have the problem entering Cuba because he was already there.[10]

A few months before William Randolph Hearst took command of it, the *New York Journal* sent Grover Flint, a thirty-one-year-old reporter who had learned Spanish during a stint in Spain. Flint went directly to Havana in late March and boarded a train to Cárdenas, a coastal city about seventy miles to the east. Flint walked off the train and headed for the interior in search of General Gómez. Flint's service with the U.S. Army in the trans–Mississippi West proved helpful as he made his way along the back roads of rural Cuba.[11]

While these three reporters were "roughing it" in search of the rebel general and his forces, most correspondents chose more comfortable accommodations in the vicinity of telegraph stations in Havana or Tampa. They gathered information from people who claimed to have witnessed recent action in Cuba. In 1895–1896, Havana's Inglaterra Hotel, with its first-class restaurant, an inviting bar, and soft mattresses, was much preferred to roughing it in the interior.[12] This commodious environment and similar hotels in Key West and Tampa were the settings for interviews and eavesdropping that formed the bases for several stories that made their way into U.S. newspapers even though they lacked not only verification from a second source but also common-sense credibility. One practice was to write up an account based on an interview in the hotel lobby and then apply a false dateline creating the impression that the reporter was in Cuba and, therefore, much closer to if not an eyewitness to the story.[13]

The classic example of a false story that grew to become a cause of trouble was one claiming Havana had fallen to the rebel forces that the Associated Press of Chicago, a small news-gathering agency with limited resources, released to its subscribing newspapers in early January 1896. The bogus account even supplied details: distraught Governor-General Martínez Campos had retreated to Spain's last refuge, Morro Castle in Havana harbor. The misguided AP of Chicago claimed that it relied on telegraphed dispatches that allegedly originated in Havana and then were sent to the town of Batabanó on Cuba's south coast (and a station for Cuba Submarine) and then, by some means not fully explained, were transmitted north to Boston, from which point the news agency circulated what it represented as a monumental scoop that instead became a monumental mistake.[14]

Such outright falsehoods spread through the wires were an embarrassment to many journalists, but there were other problems that created barriers to reliable reporting.[15] Grover Flint's frustrations exemplify how the best of intentions in the use of the electronic-communications system could go wrong. Flint eventually located Maximo Gómez and wrote a series of firsthand reports on the rebel general for the *New York Journal* from May through July 1895. Flint wrote his accounts, then split each one in half by placing every other page in a separate mailing, entrusting these multiple mailings to Cuban insurgent couriers to carry to Florida, where they would be telegraphed to *Journal* headquarters in New York. Presumably Flint also converted his copy to a coded format in case Spanish officials or a competing newspaper intercepted it. The divided, coded accounts in multiple mailings did reach the paper's editorial office, but they created such confusion that the staff of the *Journal* decided not to publish most of Flint's hard-earned reports. So while falsified and otherwise erroneous reports flowed easily through the information system, Flint's firsthand reportage became entangled and lost in the experimental system of security.[16]

Rea and Scovel, however, eventually devised a system of transmitting their accounts to New York that, at least in the early months of 1896, consistently worked. The two began an unusual collaboration. Scovel went to work for the *New York World* as a field correspondent, while Rea continued to write for the *Herald*. There was a kind of irony here because Scovel originally toiled for the *Herald* only to find that, after months in the war-torn countryside, most of his dispatches never reached the printed page because of a petulant editor in New York. Enraged that his work went for nearly nothing, Scovel left the *Herald* for the *World*, but he was willing to travel and work with Rea, the *Herald*'s man in Cuba. By late January 1896, Scovel and Rea had found a way to send their stories from rural Cuba to New York. Like Flint, they sent their reports by rebel messengers or simply by the Spanish colonial mails to their newspaper agents in Havana, who, in turn, sent the enclosures by ship to Key West, where the stories then moved by telegraph to New York. Scovel and Rea did not divide their texts into separate mailings as Flint had done, which meant that they reached their destination as coherent pieces, presumably ready to print after decoding and editorial work. The two sent firsthand reports of the fighting but reached different conclusions. Rhea became disillusioned with Maximo Gómez's avoidance of a major confrontation with the Spanish Army. Scovel, by contrast, understood guerrilla warfare and respected Gomez's hit-and-run tactics. Scovel also gave much attention to the devastating impact of Spain's policies on the rural population of Cuba. The reportage of Scovel and Rhea contributed to the negative image of the Spanish in the U.S. press. Eventually other U.S. correspondents entered the Cuban interior, and their reportage followed the themes established by this duo.[17]

CUBAN PROPAGANDA

The Cubans understood the importance of the U.S. print media, and, much more so than the Spanish, the Cubans were able to take advantage of the available opportunities. The Cuban junta in New York placed its version of events in the major dailies of that city. While there was apparently no published reference by members of the junta to the example of Venezuela's successful print media campaign, the parallel between the two efforts is indisputable. The New York press often covered these two crises in 1895 side by side. As evinced on the front pages of newspapers from New York to San Francisco, the Cubans followed the Venezuelans in making full use of the Associated Press and other telegraphic news services that constituted the electronic network in the United States.

Unlike the Venezuelans, who hired publicist-diplomat William Scruggs in their dealings with the U.S. news media and politicians, the Cubans had several prominent activists who were Cuban by ancestry but born and/or raised in North America and who had extensive knowledge of the press, politics, and popular culture in the United States. Furthermore, Cuba and periodic surges of interest in the annexation of the island were relatively well-known subjects in press coverage and political commentary in the United States. Since the middle of the nineteenth century, thousands of Cubans had traveled to the United States on a regular basis and a large number of families had settled in Key West and Tampa, Florida, and in New York City and other large cities in the northeast. Many of these immigrants and their offspring became fluent in both Spanish and English and received their education in U.S. schools and universities. Some opened businesses and practiced their professions in New York and Florida. Two individuals prominent in the independence movement were U.S. citizens. Tomás Estrada Palma was an educator who established a university preparatory school in Central Valley, New York, before he took a leadership role in the Cuban Junta in New York City in the 1890s. Gonzalo de Quesada studied at the City College of New York, New York University, and Columbia University before he became a lawyer. In 1891 he joined Estrada Palma as a member of the PRC (Cuban Revolutionary Party), which was the basis for the Cuban junta ("junta" was a convenient term in both English and Spanish for the rebels' government-in-waiting). Until he died in battle in 1895, José Martí, leader of the Cuban revolution for independence, was at the forefront of the campaign to rally support for the movement in the United States. He worked as a journalist for various Latin American newspapers while agitating for the Cuban revolution, often under the watchful eyes of Pinkerton detectives in the employ of the Spanish government. This trio, the educator, the lawyer, and the journalist made use of their familiarity with the culture and language of the United States in their roles as chief propagandists for the Cuban junta.[18]

Estrada Palma and Quesada directed their propaganda campaign from New York City and Washington. New York was the center for print media and the telegraphic network that tied together many newspapers across the United States. The junta's office in Washington furnished ready access to the Senate, House of Representatives, and the executive branch of the government. Estrada Palma and Quesada conceived and executed their propaganda campaign for news-hungry reporters and editors, the headline-hungry public, and election-sensitive politicians. They and their colleagues in the junta monitored events in Cuba, Washington, and New York, which helped them devise their propaganda messages. Their sources of information were the filibusters and other revolutionary activists who moved in and out of Cuba.[19] Occasionally the propagandists received letters directly from rebels in the field. In April 1896 Estrada Palma used a letter from the renowned Antonio

Maceo to make the case for material support of the rebel cause. Estrada Palma paraphrased the "Bronze Titan":

> The surest way of bringing hostilities to a close, says General Maceo, is to place in Cuba 20,000 rifles and 4,000,000 cartridges. It would be a great service to Cuba, he says, if the United States would not interfere with the shipment of arms and ammunition. . . . Maceo is confident of the ultimate success of the insurgents.[20]

The junta had two main roles: to generate favorable publicity, thereby gaining public support in the United States, and to fund and organize filibustering missions that supplied combatants and material for the war. The first role is the focus for this study. The junta kept its New York office open to reporters, feeding them information about Cuba in full knowledge of the nationwide influence of the major newspapers of that city. The so-called Cuban Legation in Washington supplied senators, congressmen, and employees of the executive branch as well as reporters with similar information. The impact of this propaganda campaign was most evident in the pages of newspapers in New York and Washington, but it was also apparent throughout the Midwest and the South and on the West Coast. There is further evidence of its effectiveness in the "sympathy meetings" held in large and small cities. As early as October 1, 1895, the *Chicago Tribune* reported a large gathering in that city, which was soon followed by a similarly "impressive pro-Cuba assembly" in Philadelphia in November. Mass meetings in Cincinnati, Cleveland, Columbus (Ohio), and Indianapolis (Indiana) in the Midwest and in Jacksonville (Florida) and Columbus (Georgia) in the South in 1896 testified to public interest in Cuba's fight for independence.[21]

Quesada and Estrada Palma designed their propaganda appeals for their U.S. audience. Their press releases, pamphlets, and speeches emphasized the Spanish incompetence and failure that led to fits of frustration and anger resulting in brutality and cruelty. Spain's image was that of a retrograde medieval nation. This line of negative image building found reinforcement in atrocity stories that featured General Valeriano Weyler (discussed in the next section) as the central malefactor.[22] The cruelty-to-women theme exemplified in the Evangelina Cosio y Cisneros episode was exploited by William Randolph Hearst, who employed the junta's propaganda line for his own purposes.[23]

Examples of the junta's propaganda campaign can be found in Quesada's 1896 book *The War in Cuba*. The text contains flattering biographical portraits of the heroic figures of the revolution: Martí, Gomez, Maceo, and even Quesada's compatriot in propaganda, Estrada Palma. Quesada and coauthor Henry Northrop included flattering descriptions of the city of Havana, enticing rural landscapes, and the friendliness of the Cuban people. The authors cleverly juxtaposed these scenes with a grim account of the "tyrannical" colonial rule of Spain and its roster of allegedly incompetent, corrupt, and malevolent administrative officials. A ten-page section of the book recalled the murderous tenure of General Juan Burriel, who executed fifty-two members of the *Virginius* in 1873 and sparked an international crisis within the memories of many readers in 1896.[24]

It would be an exaggeration to suggest that the junta was responsible for the entire outpouring of press and political attention in favor of the insurgents. A more judicious assessment would emphasize the organization's astuteness in playing on the themes that found widespread resonance with the press, politicians, and general public of the United States in the 1890s. Estrada Palma and Quesada understood their target audience, grasped the importance of their opportunity, and made the best of it, to the point that it is difficult to separate what they inspired in the U.S. media and public from what emerged spontaneously.

THE ARRIVAL OF WEYLER

The accounts written by Scovel and Rea reflected the difficulties that faced the administration of Governor-General Arsenio Martínez de Campos. Other reporters joined this pair in contributing to the information flow from Cuba into the outside world. Spain's main administrator in Cuba sought a peaceful resolution to the conflict during the last months of 1895 by granting concessions intended to placate the rebels. Much to his dismay, the rebel campaign intensified in the east and spread to the west, leading to fighting in the westernmost provinces of Havana and Pinar del Rio. Failure to end the rebellion became a major political issue in Spain. The Spanish press, relying in part on telegraphed reports from across the Atlantic, noted with growing alarm the successes of the rebel armies. A strong-willed, nationalistic military leader, General Valeriano Weyler, was a vocal critic of the conciliatory approach employed by Martínez de Campos. In February 1896, Weyler replaced Martínez de Campos as governor-general of Cuba.[25]

Weyler began a hard-line campaign against the insurgents. He increased the size of the Spanish military in the field and pursued the rebel contingents at every opportunity. Recognizing that rebel successes depended on the support of the peasant population, Weyler set out to divide this combination. He split Cuba into three areas by fortifying two trochas, or huge ditches, dug several feet deep and as much as two hundred yards wide. In the fall of 1896 Weyler initiated his controversial reconcentration policy under which the rural population was confined to specific towns, and all activity—including farming—outside the designated areas was prohibited. Weyler's policies had a devastating effect on rural Cuba. Starvation and epidemic disease decimated the peasants, who were forced to congregate in villages. Reconcentration deprived the rebels of much of their base of support.[26]

Weyler's efforts to suppress the rebellion included a campaign against the international press that was an effective barrier to the movement of information out of Cuba. Soon after Weyler's arrival, in February 1896, the colonial government arrested reporter Charles Michelson of the *New York Journal* and, following a ten-day incarceration in Morro Castle, deported him along with Murat Halstead, also of Hearst's *Journal*.[27] In the same month, Weyler announced his notorious reconcentration policy directed against the civilian population of Cuba and, in a separate decree, placed the press—both foreign and domestic—under military jurisdiction. Aware of the propaganda operations of the prorevolutionary junta in New York, Weyler placed severe restrictions on U.S. reporters in Cuba. The first provision in his decree of February 17, 1896, established that "those who invent or circulate by any means whatever news or information, directly or indirectly, favorable to the rebellion" were "guilty of acts against the security of the country" under martial law. These offenses were punishable by "death or imprisonment." Newspapers in the United States carried full accounts of these decrees, accompanied by critical editorial comments. The *Chicago Tribune*, for example, labeled them as "savage," and the *San Francisco Examiner* claimed that "a reign of terror is on." The *San Francisco Chronicle*, usually more moderate than Hearst's *Examiner*, condemned Weyler's press censorship and denounced his policies overall as "cruelties" enacted on the Cuban people as well as foreign journalists. The *Boston Globe* editorialized against "Weyler's draconian laws," describing them as reminiscent of "old-time Spanish cruelties" associated with the Inquisition. Nine of the metropolitan newspapers gave extensive coverage to Weyler's decrees and printed editorial and/or reportorial statements on the harm such measures did to the flow of information from Cuba.[28]

Weyler's concerns with the press were far reaching. Five of the first seven provisions in his decree dealt with the news media and communications. For example, telegraph operators were to deliver war-related messages only to "the proper officials" of the Spanish government, or, to phrase it differently, the Cuban telegraph system was for the use of the Spanish military and not journalists. Weyler also prohibited the use of the press to "belittle the prestige of Spain, the army, volunteers, firemen, or any other forces operating with the army." Another provision outlawed the transmission and/or publication of praise for the rebels.[29]

The enforcement of this decree had a chilling effect on U.S. press coverage. Charles Michelson reported difficulties in using the International Ocean/Western Union cable in Havana soon after the decree appeared. In early May, Weyler ordered the expulsion of three journalists from Cuba. Nicolas Biddle of the *New York Herald*, James Creelman of the *New York World*, and Fredrick Lawrence of the *New York Journal* arrived in New York full of stories of the "frightful" conditions in Cuba. The Spanish authorities, according to the expelled trio, had lost control of the situation and were resorting to methods of unparalleled brutality. Lawrence charged that "the full facts cannot be obtained because of the suppression by the Spanish authorities."[30] Soon thereafter, Thomas Dawley, photographer and illustrator for *Harper's*, emerged from a brief imprisonment in the now familiar Morro Castle for allegedly having attempted to take photographs of one of the Spanish trochas. Once in the United States, Dawley delivered another denunciation of Weyler's policies with the claim that the fighting between the Spanish Army and the Cuban rebels was stalemated and that "the principal war in Cuba is between General Weyler and the newspaper correspondents."[31]

The arrest of the *New York World's* prolific correspondent Sylvester Scovel in early 1897 was probably the climax of the colonial government's offensive against the press. Once in custody, Scovel faced charges based on Weyler's 1896 decree. The U.S. press speculated that Scovel faced the possibility of execution, and the State Department protested Scovel's incarceration to Spanish officials in Madrid. After nearly a month of press scrutiny and protests, Madrid ordered Scovel's release and his safe conduct out of Cuba. According to historian Joyce Milton, the Scovel incident intimidated the U.S. press corps in Cuba. Even more than in the early months of the conflict, through the remainder of 1897 reporters congregated in Havana and waited for the news to come to them.[32]

Another factor that made reporting in Cuba difficult was the risk of injury, disease, and even death. Cases of injury and illness were numerous, so much so that most reporters who ventured into the interior to interview the rebels or to witness combat were subjected to some sort of bodily harm.

At least two reporters died in the field while covering the revolutionary war from May 1895 through 1897. Twenty-three-year-old Charles Govin of the *Key West Equator-Democrat* was traveling with a small group of rebels when they were surrounded by a larger Spanish force. They failed to break out of the Spanish grasp, and Govin was hacked to death by machete on July 9, 1896. The New York newspapers gave the gruesome details of his death in the sensationalist format typical of much of the war coverage.[33] The second journalist to die was Charles Crosby of the *Chicago Record*, who fell while covering combat on March 9, 1897. Considerably older than Govin, Crosby, who had experience working and traveling in Latin America, had passed through Spanish lines some weeks earlier and managed to send some reports back to his newspaper. His colleagues on the *Chicago Record* believed that the Spanish had intercepted his correspondence and were searching for him at the time of his death.[34]

The International Ocean/Western Union cables to the United States were important in the early coverage of the revolution in Cuba, and the facilities of the British companies also contributed to the work of the correspondents. The irregular and scattered nature of the fighting made it hard for the reporters to reach the scene of battle and, by 1896, Weyler's censorship made a difficult situation worse. Like Dawley, many reporters saw Weyler as a self-avowed enemy. This blockage to the flow of information, coupled with the threat of arrest and possibility of illness, injury, and death, led most reporters to remain in the comfortable surroundings of Havana, Key West, Tampa, or even New York and to rely on sources who had recently left Cuba (or claimed as much) or simply to accept propaganda handouts from the Cuban junta.

From the foregoing, it is evident that Weyler's restrictions on the press were much more than "paper decrees" and may have contributed to the deaths of two reporters. All ten metropolitan dailies gave the decrees in-depth coverage, and their analyses indicated a unanimous rejection of censorship. However, Weyler's antipress campaign, while widely reported and deplored, did not inspire demands for U.S.-government intervention. Enforcement of freedom of the press in a foreign land did not have the same sense of urgency and broad public acceptance as the defense of the Monroe Doctrine.

THE WAR SCARE OF NOVEMBER 1896

The irregular and sometimes strangulated flow of information about Cuba, a subject in which the reading public had a considerable if fluctuating interest, moved through a new communications network that was itself at times unreliable and subject to abuse. These factors came into play during the brief war scare of November 1896. On the second Tuesday of that month, the Republican presidential nominee William McKinley won a hard-fought election contest with the Democrat-Populist nominee William Jennings Bryan.[35] The Cuban issue had moved into the background, but it had not disappeared. For example, the *New York World*'s Sylvester Scovel wrote a series of articles in November on conditions in Cuba that kept this issue close to the center of attention in the United States.[36] In the wake of the presidential election and while Scovel and other reporters continued to discuss the revolutionary struggle in Cuba, a false report concerning war between the United States and Spain suddenly captured center stage in the nation's print media and revealed the vulnerabilities of news transmitted by electronic communications.

In mid-November 1896 several New York newspapers ran a series of articles that alleged U.S. war preparations against Spain. The story spread quickly to major newspapers through the telegraphic network. The *New York Journal* claimed that, in spite of official denials and contradictory information, military and naval activity indicated that the United States was at last preparing its forces to move against Spain. The *New York World* insisted that the Spanish minister to the United States, Enrique Dupuy de Lôme, said that war between the two nations was possible and that its outcome was uncertain. The *Journal*, the *World*, and the *Sun* accepted Dupuy de Lôme's first conclusion but disagreed vehemently with the second. They predicted that the Spanish Navy faced certain destruction when the fighting broke out. The *Chicago Tribune* (tied to the *World* by a wire-service arrangement) and the *San Francisco Examiner* (still owned by Hearst) also took the war threat seriously. The *Examiner* noted "hurried preparations against Spanish invasion" and boasted that the U.S. Navy was "ready for emergencies." The *Boston Globe* represented the war threat as a serious matter, but the *Philadelphia Inquirer* judged that the excitement was the result of unfounded rumors.[37]

Officials in Madrid and Washington were quick to deny the story. Telegraphic lines that had conveyed news of the war scare to Spain via London also carried the Spanish response. Dupuy de Lôme supplied the eager press corps in Washington with a telegram from the Spanish foreign minister, the Duke of Tetuán, stating forcefully that the stories were false. Members of President Grover Cleveland's cabinet denied that war preparations were under way. One unnamed government official ridiculed press exaggerations: "The election is over; I suppose the newspapers must have something to talk about."[38] In spite of these denials, speculations about war continued in the press for several days.[39]

The precise origins of the war scare are difficult to ascertain, but the *New York Times*'s investigation into this matter yielded some interesting results. The *Times* traced the chain of events to a Chicago reporter who had inadvertently become entangled in a scheme launched by speculators in New York stock brokerage firms. Although the narrative of events was not thoroughly documented in the published account, apparently the origins of the "war crisis" were in efforts by stock speculators to cause a sell-off on Wall Street and thereby lower the price of stocks that they were obligated to purchase. The brokers or their agent used a long-distance telephone in Washington to report to New York that, according to sources in the State Department, President Cleveland had ordered the U.S. North Atlantic Squadron to Cuban waters and also instructed the army to move into ports along the Gulf of Mexico. Soon thereafter, the speculators in New York sent five or six telegrams to brokerage houses on Wall Street. From Wall Street the story quickly spread across the United States through the press and, within twenty-four hours, across the Atlantic to London and Madrid.[40]

The war scare of 1896 had only marginal importance in the relations between the United States and Spain and did not hold the attention of newspaper editors, even the excitable editors of the *San Francisco Examiner* and *Chicago Tribune*. Of the ten metropolitan dailies under scrutiny here, only one gave the matter more than two days' attention, and that was the *New York Times*, which published a third article to investigate the origins of the false story. The *Times* labeled the story "ridiculous" and the *New Orleans Picayune* found it to be absurd.[41] The sensationalist urge had been curbed.

WEYLER THE "BUTCHER"

While the war scare quickly evaporated, the interest in General Valeriano Weyler did not, and he was soon being described as having descended into depths of villainy and barbarism. An experienced officer in the Spanish Army who had served his nation as military attaché in Washington during the U.S. Civil War and who had also played an active role in Spain's previous military campaigns in the Caribbean, Weyler was convinced that attacks on the rebels and their bases of support within the rural civilian population would bring an end to the uprising. His military thrusts brought some gains, but his reconcentration policy had unfortunate results. Forcing the peasant population into enclosures to separate them from the insurgents led to abandoned fields and the loss of crops and food supplies. The rural folk, herded together into cramped living quarters, suffered from malnutrition and diseases that flourished amid unsanitary health conditions. The death rate in rural Cuba rose rapidly. For a leader who was sensitive to the foreign press, as his decree of February 1896 made clear, his reconcentration policy was a calculated risk that gave eager, inquisitive reporters who were already angered by his censorship a story that would answer the predispositions of editors and readers in the United States to judge the Spanish as brutal throwbacks to medieval cruelty.[42]

And such stories began to flow from Cuba by way of Key West, Tampa, and New York onto the front pages of newspapers across the United States. The network created by the telegraph lines of Western Union and the wire services such as the Associated Press and the news-service branch of the *New York Journal* connected cities across the United States and made possible the rapid dissemination of the stories. Marcus Wilkerson, Joseph Wisan, Charles Brown, Joyce Milton, and other historians have covered this theme in depth so that it needs only brief recounting here. Both Wisan and Wilkerson observe that Hearst's *Journal* and Pulitzer's *World* prepared the way with atrocity reports that alleged Spanish police fed their prisoners to the sharks off Havana and that drunken army officers went on a killing spree among helpless Cuban civilians. These two newspapers and the U.S. press in general recognized the consequences of Weyler's imposition of reconcentration. News coverage and editorial opinion ran along the lines expressed by the *Journal:* "The rural population are given the choice of being murdered by Spanish soldiers or of dying of starvation in the towns. . . . This is wholesale murder, not war."[43] Reporter James Creelman's words in the *World* voiced a similarly grim message: "The horrors of a barbarous struggle for the extermination of the native population are witnessed in all parts of the country."[44] The detailed portrayal of harsh Spanish policies appeared prominently and simultaneously in the news sections and on the editorial pages of papers such as the *Chicago Times-Herald*, the *Boston Herald*, the *Charleston News-Courier*, the *Cincinnati Commercial-Gazette*, and the *New Orleans Times-Democrat*. *Leslie's Weekly* devoted one entire issue to the conflict in Cuba.[45]

According to Wilkerson, Pulitzer's *New York World* was at the forefront in the verbal assault on Weyler and Spanish rule in Cuba, and the circulation figures for this newspaper reflected public interest in this approach to the crisis. The daily circulation for the *World* was about 400,000 in early

1895. By April 1896, the morning and evening editions registered a total of 742,673 per day and, by early 1898, the total circulation exceeded 800,000.[46] The impact of these atrocity stories was further intensified in the speeches of senators, congressmen, and members of the executive branch, who often cited the *New York World* as their source.[47]

Detailed portrayals of harsh Spanish policies appeared in the news sections and editorials of a wide range of newspapers across the United States, not only in the pages of the *World* and *Journal* of New York. General Valeriano Weyler in his capacity as the commanding officer of the Spanish forces in Cuba became the symbol of Madrid's repressive policies. The sensationalist newspapers of New York placed Weyler at the center of the press agenda, and other more responsible publications followed this story line. U.S. press coverage of Weyler was critical right from the early days of his command. Although it is difficult to pinpoint an event that served as the main catalyst for the emergence of a negative image of the Spanish general, his announcement in February 1896 of the new policies he intended to pursue brought an outpouring of press criticism. These decrees established his executive authority in Cuba and helped to create his image as a tyrant in the United States. Even the moderate *New York Times* labeled these laws as "draconian," and the typically strident *Chicago Tribune* agreed.[48] All ten metropolitan newspapers characterized these decrees as harsh. Except for the *Chicago Tribune* and the *San Francisco Examiner,* their rhetoric did not match the shrillness found in the New York papers of Pulitzer and Hearst, but the coverage of the decrees was similar: they all printed extensive verbatim quotes of translations of the proclamations that placed severe limits on the movement of the civilian population and established substantial restrictions on freedom of the press and the activities of journalists. Newspapers editors and reporters across the United States found the decree relating to the press to be especially onerous.[49]

While there were periods during which Weyler's name did not appear in the headlines, the Spanish commander in Cuba found himself cast in the role of the primary villain in what amounted to a Victorian morality play. For example, he returned to Havana after rebel general Antonio Maceo's death in December 1896 as a conquering hero for Spanish loyalists on the island. He boasted, according to Whitelaw Reid's *New York Tribune,* that suppression of the rebellion was "an accomplished fact." This boast, made only days after the confirmation of Maceo's death, offended many readers in the United States, who shared the insurgent Cubans' anguish following the rebel leader's demise. Weyler appeared to be bragging about what some readers saw as an assassination. Rightly or wrongly, Weyler's celebration of his battlefield triumphs helped to cement his reputation as an ignoble, brutal commander.[50] Similar stories continued to flow through the cables from Havana to New York and then across the national wire network. The *New York Times* led the way on coverage of what its editors saw as Weyler's imperial tyranny.[51] Most devastating was the effect of the reconcentration policy that, by the summer of 1897, began to dominate the news from Cuba. Judge William Calhoun went to Cuba as a representative of President McKinley to report on the medical and social conditions. Calhoun's comments that he made on his return to the United States in June 1897 received coverage in all ten metropolitan dailies. This indictment of Weyler was not a momentary story that faded after a day or two. Interest in him was sustained across the duration of his tenure in Cuba, and his tarnished image affected the image of the Spanish regime after his departure.[52]

While this steady barrage of anti-Weyler and anti-Spanish opinion and information became typical of the coverage in the press, the usually antijingo *New York Times* provided flattering portraits of Tomás Estrada Palma and Gonzalo de Quesada, two of the leaders of the Cuban junta in New York City. The junta was the main propaganda and diplomacy agency for the Cuban rebels, and the *Times* lauded its conduct in the United States: "The work of the Junta in this country has been one of delicacy and difficulty, but it has been performed in the main with such discretion as to win for Gen. Palma and his associates the respect of the American people."[53]

The expressions of moral outrage in the U.S. press tended to obscure the complexity of the Spanish predicament. Weyler's reconcentration policy coupled with the dispersal of the rebels gave the Spanish an advantage in the struggle in western Cuba, at least temporarily. Havana was secure, and the governor-general built up his army with reinforcements to make it even more formidable. But these military gains carried a high cost. Even revisionist studies of Weyler concede that reconstruction was a cruel policy that played into the hands of Estrada Palma and the New York junta. In this sense the U.S. press had become a part of the revolutionary process. Weyler revealed frustration with his image as "the butcher" in the United States at the same time that his military campaign against the rebels seemed to be succeeding. So long as this image was dominant in U.S. newspapers, the Cuban junta enjoyed success in the recruitment of funds, soldiers, and public support in the United States.[54]

THE OLIVETTE AFFAIR

"Butcher" Weyler, the atrocities associated with reconcentration, and skirmishes in the Cuban outback proved to be the source of "good copy" that boosted circulation for New York newspapers, but Hearst wanted to surpass rival Pulitzer and his stable of reporters. The editor of the *Journal* offered the nation's best-known journalist, Richard Harding Davis, $3,000 plus expenses for a month of reportage from Cuba. Hearst teamed Davis with Frederick Remington, the renowned illustrator, for a series of illustrated articles from the war-torn island. Soon after arriving in Havana, Davis made an appointment with the notorious Weyler for an interview. Much to the reporter's surprise, the Spanish general was a polite, considerate host. Weyler offered Davis the opportunity to travel with his army on its next expedition to the interior, an invitation that Davis accepted. Davis's contract with Hearst was for only one month, however, and when it became apparent that Weyler would not return to the field for several weeks, Davis and Remington sought other options. In spite of his unexpectedly pleasant meeting with Weyler, Davis did not emphasize this side of the general's nature in his published writing.[55]

With no immediate prospects of moving outside the safe environs of Havana and nearby communities, Davis and Remington grew frustrated, as they paced in the frequently trod tracks of other journalists. Much to his dismay, Davis found that Remington was a poor traveling companion, who complained frequently of the many unavoidable discomforts. Furthermore, his blatant racism made Davis uncomfortable in Cuba's multiracial society. The legendary telegraphic exchange in which Remington informed Hearst that "there will be no war," to which the jingoistic editor replied, "Please remain. You furnish the pictures and I'll furnish the war," may or may not be accurate, but it is highly probable that the war of which they spoke was not the Spanish-American War of April–July 1898, but the insurgent struggle against the Spanish colonial government. In any case, Remington soon left, and Davis joined up with Scovel and Rea in an attempt to reach a battle zone. The trio's plans were scuttled by none other than the *New York Journal* itself, which informed its readers, including the staff of General Weyler, that Davis and Remington were riding with the rebel army. This falsified account placed Davis in jeopardy and caused him to avoid any combat zone for fear of being fired on by government forces. Scovel and Rea regretted that they had to leave him on the sidelines of the war, but his difficulty was the result of the falsehood concocted by his colleagues on the *Journal*.[56]

With Davis's usefulness in Cuba compromised, the reporter booked passage on the steamer *Olivette* for New York only to encounter the largest and most controversial story of his brief time in Hearst's employ. His dinner companion one evening was Clemencia Arango, the sister of a rebel officer. Davis found her to be an articulate, attractive young woman and was appalled at her description of her detention by Spanish officials in Havana before she boarded the *Olivette*. According to her account, she was ordered to remove her clothing so the Spanish could determine if she was carrying a message or other items that would benefit the rebel cause. The official found no such items, and she departed on her scheduled voyage.

Davis, outraged by this incident, wrote a highly charged version of these events tinged with patriotic sentiment and sent the copy by telegraph from Tampa to New York, which the *Journal* published under the headline: "Does Our Flag Shield Women?" Hearst was pleased with the story of a young, aristocratic woman whose dignity and privacy were violated by Spanish officials. Remington dashed off an illustration of his perception of the sordid event that featured three Spanish males standing around Miss Arango with all of her clothes removed. The headline, story, and illustration created the effect that Hearst sought.[57]

One large problem with this piece of journalism was that the reporter and illustrator were separated by two thousand miles. The telegraph enabled Davis to take full advantage of his sense of outrage and gain a scoop, but Remington's illustration contained a basic factual error. Arango was searched by matrons in the service of Spanish officials, not by men. Davis did not make this clear in his article, and Remington's assumption that lascivious male police officials were the malefactors was natural, given the sensationalist, anti-Spanish mood prevailing in the editorial office of the *New York Journal*. The deception became an issue when Pulitzer's reporters interviewed Arango on her arrival in New York and discovered that the impression conveyed by the Davis-Remington piece was indeed false. Davis was angered by the illustration and vowed never to work again for Hearst.[58] While Hearst and Remington expressed few if any regrets in the matter, the episode made clear the difficulties of reporting distant events by telegraph. Davis did not state who searched Arango, but a few minor editorial changes could have clarified this point. Such editorial work, however, was not possible via telegraph over such a distance, and the imperative of securing the scoop forced editors—especially on the Hearst staff—to move quickly. Under the circumstances, the speed and convenience of the telegraph were outweighed by the lack of precision in the prose and the precipitate decisions in the editorial offices.

More importantly, the *Olivette* story did not generate interest in the metropolitan newspapers outside Hearst's control. The *San Francisco Examiner* ran the story on its front page for three days, but the *San Francisco Chronicle* ignored it and instead featured an article on Weyler's campaign against the insurgents. The usually excitable *Chicago Tribune* gave it one day on the front page and then dropped it, as did the *Washington Post*. The other metropolitan newspapers gave it little or no attention.[59] Hearst's plans for a Davis-Remington collaboration with a nationwide impact turned into a fiasco.

THE CREATION OF NEWS: THE CISNEROS AFFAIR

Hearst was not troubled by distortion and deception in the pages of his newspaper. This predisposition came to the forefront again in the summer of 1897 when press coverage of Cuba hit a lull to some extent because of Weyler's intimidation of roving journalists and the summer doldrums in the insurgent campaign. Hearst was searching for a story and found one, almost by accident, if James Creelman's account is correct. According to Creelman, a copyboy brought into Hearst's office a telegram from the *Journal*'s George Eugene Bryson that read simply "Evangelina Cisneros, a pretty girl of seventeen years, related to the president of the Cuban Republic, is to be imprisoned for twenty years on the African coast for having taken part in uprising Cuban political prisoners on the Isle of Pines." According to Creelman, Hearst read the telegram, slapped his knee, and shouted, "We've got Spain now."[60]

The dapper Karl Decker and the disguised Evangelina Cisneros (with cigar) posed for this publicity photograph as part of Hearst's campaign to build public sentiment against Spain (Evangelina Cisneros, *The Story of Evangelina Cisneros Told by Herself* [New York: Continental Publishing, 1897], 201).

What followed pushed journalistic standards to a new low. The image of Weyler as a "butcher" was an exaggeration, but his campaign against the press and his reconcentration policy inflamed the anger and imaginations of even reasonable correspondents and editors. The *Olivette* affair was the result of a hurriedly written account of a small incident blown out of proportion by inadequate communication among reporter, illustrator, and editor. But the story of Evangelina Cossío Cisneros was more manipulation and fabrication of an event than distorted reporting. Hearst literally seized on an isolated, largely irrelevant circumstance and turned it into an international crusade far removed from the actual issues in the conflict between the Spanish and the Cubans.

Within a few hours after his initial inspiration, the energetic editor ordered the transmission of hundreds of telegrams across the United States and through submarine cable to Europe in an effort to launch a public-relations campaign ostensibly to save young Miss Cisneros from the clutches of, in the *Journal*'s view, the evil and brutish Spanish but in actuality to promote his newspaper, himself, and the Cuban issue as the centerpieces in a chivalric-type drama worthy of the attention of readers on both sides of the Atlantic. Creelman's memoir placed this statement from Hearst to his staff in quotation marks:

> Telegraph to our correspondent in Havana to wire every detail of this case. Get up a petition to the Queen Regent of Spain for this girl's pardon. Enlist the women of America. Have them sign the petition. Wake up our correspondents all over the country. Have distinguished women sign first. Cable the petitions and the names to the Queen Regent. Notify our minister in Madrid. We can make a national issue of this case. It will do more to open the eyes of the country than a thousand editorials or political speeches. The Spanish minister can attack our correspondents, but we'll see if he can face the women of America when they take up the fight. That girl must be saved if we have to take her out of prison by force or send a steamer to meet the vessel that carries her away—but that would be piracy, wouldn't it?[61]

Hearst's electronic publicity campaign to liberate Weyler's young female prisoner was successful in eliciting responses from celebrities. Telegraphed responses poured into the *Journal*'s offices and were promptly reprinted in the next day's edition. Julia Ward Howe, President McKinley's mother and the widow of Jefferson Davis, and Clara Barton, among others, in effect endorsed Hearst's campaign. Thousands of women signed the petitions that circulated in the United States and Great Britain. Within a few days Hearst had assembled a triumph in the amorphous arena of international public opinion.[62]

The story of the release of Cisneros was more complicated though no less deceptive. Again the telegraph played a large role in Hearst's operations because quick action was essential if he was to exploit the emotionalism he had aroused in the newspaper reading public. Letters, petitions, and statements by baffled diplomats and swaggering anti-Spanish patriots created a circus-like atmosphere from late August into September, but soon public interest began to decline. Concerned about the decline in newspaper sales, Hearst sent an agent to Cuba, Karl Decker, who organized the apparent jailbreak. Decker and two associates did actually saw though the bars on the window of Cisnero's cell while perched on a ladder laid horizontally across the space between the jail and another building, the street menacing thirty-five feet below. The outcome of this adventure was assured, however, because Decker had bribed the Spanish jailer with Hearst's money.

The culmination of the Cisneros episode was not her escape from the Spanish prison but her reception in the United States. By early October coverage of the Cisneros story spread through the telegraph network to appear on front pages across the entire nation from New York to Charleston, South Carolina, New Orleans, Indianapolis, and San Francisco. The debut of this seventeen-year-old into society as arranged by Hearst put New York debutantes to shame. Hearst turned to the telegraph and the influence of his newspaper to organize a series of public events in New York and Washington. George Bryson, who had originally found the Cisneros story and then found himself expelled from Cuba by Weyler, greeted her in the New York harbor and, with other members of the *Journal*'s staff, escorted her along streets lined with seventy-five thousand spectators to a suite of rooms in the posh Waldorf-Astoria Hotel. Fortunately for Hearst's plans, Cisnero's physical appearance coincided with the widely held notions of the day concerning female beauty. Her presence in front of large audiences of the excited and curious further reinforced Hearst's adolescent assumption that he was a medieval knight who had overseen the rescue of a beautiful young maiden from the clutches of a dastardly villain. The power of the press seemed to know few limits in the United States. The *Journal* devoted 375 columns of coverage to her story, much of it on the front page. On October 12 the *Journal* boasted that "international complications may result from the Cisneros rescue." President William McKinley, Secretary of State John Sherman, the widow of former president Ulysses S. Grant, and political and civic leaders expressed their admiration for the derring-do of Decker and acknowledged the symbolic importance of the liberation of the young woman from the Cuban prison. No one in the United States seemed to notice that Decker's actions had violated diplomatic procedure and Spanish law.[63]

Hearst's contrived media event drew into its web the president, other political dignitaries, and many newspapers from coast to coast, but it also met a cool reception among editors who rejected the *Journal*'s sensationalist methods. Six of the ten metropolitan newspapers under study here gave the Cisneros affair little attention. Whitelaw Reid's *New York Tribune* apparently ignored it altogether, and the *San Francisco Chronicle* and the *New Orleans Picayune* gave it only brief mention. The *Washington Post* gave the rescue six columns, two of which were devoted to Karl Decker's contrived mission. The *World* limited coverage of the affair to twelve columns, and the *Times* gave it ten. In brief, Hearst's publicity efforts, in spite of celebrity participation, fell far short of becoming a significant part of the coverage of the Cuban issue in the nation's information system.[64] The *New York Times* described Cisneros's reception in New York City, but its editorial skewered Hearst, Decker, and the entire "rescue" operation as an illegal "invasion of Havana by our contemporary's agent." The *Times* argued that breaking into a government jail in a foreign country could only provoke a diplomatic incident that in this case would further worsen Spanish-U.S. relations.[65] Such diplomatic complications did not arise, but the *Times* and the other three newspapers that minimized their coverage of this story made an important point. They did not practice this form of sensationalism and rejected Hearst's manipulation of events.

CONCLUSIONS

By the end of 1897 a pattern had emerged in the metropolitan press coverage of the Cuban revolution. Those papers that followed the Hearst-Pulitzer approach (*San Francisco Examiner, Chicago Tribune,* and, to a lesser extent, *Washington Post* and *Atlanta Constitution*) had attempted to evoke press and public concerns about the war scare of 1896, the *Olivette* affair, and the Cisneros affair. Only the Cisneros story held a prominent position on the front pages across the country for more than a week, and that level of coverage was to be found in only four of the ten metropolitan dailies. In contrast to the Chilean crisis of 1891–1892 and the Venezuelan episode of 1895–1896, the coverage of Cuba did not inspire a significant nationwide demand for war.

The themes developed by Scovel and Rhea in their 1895–1896 coverage were revisited in many newspapers, but these detailed accounts of skirmishes and guerrilla raids did not capture public attention. Although the sensationalist press adopted a potentially inflammatory slant, the mainstream press did not follow this approach. The *New York Times* conducted an investigation to determine the origins of the war scare of 1896, and several newspapers printed revealing versions of Karl Decker's role in the Cisneros "escape" from a Spanish prison. As press historian W. Joseph Campbell has noted, there was a substantial campaign against the yellow press that included a boycott (or banning) of the *New York Journal* and the *World* from libraries in New York City and across the nation. The campaign seemed to reach its peak in 1897 and diminished thereafter.[66]

The cumulative effect of these stories from Cuba and the related controversies, however, was large. In spite of the anti-yellow-press campaign and the failure of particular episodes to capture a broad readership, the general press coverage of the revolution in Cuba was persistent and—allowing for differences in topics, timing, and perspective—pervasive. Most importantly there were topics that transcended the sensational-versus-responsible editorial divide, demonstrating lasting interest throughout 1897: the policies and personality of Valeriano Weyler. The relatively calm coverage in the *New York Times*, the *New York Tribune*, the *New Orleans Picayune*, and the *San Francisco Chronicle* conveyed essentially the same information and similar reportorial and editorial judgments on the Spanish general as other newspapers. The coverage of Weyler fed the jingoistic mood of the nation and set the stage for the dramatic events of early 1898. President McKinley, Secretary of State Sherman, and other members of the administration had been able to ignore the stridency, stunts, and falsifications of the sensational press, as their concoctions appeared and faded from memory from 1895 through 1897.[67] The image of Weyler and the lingering effects of the reconcentration policy, however, became consistently and widely covered stories for both jingo and nonjingo papers and could not be avoided by the president, his government, or politicians of either party. It was the combination of jingo and nonjingo coverage of Weyler's policies that symbolized the Cuban situation and made it a prominent issue in the United States information system through 1897.

CHAPTER TEN

Diplomacy Under Stress

Washington, Havana, and Madrid

JOHN LONG, SECRETARY OF THE NAVY, found himself awakened by his daughter at 1:30 a.m. on February 16, 1898. She had just returned from a ball, the highlight social event in Washington on that evening, but such pleasantries were quickly forgotten when Secretary Long read the contents of the telegram that she handed him. The message from Captain Charles Sigsbee of the USS *Maine* informed Long that this battleship had been destroyed by a massive explosion at about 9:40 p.m. in Havana harbor with a large loss of life. Long quickly sent Sigsbee's telegram to President McKinley whose burden of leadership became much heavier as news of the disaster spread throughout the country. Sigsbee, like Long and McKinley, understood the volatility of U.S. public opinion and its impact on the tense relations between the United States and Spain. He included in his dispatch to Long the admonition that "public opinion should be suspended until further report."[1] Given the recent events in Cuba and the eagerness of the nation's newspapers for dramatic news, the suspension of public opinion was impossible.

There was an irony in Sigsbee's expression of hope that the public response to the explosion of the *Maine* could be limited, given his familiarity with and trust of certain members of the press. Soon after composing this message and a second brief telegram to Admiral Montgomery Sicard at Key West, Sigsbee saw George Bronson Rea of the *New York World* among the correspondents milling about Havana harbor. Sigsbee, distraught from the shock of events, called Rea aside and asked him to take his telegrams to the cable office for transmission to the United States. Rea honored this request, but, after reading Sigsbee's urgent message to the secretary of the navy (composed in plain English and not naval code), the reporter included it in his account of these events the next day, thereby claiming a scoop for Pulitzer's *World*.[2]

Sigsbee's reliance on Rea was indicative of the open and sometimes friendly relationship between U.S. government officials and correspondents in the 1890s, but it also revealed that the naval officer did not grasp the competitive and often diametrically opposed roles of the press and the administration in the use of the information system. While diplomats and politicians in Washington and Madrid used transatlantic cables in their struggle to find a way to avoid what historian John Offner has aptly termed "an unwanted war," contributors to the print media fed their readers' mania for exciting stories based on tension and strife. The cable system served both government and press, but it was the reporter not the government official who would exploit the system to full advantage. By early March 1898 the notorious yellow press led by Hearst and Pulitzer shared the vocabulary of moral outrage and patriotism with many respectable newspapers across the country. The rhetoric of warfare seemed to boost circulation. The new media network based on the telegraph made this approach to journalism a nationwide phenomenon.

UPGRADED CABLES, ENHANCED WORDS AND IMAGES

The network of cables and the format of front pages underwent changes that by 1898 made the news more readily available and more exciting than ever before. The movement of dispatches became more extensive over the course of the Cuban crisis because of technical improvements in the electronic connections between and among the United States, Spain, and Cuba. At the same time, editors began to use larger and more pointed illustrations in the form of editorial cartoons and adopted more frequent banner headlines in larger print. While diplomats and political leaders tried to minimize the discord between Spain and the United States, the coverage of the revolution took on a faster pace, and the style of presentation became more dramatic.

The most significant improvements in the cable network took place in the North Atlantic, where the number of working transatlantic cables grew from one in 1866 to twelve by the end of the century. Private corporations in Britain, France, and the United States employed cable-laying ships that by the 1880s became proficient in this task. For example, U.S. corporate giant Western Union, which had acquired Scrymser's International Ocean, laid two cables between Cornwall and Nova Scotia in 1881. Entering into joint-purse agreements, European and U.S. competitors minimized competition by setting rates and sharing revenues based on their respective proportions of the messages sent.[3] By the late 1890s the North Atlantic system was highly reliable. Diplomats and journalists could count on the delivery of messages made even quicker with the advent of a new practice called duplex telegraphy that enabled two messages to pass through the same cable at the same time in opposite directions. This doubled the capacity of most existing cables. Automatic retransmission equipment made it possible to relay (or pass on) messages by means of a mechanical device that replaced human operators and thereby eliminated the chance of human error in the process.[4]

One set of technical changes pertained to the long-running struggles of the cable companies in the face of undersea interruptions of the electronic messages. The floor of the Caribbean was perhaps the most difficult area in the world for submarine lines. From its outset, West India and Panama had to contend with rocks and coral reefs. The company reported numerous cable failures in the 1870s and 1880s. The company purchased two cable repair ships, the *Grappler* and the *Duchess of Marlboro*, and kept them busy. The company's board of directors described them as "continuously at work" in 1881.[5] The *Grappler* reported its frustrations with a cable break between St. Kitts and St. Thomas: "The great inequalities in the depths of the water, and the rocky nature of the bottom rendered it impossible [to repair the cable], . . . and eventually we had to abandon the old cable, and expend 106 knots of new cable in effecting the repair."[6] The *Grappler* encountered similar problems with the cable that connected Santiago, Cuba, and Kingston, Jamaica in 1885.[7]

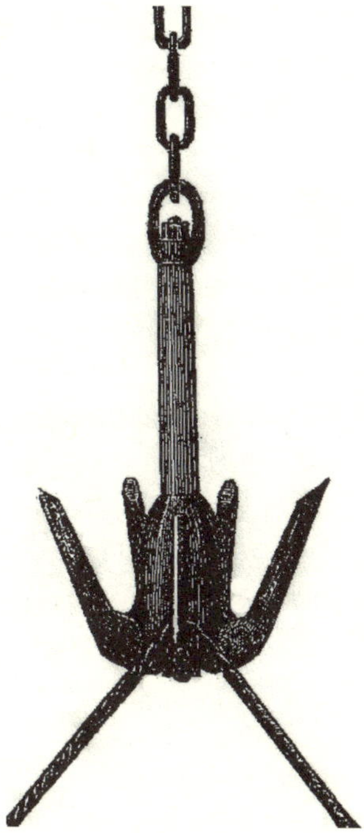

The grapnel was essential in the recovery of damaged cables from the ocean floor (William Maver, "Ocean Telegraphy," pt. 1, *Electrical World*, May 10, 1885, 153).

These cable repairs were difficult under the best of circumstances, but eventually West Indian and Panama gained control of the situation. The repair ship lowered a device known as a grapnel that resembled an anchor except it had multiple arms that turned upward at their tips to hold the cable. The ship dragged the grapnel along the sea floor until it snagged the cable. One of the main problems was finding the cable, which could be submerged in great depths on irregular, uncharted bottoms.[8] Another improvement in the communications system came with the placement of a second or duplicate cable alongside the original. If the first still functioned, the duplicate added not only reliability but also capacity. In the 1880s West Indian and Panama experienced fewer interruptions, and by the 1890s the standard company evaluation simply stated "no interruptions to service."[9]

West Indian and Panama cables were essential for journalists sending their accounts of events in Cuba to editorial offices in the United States. The unreliable lines of the 1870s would have made reportage from Cuba irregular and fragmentary. Instead, newspapers received a steady stream of coverage that found its way onto the front pages across the United States. The Spanish government's censorship in early 1896 reduced the movement of information via the Havana–Key West cable. Some reporters used the West Indian and Panama link to Jamaica and, after the onset of war, many made the difficult trip by boat to Jamaica to access the international network through the West Indian and Panama line between Jamaica and Panama.

The newspapers that received these on-the-spot reports deployed innovations that paralleled the improvements in submarine cables. By the early 1890s, over 2,800 linotype machines had been installed in newspaper print shops.[10] The layout of the front page of many newspapers changed. Stories began to appear new and attention-grabbing formats. Headlines became larger. Banner headlines stretched across two or three columns and by 1897–1898 sometimes reached across the entire front page. Bigger headlines and fonts meant fewer words per page, but this layout had a greater dramatic impact.[11] Front-page coverage of the threats of war reflected the evolution of this trend in the 1890s. The possibility of a war between the United States and Chile received front-page coverage in January 1892. All ten metropolitan newspapers devoted one column to this story. Stories concerning tensions between the United States and Britain over the Venezuela–British Guiana boundary in December 1895 also appeared on the front pages of all ten newspapers, the stories averaging nearly two columns. The controversy surrounding Spain's diplomatic representative to the United States Enrique Dupuy de

Lôme was on all ten front pages in February 1898, taking up an average of just over four columns under large headlines. This controversy broke less than a week before the *Maine* exploded in Havana harbor on February 15. The *Maine* incident, unmatched for its tragic loss of life and dramatic quality, received just over four columns under several banner headlines. There were no banner headlines in 1892, but in 1898 they were commonplace, with two exceptions: the *New York Times* and the *New York Tribune*.[12]

At the same time editors began to tie images to words. Editorial cartoons became larger and more numerous. As was the tradition dating back to the early years of the nineteenth century, these cartoons often carried a point forcefully made, but in the context of the Cuban issue and deteriorating relation with Spain, these cartoons acquired international importance. For example, cartoon depictions of the ill-fated diplomat Dupuy de Lôme and General Valeriano Weyler, Spain's military commander in Cuba, were derogatory to the point of mockery. In general, as media historian David Spencer has noted, the newspaper cartoons of the late nineteenth century supported U.S. imperial adventures in the Caribbean and Latin America.[13] Slanted editorial cartoons coupled with dramatic stories from Cuba and emphatic banner headlines created a charged political environment in which diplomacy appeared tedious and tenuous.

THE SPANISH POINT OF VIEW ON THE U.S. PRESS

The coverage of the plight of Sigsbee and the *Maine* constituted a phase of the trend that had been present in the U.S. press for some time. Only four months earlier Hearst's *New York Journal* converted a contrived event into a national pro-Cuban celebration in the case of Evangelina Cossío Cisneros, an event that had had a large impact on officials in the Spanish government. Evangelina's "escape" from a Spanish jail and her reception in New York and Washington was in many ways the epitome of the sensationalist coverage of the Cuban revolution against Spain. Spanish prime minister Antonio Cánovas del Castillo observed this spectacle with both disbelief and distress. The comment he made in an interview with James Creelman that "the newspapers in your country seem to be more powerful than the government" was based not only on the Cisneros affair but three years of watching Hearst and the other magnates of the press in the United States publish exaggerations, invented stories, and distorted editorial judgments that invariably represented Spain and its officials as one-dimensional monsters with few redeeming virtues.[14] The fact that the Cisneros story and similar inventions of the yellow press had little impact on most metropolitan newspapers did not lessen the anxieties of Spanish officials.

Spain's minister in Washington was Enrique Dupuy de Lôme, an experienced diplomat, who, like his colleagues in Madrid, was often disconcerted by the tone and content of such newspaper stories. A sensitive observer with a Spanish perspective on events in the United States, he issued statements to the U.S. press in an effort to correct falsehoods on the pages of the nation's newspapers. For example, on March 19, 1896, Dupuy de Lôme countered the accumulation of hostile criticism of General Weyler that had led some newspapers to conclude that the general would resign. The Spanish minister cabled Weyler in Havana and then released to the press in Washington the general's emphatic telegraphic denial.[15] At about the same time the U.S. Senate was debating the Cuban situation, using the press to obtain much of its information. Dupuy de Lôme pointed out the flaws in the senators' grasp of the alleged facts drawn from unreliable sources.[16] The Spanish minister also denied the threat of war in November 1896 and strongly urged Madrid to order the release of the captured Sylvester Scovel the following March because the nationwide newspaper coverage of his incarceration had had a negative impact on "public sentiment."[17] Early in the Cisneros drama, Dupuy de Lôme wrote a highly publicized letter to Mrs. Jefferson Davis, who had taken a prominent place in the Hearst-inspired campaign to free Cisneros. Dupuy de Lôme's response on behalf of the queen regent of Spain encapsulated his understanding of Hearst's machinations: "The information received from Cuba by the Spanish Government and laid before her majesty, and that has been transmitted to me by cable, shows, in my opinion, that a shameless conspiration to promote the interest of one or more sensational papers is at the bottom of the romance that has touched your good heart."[18]

The Spanish official most directly affected by the flow of information from Cuba and Spain to the United States and the augmentation and distortion of this information within the U.S. print media was General Weyler. The exaggerations of his policies and the distortions of his personality appeared in both the yellow press and in the more moderate metropolitan dailies. He denied that his troops had committed "outrages" against noncombatant Cubans, as claimed in the U.S. press in April 1896.[19] Later that same year, Weyler judged it necessary to explain that Antonio Maceo's death came in battle and not in an ambush on the way to a peace conference with Spanish officials.[20] Weyler was convinced that the Cuban junta was able to place its propagandistic stories in the New York press at will and that by 1897 William Randolph Hearst was, in effect, a propagandist for the rebel movement.[21]

Weyler's use of censorship and his arrests and deportations of U.S. reporters offer further evidence of his concern about the work of the press. James Creelman, at the time on the staff of the *New York World*, wrote an account of his meeting with Weyler that is consistent with the general's vaunted hostility to U.S. reporters but, at the same time, reveals yellow journalists' penchant for embellishing their reportage with cultural and personal innuendo. Creelman's description of Weyler was an indictment based on physical appearance and body language. Weyler's mouth and jaw projected "a singular expression of brutal determination." His eyes were "gray and cold." In speaking, "he jerked his words out in the curt manner of a man accustomed to absolute authority." Weyler's face was "smileless, cruel . . . with just a suggestion of treachery in the crows feet about the eyes."[22]

With the main thrust of the interview already established, Creelman proceeded to describe Weyler's expressions of hostility toward the press. Creelman asked about the resilience of the Cuban insurgents, and the general replied "with a sudden passion" that "the American newspapers are responsible. . . . They poison everything with falsehood. They should be suppressed. . . . Men like you!" Weyler accused Creelman and his colleagues of exciting "rebellion everywhere," calling them all "meddlesome scribblers." Then the general threatened Creelman with constant surveillance and even death if he should have further involvement with the insurgents.[23] In his slanted version of the interview, Creelman used several rhetorical devices to place Weyler in a most unfavorable light, but it is reasonable to conclude the reporter's description of the general's expression of frustration with the U.S. press was genuine. Not only Weyler but also Dupuy de Lôme and the government in Madrid were convinced that the press in the United States was a significant factor in the formulation of policy toward Cuba and Spain.

PUBLIC OPINION AS A DIPLOMATIC WEAPON: MCKINLEY, WOODFORD, AND THE SPANISH

The propaganda work of the Cuba junta in New York City and the preference of editors for exciting stories from Cuba provided newspaper readers in the United States with a large if irregular flow of coverage and comment that damned Spain while praising the rebels. The presidential conduct of diplomacy in the United States before the 1890s rarely included such expressions of public opinion as factors in decision making. President Grover Cleveland

and Secretary of State Richard Olney, however, had a transformational experience in the Venezuelan boundary controversy. The dramatic outpouring of bellicose condemnations of the British in the U.S. press following Cleveland's challenge to the Salisbury government in December 1895 indicated that the print media and their readers could be roused to warlike intensity, a fact that had obvious implications for voting in the next election.

That the relationship between the McKinley administration and the press was tense is somewhat ironic because the president and his staff worked hard to build a good rapport with journalists. McKinley gave more attention to the press than his two immediate predecessors, Harrison and Cleveland. He and his staff were proactive—they created an environment in the White House that made the job of reporters easier. Journalists found a larger and more convenient work area in the White House, and McKinley's chief of staff, John Addison Porter, met with journalists twice a day—at noon and four in the afternoon—to discuss the day's events. McKinley's assistant secretary, George Cortelyou, scanned several newspapers and kept a "scrapbook" of clippings on current issues for the review of the president. McKinley himself, in spite of his reserved demeanor, was friendly with reporters at least during the first year of his administration. He initiated informal conversations with reporters, and the press response was favorable.

Historians disagree on the weight given to press opinion in the McKinley government's formulation of foreign policy. John Maxwell Hamilton, Robert Hilderbrand, and Stephen Ponder claim that McKinley and his advisors saw much importance in press opinion and, furthermore, developed a systematic approach to the management of public opinion. Richard F. Hamilton and W. Joseph Campbell, on the other hand, find that McKinley and his advisors read but largely ignored the sensationalist newspapers and for the most part did not let editorial opinions influence their policy making. There is no attempt here to resolve this difference, but it is clear that McKinley was deeply affected by press opinion. While he may have dismissed the exaggerations of the yellow press, he believed that the more responsible newspapers reflected the mood of the public and that the mood of the public was important in the conduct of governmental responsibilities. Media scholar John Maxwell Hamilton and a team of researchers found in their sampling of ten metropolitan newspapers that even the more conservative (tending to favor U.S. neutrality) came to accept intervention in Spanish Cuba. This finding is consistent with the analysis of the metropolitan press discussed in this study. Furthermore, historian Louis Pérez has documented a persistent interest in Cuba among U.S. political leaders throughout the nineteenth century, including an enthusiasm for the idea of annexing the island. The excitement in the press on the Cuban issue and the long-term interest of the United States in controlling the island combined to create anxieties in Madrid. McKinley and officials in the State Department used evidence of growing hostility toward Spain that was being expressed in editorials, news coverage, and letters in U.S. newspapers to impress on the Spanish the need to end the strife in Cuba. The McKinley administration gave the impression that the options the Spanish had were limited by the demands emanating from the U.S. public as expressed in the press and in the speeches in the House and Senate.[24]

Secretary of State John Sherman made one of the first pointed references to public opinion in a letter to Dupuy de Lôme on June 26, 1897. Recalling Weyler's military campaign, Sherman reminded the Spanish diplomat that these events had "deeply affected the sensibilities of the American people [and] . . . painfully impressed their government."[25] A few weeks later Sherman and McKinley instructed the new U.S. minister to Spain, Stewart Woodford, to be mindful of "the sentiment of the American people." Woodford was to make clear to officials in Madrid that the United States was an "onlooking nation," profoundly concerned about affairs in Cuba, that was getting most of its information, by implication, from the U.S. press. The new minister was acutely sensitive to press coverage. Much of his anxiety came from the constant pressure for interviews. He was careful in his statements to reporters and wrote to McKinley of his refusal to respond to a telegram from Hearst regarding the Evangelina Cisneros affair.[26]

Woodford read his instructions to the Spanish minister of state, the Duke of Tetuán, in September of 1897. Soon thereafter, however, Woodford became frustrated by the lack of progress in the achievement of peace in Cuba and the growing evidence of a Spanish diplomatic offensive in Europe to pressure Washington to back away from the crisis. He began to explore the possibility of going to the U.S. press with selected information, especially his instructions from McKinley and Sherman, to counter the Spanish diplomatic offensive. Woodford tried to persuade Sherman and McKinley of the value of a public-relations effort: "The bankers, the merchants, the scholars, and the Governments of the world are interested in knowing the exact facts and in some way they will either get the truth or they will be misled and will suffer because of misinformation." McKinley revealed his sensitivity to the press when he reprimanded Woodford for his initiatives and ordered an end to the effort. Tetuán, when informed of Woodford's request to provide the press with portions of his instructions, also objected. Woodford dropped the matter.[27]

The importance of the U.S. press and public opinion emerged again in a conversation between Woodford and Spain's queen regent. She surprised Woodford with her personal and confidential request that the McKinley administration break up the Cuban junta. Her argument was that the continuation of the conflict owed primarily to this organization because it supplied material support and recruits for the rebels. Woodford replied that in the United States, the president had to respond to the will of the people. The public placed the blame for the fighting on Madrid and saw the junta as depicted in newspapers across the country as a legitimate organization entitled to the freedoms guaranteed in the U.S. constitution (members of the junta, including Estrada Palma, were U.S. citizens). Furthermore, Woodford affirmed that "the majority—the large majority of the American people think that the war in Cuba should be stopped at once and that the president should do all that may be necessary to stop it at once."[28]

Woodford and McKinley communicated regularly about the diplomacy in Spain and the fighting in Cuba. After his judicious analysis of these exchanges, historian Offner concludes that "the president of the United States could not join with Spain in crushing the Junta in the face of overwhelming public sentiment in its favor. American public opinion, which included important politicians and civic leaders, championed Cuban freedom."[29] And in the late nineteenth century, government officials often equated (or saw a close correlation between) press coverage and public opinion. McKinley and Woodford made this point emphatically and repeatedly to the government in Madrid. Woodford also made it known throughout Europe (although not via the press) that the United States saw public opinion in favor of Cuban freedom as determinative. Woodford's success in conveying this message was indicated by a Russian diplomat's advice to the French to avoid involvement in the issue because McKinley was "restraining the American public" on a matter of great sensitivity within the United States.[30]

THE DOWNFALL OF DUPUY DE LÔME

Enrique Dupuy de Lôme found himself in an extremely difficult position. As Spain's chief diplomat in Washington, he had to work for cordial relations between the two countries and at the same time report to Madrid on the growing hostility in the press and public of the United States and the increasingly heated invective in the U.S. Congress. As he interacted with President McKinley and Secretary Sherman, Dupuy de Lôme saw evidence of what he concluded was weak, intimidated leadership adjusting to unpredictable and often irrational outbursts of press opinion.

Dupuy de Lôme frequently commented on the close relationship between press opinion and policy. As early as April 1896, he advised Madrid of the prejudicial influence of the Cuban junta on public opinion in the United States. Dupuy de Lôme's reports to Madrid reflected the seemingly changing mood of the U.S. public. Sensitive to the possibility of intervention, he shared the chagrin of most Spanish officials over the publication in November 1897 of an essay by the former U.S. minister to Madrid, Hannis Taylor. This essay assumed that Cuban victory was inevitable. Even though Taylor had left his diplomatic post several months before his essay appeared in print and had little if any contact with the McKinley administration, Spain's government interpreted his piece as representing an important perspective from an influential figure in U.S. diplomacy. Taylor argued that Spain had lost control of Cuba and had no prospect of regaining its colonial mastery. He insisted that the only way to end the fighting was by direct U.S. intervention.[31]

Spain made concessions with its implementation of reforms in the colonial government of Cuba on January 1, 1898, but those reforms brought the opposite of what Madrid intended—an intensification of press and political criticism in the United States. Continued stories of displaced Cuban war refugees in the U.S. press damaged Spain's image.[32] The recently enacted reforms were controversial among Spanish nationalists in both Cuba and Spain and provoked violence in the streets of Havana. These reforms included provision for the autonomy of Cuba within the Spanish Empire, approximating the position of Canada within the British Empire. Anti-autonomy riots broke out in Havana in mid-January, and news of these disturbances flashed across the front pages of newspapers in the United States. The *New York Times* described "bloody battles in Cuba," the *Chicago Tribune* headlined "wild mobs" in the streets of Havana, and the *New Orleans Picayune* reported "great tumult." Only McKinley loyalist Whitelaw Reid's *New York Tribune* offered low-key coverage, claiming that there was "no anxiety about Cuba" in Washington. The *New York Times* editorial captured the prevailing mood: Spanish rule on the island had degenerated into a "farce." It was time for "further action by the United States." Disturbed by the press frenzy, Republican representative Robert Adams (chair of the House subcommittee on Cuba) echoed the concerns of Congressman James Blount during the Chilean crisis of 1891–1892. He cautioned the government not to take action on the basis of newspaper coverage and urged it to wait for more reliable consular reports. Dupuy de Lôme feared that Adams's caution was the exception and warned Madrid that "the change of sentiment [has] been so abrupt and our enemies so influenced by it [and] so numerous" that additional controversies could impact the policy of the government in Washington.[33] The arrival of the U.S. battleship *Maine* in Havana in January 1898 emphasized that point.[34]

Dupuy de Lôme not only had to endure political pressure and the fluctuations in press coverage in the United States but also criticism from the press in Spain. Transatlantic cables furnished the Spanish press with news from Cuba and the United States. Spanish newspapers and periodicals did not have a mass readership to the extent that comparable publications had in the United States, but these organs of mass communication began to have an impact in Madrid and other cities in Spain. Political activists organized public demonstrations in response to press coverage. In early 1896 the return of the disgraced Martínez Campos from Cuba sparked inflammatory antigovernment comments in the urban press and disorderly demonstrations as well. Police handled some demonstrators roughly, and shot and killed one protester. *El País*, a leftist newspaper in Madrid, called for another public demonstration, and over twenty-five thousand protesters filled the streets of the capital city to accompany the cortege of the fallen activist. Patriotism intensified as neither Weyler nor his successors were able to end the revolution. The press reflected the concerns of Spanish families whose sons served in Cuba under the threats of death in combat and death by disease.[35] Dupuy de Lôme faced pressure both from Madrid and Washington.

El País was joined by *El Imparcial*, *El Liberal*, and *El Heraldo* as voices of nationalist enthusiasm.[36] The magazine *La ilustración espanola y americana* offered similar coverage of the Cuban crisis. It did not limit itself to current press reports but also combined retrospective articles with recent photographs and engravings. *La ilustración* was nationalist in its own way, featuring both words and pictures that flattered Spanish officers and troops while denigrating Cuban rebels for their disheveled appearance. The looming threat of the United States came through in an illustration of the battleship *Maine* "in all of its splendor" anchored in Havana harbor.[37]

Buffeted by press opinion in Spain and the United States, by the shifting diplomacy of the McKinley administration, by political instability in

Madrid, and by dismal reports from the fields of battle in Cuba, Dupuy de Lôme lapsed into intemperate expressions of anguish in a letter to a trusted friend, José Canalejas, the owner and editor of *El Heraldo.* Dupuy de Lôme was aware of the disastrous results that followed the coverage of any diplomatic mistake in the U.S. press. He was dismayed to see on the front page of the *New York Journal* of February 9 the full text of his critique of McKinley's apparent catering to the press and public opinion. The beleaguered Spanish diplomat concluded in supposed confidence to his friend that "it once more shows that McKinley is weak" and that he was "bidd[ing] for the admiration of the crowd" in order to stay "on good terms with the jingoes of his party."[38]

Historians tend to emphasize this critical assessment of McKinley more than Dupuy de Lôme's plans, expressed in the same letter, to generate favorable publicity in the United States. The "crowd" that McKinley sought to win over was stirred by the newspapers' daily reports from Spain and Cuba. The minister's awareness of the importance of propaganda was spelled out in a recommendation to Canalejas that Spain publicize an effort to improve commercial relations between the two countries "even if only for effect." The plan included inviting a leading Spanish businessman to visit, so that, according to Dupuy de Lôme, "I may make use of him here to carry on a propaganda [campaign] among the Senators and others in opposition to the Junta."[39]

The extent of his frustration with the press and public opinion in the United States came through in the same letter. Dupuy de Lôme saw them as more important than the McKinley administration in an ominous situation: "The danger of encouragement [is] being given to the insurgents by a part of public opinion if not by the government."[40]

Dupuy de Lôme's hopes for a productive Spanish propaganda campaign disappeared with the publication of his letter on the front page of the *New York Journal.* The story flashed quickly through the telegraph system, subsequently appearing in major newspapers across the country. The *Washington Post* intruded on the *New York Journal's* scoop by publishing a version of the letter, also on February 9, beneath headlines that indicated the possibility of forgery and identifying the New York junta as the source. The *Post,* however, joined with other newspapers in attesting to Dupuy de Lôme's authorship of the letter the next day, and headlines in Boston, Philadelphia, Chicago, San Francisco, and New Orleans enticed readers with essentially the same story. The *San Francisco Examiner* rivaled the *New York Journal* with unabashed hyperbole by calling the Spanish minister "abusive" and characterizing his plans for a public-relations campaign as evidence of "deceitfulness." Furthermore, the *Examiner* caricatured Dupuy de Lôme in an editorial cartoon that featured the Spaniard twisting an exaggeratedly long mustache. Even the reluctant *New Orleans Picayune* presented Dupuy de Lôme in a highly unflattering light. Disgraced and discredited, he left Washington to return to Spain. Woodford again instructed the Spanish on the importance of the U.S. press when he demanded an official apology for Dupuy de Lôme's letter. Madrid complied, but U.S.-Spanish relations were at a new low, and, as many experienced Washington diplomats grasped, the departure of Dupuy de Lôme deprived this relationship of a competent if at times arrogant diplomat who understood the nuances of diplomacy in a context susceptible to the whims of newspaper coverage and public opinion. It was an extraordinary irony that his own words would contribute to the crisis and remove him from the scene. The ensuing controversy "was a public relations triumph for the [Cuban] Junta."[41] All ten of the metropolitan newspapers drew attention to the letter with an intensity not unlike that of the yellow press.[42]

THE DESTRUCTION OF THE *MAINE*

Fitzhugh Lee had seen conditions in Havana deteriorate to the point that by January 1898 he had begun carrying a pistol in a holster on his hip most days. Serving as Washington's consul general in Cuba, Lee was the ranking U.S. representative on the island. A veteran of the Confederate cavalry in the Civil War, the nephew of General Robert E. Lee, and the former governor of Virginia, Lee was experienced in both military and political affairs. Although lacking background in Cuba (or elsewhere in Latin America), Lee had been in charge of the consulate since early 1896.[43] His understanding of Spain's colonial problems had deepened during his tenure, and he was disturbed by the antiautonomy riots of January 12, 1898. Spanish defenders of recently dismissed General Weyler, many of whom were soldiers, had taken to the streets to attack the offices of newspapers supportive of Cuban autonomy within the Spanish Empire. The electronic coverage of Cuban affairs in the U.S. press produced headlines such as "The Spectacle of Army Officers Wrecking a Building" in the *New Orleans Picayune*.[44] The *San Francisco Examiner* saw serious consequences: "Havana Mobs May Give M'Kinley Cause for Intervention."[45] An editorial in the *Examiner* concluded that Spain must be removed from Cuba, by force if necessary.[46]

Because of the instability in Havana, supposedly the most secure city on the war-torn island, Lee faced the difficult decision of whether to ask

Washington to send help for the protection of U.S. citizens. The typical form of assistance in such situations was the dispatch of a warship to standby in a port near U.S. government representatives, businessmen, and their families. Havana had a large harbor, and the McKinley administration assigned the battleship *Maine* to Key West to await the call to proceed to Havana. The *San Francisco Examiner* placed a sketch of the *Maine* on its front page of January 13, accompanied by a caption that underscored the importance of telegraphic communication in naval preparations as well as journalism. The caption was based on a telegraphed report from Key West datelined January 12 and informed readers that the *Maine* was under orders that arrived from Washington "by telegraph to-night. In compliance with these orders, Captain C. D. Sigsbee, commanding the battleship, has made all arrangements to get under way without delay. He had been placed in communication with (Consul-) General Lee, who will cable him whether or not it is necessary for him to go to Havana."

For several weeks Fitzhugh Lee was in telegraphic communication with Captain Charles Sigsbee of the *Maine* in Key West, which was only a six-hour voyage from Havana. Lee and Sigsbee had devised a special telegraphic code to summon the *Maine*, but the consul general struggled with the question of whether to call for the *Maine* (which he was empowered to do) or to give Governor-General Ramón Blanco more time to establish order. Lee was especially concerned that the arrival of the *Maine* would result in more anti-U.S. street demonstrations and make Blanco's task even more difficult. Therefore, on January 25, Lee cabled Washington to delay the sending of the *Maine* until conditions in Havana were calmer. This cable was too late. On the previous day, McKinley and his advisors decided to send the *Maine* to Havana. Sigsbee's ship reached Havana on the morning of January 25. John Caldwell, a correspondent of the *New York Herald*, received a coded message from his editors that the *Maine* was en route. Caldwell lived in the Hotel Inglaterra, where Lee also resided. Caldwell rushed to Lee's rooms with this information. The consul general apparently retorted, "Nonsense!" Then both men looked out the window of the hotel to see the battleship gliding into Havana harbor.[47] As was often the case, the journalist was a few steps ahead of the government official in the use of the cable system.

The *Maine* anchored in the harbor for three weeks before it was torn apart by a huge explosion that took the lives of over 260 sailors and officers. Tensions had abated in Havana somewhat during this period, but several reporters remained alert to signs of trouble that might lead to a newsworthy story. As the sound of the massive explosion reverberated through the streets of Havana, two of the most experienced journalists in Cuba, Sylvester Scovel and Charles Rea, were enjoying an evening of carnival celebrations with Mrs. Scovel in a café not far from the harbor. They were shaken in their seats as the impact of the explosion hit. Rea looked toward the harbor and saw the night sky "lit up with an intense light, and above it all could be seen innumerable colored lights resembling rockets." He and Scovel dashed into the street and quickly determined to make haste to the harbor after escorting Mrs. Scovel to a place of security. The two arrived in time to witness the death throes of the *Maine*, fires still burning and explosions from rounds of ammunition occasionally bursting forth. The two mingled with other reporters. Given their profession's obsession with the "scoop," the attention that they first focused on the *Maine* and human victims soon shifted to the problem of how to get their accounts to their editorial offices.[48]

GETTING THE STORY THROUGH

Reporters rushed to the Havana cable office by horse-drawn cab to face what they knew would be an unreceptive censor. Only four messages passed between Havana and Key West along the submerged lines of International Ocean/Western Union. Both of Sigsbee's messages—the report to Secretary Long and the shorter telegram to Admiral Sicard—passed through the censor. The ever-resourceful Scovel was carrying in his possession a blank sheet with the censor's stamp of approval already affixed, so he penned a sixty-four-word story that reached New York in time to make the 5 a.m. edition of the *World* on February 16. F. J. Hilgert's dispatch for the Associated Press was actually the first to survive the censor and enter the telegraphic network in the United States. He had prepared a lengthy account, but only the first hundred words left Havana. Hilgert's story reached several newspapers a few hours before Scovel's lines reached the *New York Journal*.[49]

To avoid the censor, Walter Meriwether of the *New York Herald* took the steamer *Olivette* from Havana at noon on February 16 to Key West. He was fortunate to accompany several of the surviving members of the crew of the *Maine* for the six-hour voyage. Just before his departure, Meriwether, who also had a prestamped censor's sheet, used that document to send a simple, innocuous message to the *Herald*'s agent in Key West to hold an open key to the underwater line to Punta Rassa and points north. On arrival, Meriwether had a twelve-thousand-word story replete with interviews of the survivors. It reached New York by the evening of the sixteenth.[50]

Meriwether's use of the *Olivette* to avoid the Spanish-controlled telegraph office in Havana pointed to what became the most reliable way to supply editors with copy from Havana. The stunning news that 260 sailors and officers perished in the explosion and that the once-proud *Maine* was reduced to twisted, smoldering wreckage with a flight of buzzards spread across the masthead shocked readers in the United States, Spain, and other European and Latin American nations as well. The dramatic, violent, and tragic elements in this story were sensationalist enough for both editors and readers of the yellow press without any embellishment, but, for William Randolph Hearst and Joseph Pulitzer, this disaster offered additional opportunities they could not resist. Both Hearst and Pulitzer followed Meriwether's example. Hearst procured a three-vessel charter fleet and Pulitzer secured the services of two former filibuster boats. The Associated Press fell behind these New York newspapers. Charles Diehl, the AP's assistant manager in charge of Cuba coverage, complained that general manager Melville Stone was too slow to respond to the need for oceangoing messenger ships. Stone did not approve funding for an AP dispatch boat until the last part of April, after the New York papers had established a lead in covering Cuba.[51]

The editorial offices in New York were soon flooded with telegraphed accounts of the explosion and its aftermath. Hearst set the tone for his newspaper's coverage of the event early on February 16 when, in a telephone conversation with the *Journal*'s night editor, the master of sensationalism asked, "Have you put anything else on the front page?" The editor replied, "Only the other big news." Hearst responded, "There is not any other big news. Please spread the story all over the page. This means war."[52]

The government in Washington responded to the outpouring of questions concerning the identities of the victims, the well-being of the survivors, and the causes of the explosion. These responses, no matter how carefully phrased, elicited grief and anger. The Navy Department reported "a great demand for the news" concerning the fate of the crew. Overwhelmed by the volume of this demand, naval officials "answered with a blood-curdling official brevity." At the same time, the press reported that Consul General Lee's "secret dispatches" from Havana indicated that the explosion was the work of "hostile hands."[53]

Hearst, his chief editor Arthur Brisbane, their Havana-based reporters, and New York copyeditors filled their paper with information, speculation, and rumors. From February 17 to 24, the *Journal* devoted an average of eight and a half pages to the *Maine* story and its repercussions. Serving as investigative body, judge, and jury, the *Journal* concluded that the explosion resulted from the actions of "criminals" who deserved "conviction" and offered a reward for evidence that would reinforce their premature verdict.[54] On February 17 the *Journal* summed up the case for the Spanish as the conspiratorial evildoers: "The warship *Maine* was split in two by an enemy's secret infernal machine." As Joseph Wisan observes in his pioneering study of the New York press in this period, "The news columns of Hearst's paper seemed deliberately intended to inflame the public."[55] An inflamed public was certainly an interested public. The *New York Journal* claimed that its circulation rose sharply from a daily average of 416,885 for the week following January 9 to 1,025,624 on February 17 and 1,036,140 the following day.[56]

Pulitzer's *New York World* used a similar if not quite so strident approach, and its editorials were only slightly less warlike in tone than those of the *Journal*.[57] By contrast, the *New York Times* seemed to accept the theory that the explosion resulted from an accident and warned against precipitous action against Spain. The *Times* criticized the sensationalist extremes of the *Journal* and the *World* and advised patience in the determination of the cause of the explosion. In the immediate aftermath of the destruction of the *Maine*, the *Sun* and *Evening Post* of New York joined the *Times* and *Tribune* of that city in advising caution in the policy toward Spain. Seven of the other eight metropolitan newspapers under study here also manifested a measure of caution, but their coverage in general included speculative interviews with Captain Sigsbee and several less qualified individuals. Hearst's *San Francisco Examiner*, however, was the glaring exception, issuing numerous heated accusations against Spain.[58] The organizations that devoted the largest amount of money to covering the story and employed the most modern forms of technology in order to gain access to information from the scene of the event provided the reading public with the least reliable reportage. Hearst and Pulitzer sent teams of experienced reporters to Cuba, chartered dispatch boats for the Havana–Key West run, and held cable connections open to handle their voluminous and expensive coverage. This technology and expertise, however, did not insure factual accuracy and dispassionate analysis. It did contribute to higher circulation figures and more revenue.

ANOTHER NEWS VACUUM: WAITING FOR MCKINLEY'S DECISION

The destruction of the *Maine* and its extensive press coverage posed a serious concern for President McKinley, yet he and his foreign-policy advisors maintained a grasp of the situation and avoided hasty action. Coming only a week after the publication of the Dupuy de Lôme letter, the *Maine* affair could have produced an intense crisis in the relations between Washington and Madrid, but diplomats of both countries "handled the dramas well." Within a few days, the unanswered questions of the *Maine* crisis entered "routine investigative channels."[59]

Furthermore, McKinley was quick to establish means of coping with the press and the public. The president instructed Secretary Long to inform reporters that preliminary investigation confirmed the explosion was probably an accident and that there was no "cause for alarm." McKinley also authorized his executive secretaries to release copies of the telegrams from Havana so as to prove that the situation was being handled in a routine manner.[60] In spite of this calm exterior, the McKinley administration was making plans for war. It was clear that if the government in Madrid did not make major concessions, the political pressure in Congress and the public mood throughout the nation would be such as to make war hard to avoid.[61]

President McKinley was concerned about expressions of public opinion—especially on the editorial pages and front pages of major newspapers. As early as mid-1897, the president was regularly perusing these publications, and his staff maintained large books of clippings for reference. McKinley also increased the number of his staffers from six to eighteen in 1897 to help in the monitoring of the press. McKinley's personal relationship with the press was ambiguous, however. A reticent yet friendly man, he was on good terms with many reporters, but he was also a "lackluster" subject with little talent for publicity. While historians are not in agreement on the effectiveness of his press strategy, it is clear that McKinley and his administration were following the portrayal of the Cuban issue in U.S. newspapers by late 1897, and, after the *Maine* disaster, the way it was represented became a daily priority in the White House.[62]

The front pages of U.S. newspapers were filled with details about the destruction of the *Maine* for several days. To some extent, the exuberant Hearst was right; there was no other news, or, at least, no other event that sparked so much public interest. A news vacuum appeared like the one that appeared in the last half of 1895, during which tensions that arose from the uncertainties surrounding relations between Great Britain and the United States over the British Guiana–Venezuela boundary intensified to the point of becoming a war scare. Reporters and editors understood that the fate of the *Maine* created the threat of war with Spain, but neither Washington, nor Madrid, nor Havana produced information of sufficient importance and dramatic impact to satisfy the public demand.[63]

Inventive reporters and imaginative editors went to work to supply their readers with interesting but not always accurate stories in an effort to maintain high circulation figures. New York was the communications center of the nation, and reportage that originated in this city rapidly spread across the nation. Hearst and Pulitzer had innovated in the appearance of the front pages of their newspapers by introducing heavy-type, large-letter headlines strung out across several columns—or in certain cases even across the entire front page.[64] The coverage ran the gamut from maudlin trivia to nonsensical braggadocio and from manipulative expressions of grief to gruesome details of mutilation and death. The press played to the emotions of its readers.[65] Captain Sigsbee's dog, Peggy, revealed symptoms of trauma long after she survived the explosion. While Peggy was fearful and anxious, leading sports figures in the United States apparently were not. Boxing champions John L. Sullivan and James J. Corbett, baseball great Cap Anson, and hammer thrower Jim Mitchell were willing to join a cadre of athletes and other males with impressive physiques and physical skills to go to Cuba and, according to Hearst's *Journal,* intimidate the Spanish Army into a quick surrender.[66] The same newspaper blamed the death of Senator John Thurston's wife on the emotions aroused by a brief trip to Cuba in early 1898.[67] Even the respected Sylvester Scovel played to the emotions of pathos and patriotism in the pages of the *New York World* with his description of vultures devouring the flesh of a deceased sailor whose body randomly surfaced from within the twisted remains of the ship.[68]

The appeals to raw emotion took another path that made use of existing prejudices and stereotypes to reinforce the public's fascination with conspiracies. The presumption that somehow the Spanish were responsible for the deaths and the destruction was widespread in the press and among politicians, but it was not clear to anyone who planned and implemented the terrible deed. In late February, a conspiracy theory that tied the much-despised and recently dismissed Weyler and his angry followers to the *Maine* disaster surfaced in U.S. newspapers. U.S. minister Stewart Woodford conveyed a report to McKinley that claimed correspondents in Havana "are doing all they can to raise a war scare between America and Spain, spreading no end of lies and succeeding in exciting a bad feeling." The Hearst papers and the Associated Press used the telegraph system to broadcast these stories throughout the country. Hearst's *San Francisco Examiner* led with the headlines "Weylerites Plotted the Deed" and "Upon the Shoulders of the 'Butcher' Is Laid the Crime."[69] The *Chicago Tribune* carried the same story on the same day and expressed it in language designed to elicit deep resentment: "Weyler gave his order and the *Maine* went down. . . . Not alone do the Weyler people tell it as they joyfully congratulate each other over the blow to the Yankee, but the Blanco party has the same story."[70]

Other lesser-known conspiracy theories appeared in big-city newspapers. The *Chicago Tribune* and the *New York World* carried Scovel's observations about angry pro-Spanish "volunteers" who were "recruited from the worst elements of the population, . . . the most bigoted and ignorant class," in Havana to menace U.S. citizens who remained in the Cuban capital. Not to be outdone, Karl Decker of the *New York Journal* claimed that he had information from a "trustworthy source" that tied the destruction of the *Maine* to three Spanish merchants with stores in the Havana business district.[71] Even the respectable *New York Times* joined the conspiracy game by featuring an interview with an unnamed Englishman identified only as a "seeming electrical engineer," who stated that he had sold several submarine mines to Spanish officials who apparently had placed eight to ten of them in Havana harbor.[72]

With such opinions and stories spreading through the wire services and newspapers in the United States, Governor-General Blanco revived the anti-press policies of previous administrations in Cuba. The first journalist to be expelled was the French-born Honoré Lainé, a reporter for the *New York Journal* and several other newspapers. The Spanish colonial administration accused Lainé of consorting with the rebels, a charge that had been directed against him once before after a trip to the Cuban backlands. Police officials arrested him, placed him in prison briefly, and then expelled him from the island. Thanks to the telegraphic news network, descriptions of Lainé's plight appeared in U.S. newspapers on both coasts on the same day.[73] Within the next few weeks three other journalists—Freeman Halstead of the *New York Herald*, Charles Rea of the *New York World*, and Charles Musgrave of the *New York Journal*—all were either harassed, pursued, or arrested before departing Cuba.[74]

Lainé's expulsion in early March was not only a blow against the *New York Journal* but also a threat to the U.S. press in general. By that time newspapers across the country were advocating caution less and beginning to emphasize war preparations more. The moderate Republican *New York*

Tribune headlined, "Administration On Guard . . . Not to Be Caught Unprepared If Emergency Arises."[75] Owner Whitelaw Reid of the *New York Tribune* was an influential Republican Party leader who had run as the vice presidential candidate with Benjamin Harrison in the unsuccessful campaign of 1892. Reid wrote a revealing letter to McKinley on March 8 that testified to the moderate Republican's growing support for war while at the same time reiterating his often-stated condemnation of sensationalism. "The more intelligent classes are not greatly affected by the sensational press," he suggested but then went on to say that he saw a strong inclination in the public to accept war as the necessary policy: the people of the United States "really seem to feel, as every patriot must, in this crisis, a readiness to hold themselves subject to call when and where the Country needs. Conservative public sentiment will sustain purchases of ammunition and even of war-ships." The day before he wrote this letter to McKinley, Reid instructed the *New York Tribune* office by coded telegram to make "thorough arrangements" for "war correspondents."[76]

The *San Francisco Chronicle* echoed the same message in the headline "Men and Guns Will Be Ready" on March 6, and two days it later warned Spain in an editorial that military preparations left the Spanish with a simple choice—either give in to U.S. demands or expect armed conflict.[77] The *Atlanta Constitution* placed an illustration of what it described as a "ten-inch bore, 67,000 pound rifle" at the top center of its front page on March 7. This large weapon made a stop in Atlanta on its way to Pensacola. Its imposing presence inspired the headline "This Big Gun Booms War Talk; Gave Color to Many Rumors Afloat," a headline that suggested similarities with the press coverage of another large naval gun on its way across the Rocky Mountains to California during the height of the U.S.-Chilean war scare of 1891–1892.[78]

Unlike the big-city press from New York to Chicago to San Francisco, which trumpeted its suspicions about Spain and increasingly voiced a prowar sentiment, the daily and weekly papers in smaller cities and rural communities across the South, Midwest, and Great Plains were hesitant to call for war at first, but by late February and early March joined their metropolitan counterparts in expressing the "martial spirit." For example, the *Indianapolis Journal* was skeptical of the yellow press of New York and its propensity to condemn Spain, but by February this Indiana paper accepted war as necessary. Indiana newspapers in Wabash, Madison, and South Bend agreed.[79] Publications in Iowa followed a similar pattern. Press organs in

Dubuque, Davenport, and Council Bluffs were in accord with the national mood of belligerency by early to middle March. Newspapers in North Carolina and Texas set aside their skepticism about the yellow press to express support for war as a solution to the Cuban crisis. On March 29 the *Raleigh News and Observer* was emphatic: "North Carolina is for war—if war be the only thing by which the Cuban butcheries can be stopped and the island freed." Historian Marvin Olasky summarizes the mood in Texas in words that applied to many newspapers outside the major cities: "Texas editors began to suggest that basic animosity toward New York newspapers was no reason to disregard evidence indicating that, for once, the yellow journals might have it partially right." In Ohio the *Press-Post* of Columbus (the largest circulation newspaper in the state) was receptive to the Cuban junta's propaganda and an advocate for war as early as February, while the moderate *State Journal* supported McKinley's diplomatic approach. By mid-March both newspapers were carrying extensive coverage of war preparations and patriotic statements made by politicians in Washington and the jingo press in New York.[80]

Whitelaw Reid's journey across the United States from the Pacific coast to New York in mid-March brought him to an awareness of the same phenomenon reflected in the contradiction between editorials and news coverage. In a letter to Donald Nicholson, the *New York Tribune*'s editor, the staunch antisensationalist commented on the powerful appeal of "the frightful sensationalism" in news coverage from the Pacific coast across the heartland of the country: "It was an actual amazement to me, in crossing the continent, to notice how extensively the papers which have so denounced what is termed 'yellow journalism' on the editorial pages, were nevertheless repeating, with job type display, the *Journal*'s most sensational news. It is a queer world."[81]

The two and a half months between the destruction of the *Maine* and the declaration of war saw some news events of genuine importance in addition to the flow of inflated minor incidents, invented conspiracies, and political posturing. For example, the unanimous congressional approval of McKinley's request for $50 million to augment the nation's armed forces created reverberations of patriotic excitement, as the press circulated the news throughout the country.[82] Soon thereafter, on March 18, the moderate, nonpolemical senator from Vermont, Redfield Proctor, delivered his report on his recent trip to Cuba, which added credibility to the rebel cause by its indictment of Spanish colonial policy—especially reconcentration. Proctor projected an unusual image for the late 1890s. With his understated manner and objective prose, he drove home a moral judgment that the inept, heavy-handed Spanish must be removed from Cuba. The national press featured Proctor as a protagonist of intervention on the grounds that the Spanish had lost control of the island and not just on account of the *Maine*.[83] The juxtaposition of the news of a huge budget for war preparation with Proctor's speech carried the nation's press and politicians in the direction that the jingoes wanted—war with Spain.

THE DECISION FOR WAR

The press and public eagerly awaited the report of the Sampson court of inquiry into the causes of the explosions on the *Maine*. When President McKinley released this report on March 28, the heavy pressure of public and political opinion bore down on the anxious chief executive. The drive for a "scoop" pushed the Associated Press to publish a summary of the report a day before McKinley made his presentation.[84] Already upstaged by this news leak, McKinley also suffered in comparison with the navy, which had concluded that the explosions originated outside the ship. McKinley's message seemed to indicate uncertain leadership at the time that the press, the public, and politicians in both the Democratic Party and the president's own Republican Party joined in the "impassioned outcry for immediate action."[85]

Factual reporting mixed with rumors and conspiracy theories, but the reporters and editors behind them generally shared a single conclusion: the United States must prepare for war with Spain. Descriptions of naval and military preparations and the depth of the "war spirit" appeared in both sensationalist and nonsensationalist papers. In

Whitelaw Reid, editor of the *New York Tribune* and a perceptive observer of the U.S. press coverage of events in Cuba (Stanley Waterloo, ed., *Our Living Leaders* [Chicago: Monarch Book Company, 1896], 383).

early April the *New York World* issued its own ultimatum to Spain, which was supported by a poll that claimed to cover every congressional district in the country. Pulitzer's editorial conclusion used contorted logic to reject further diplomacy: "Stop the nonsense! Let us have peace even if we must fight a battle or two to compel it!" The nonsensationalist *New York Herald* had been consistently opposed to war, but on March 30 it shifted to a prowar editorial position.[86]

McKinley's seeming inaction brought on a revolt within his own party. On the night of March 29 and the following day, approximately one hundred Republicans in the House of Representatives (over half the party's membership in that body) agreed to the establishment of a deadline by which McKinley had to take executive action or else they would join Democrats to vote for war with Spain. The leaders of this Republican schism came from Illinois, Iowa, and Nebraska and there was significant participation from the delegations of New York, Pennsylvania, Michigan, Ohio, Wisconsin, and West Virginia. Their most probable motive, in the analysis of John Offner, was concern about the off-year congressional election in the fall of 1898.[87] Political survival was central in the revolt. Newspapers throughout the country, with their connections to the electronic information system, played an essential role in elevating the Cuban question to the status of a nationwide issue with large political ramifications.

Outside the arena of press scrutiny, the McKinley administration engaged in tough diplomacy with Madrid. On March 29 the president instructed U.S. minister Stewart Woodford to present Washington's demands (including an armistice in Cuba) to the Spanish government, and, echoing Harrison's aggressive approach to Chile in 1892, declared that the government of Prime Minister Praxedes Sagasta should reply by cable within forty-eight hours. The Spanish replied within the time limit of the ultimatum but rejected one of the main requirements—the demand for an armistice.[88] McKinley used assertive diplomacy, but his administration apparently chose not to publicize his firmness and thereby missed an opportunity to bolster his public image.

While McKinley agonized in the White House, Fitzhugh Lee, U.S. consul general in Cuba and an outspoken critic of the Spanish, selected a propitious time, the second week in April, to return to Washington. Spanish minister of colonies Segismundo Moret y Prendergast had identified Lee as a rebel sympathizer, leading to a tumultuous reception in the United States that was bad news for Madrid as well as the McKinley administration. His trip by rail from Florida through the states of the old Confederacy (his uncle was General Robert E. Lee) on his way to the nation's capital elicited press descriptions such as "the triumphant tour," and Lee himself became "the hero of the hour" in front-page stories in Washington, New York, Boston, Philadelphia, Chicago, Atlanta, and New Orleans. These papers carried detailed accounts of Lee's rail stops along the way to Washington that revealed a surge in national unity that transcended the divisions of the Civil War era. Press coverage of these public outpourings was as extensive in the North as the South. Both the *Philadelphia Inquirer* (Republican) and the *Boston Globe* (Democratic) reported that Lee drew ovations from the massive assemblages as the universally admired "conquering hero." These public demonstrations of support reinforced Lee's criticisms of Spain and indicated spontaneous sentiment in favor of action. A crowd of five thousand greeted Lee in Savannah, while reports from Richmond estimated ten thousand at the train station in that Virginia city. The *New York Times* observed that five thousand well-wishers gathered in the streets around Washington's Shoreham Hotel where Lee was staying.[89]

The *New Orleans Picayune* captured the mood of the Washington reception: "'Now there's a fighting man in town' yelled an enthusiastic spectator.' . . . The crowd cheered lustily." The same newspaper reported from Waycross, Georgia, a day earlier that one aged veteran had said, "I am 62, but all I want you to do is sound the bugle, and I will be with you, saddled and booted. Don't put it off too long, general. Let us go to haul down the Spanish flag." This sentiment was common all along the line.[90] These belligerent statements found support in the release of U.S. consular reports to the press that provided details about Weyler's reconcentration policy on Cuban civilians. Lee's testimony before the Senate Foreign Relations Committee covered the destruction of the *Maine*. Lee concluded that the known evidence pointed to Weyler and his loyalists as the likely culprits.[91]

Fitzhugh Lee's emphatic condemnation of Spain undercut McKinley's persistent efforts to find a diplomatic solution. The president reinforced his image as an indecisive leader by postponing his much-awaited address to Congress on Cuba from April 4 to April 6 and then again to April 11. The statement itself made matters worse for the administration. It was "legalistic, stilted, and florid. . . . [M]any were disappointed by the president's failure to capture the emotion of the occasion."[92]

McKinley's public-relations blunder was ironic given that he and his staff had exerted so much effort to win the allegiance of the press and the public in the weeks after the February 15 destruction of the *Maine*. By early March

McKinley's staff had begun distributing written statements to White House reporters, an innovation in president-press relations. These press releases gave newspapers a convenient summary of information that was intended to present the administration in a favorable light. Throughout the crisis the president's staff clipped articles and editorials from newspapers.[93] Although McKinley generally ignored the Pulitzer-Hearst yellow press, he paid close attention to other newspapers—especially the New York papers, those of other large metropolitan centers, and his hometown's *Canton Repository*.[94]

Those who worked with McKinley confirmed his sensitivity to press opinion. Chauncey Depew, a railroad executive and a leader in the Republican Party, stated that the president "studied so constantly public opinion that he became almost infallible in its interpretation" and that his "faith in the public intelligence and conscience was supreme."[95] Henry Pritchett, superintendant of the Coast and Geodetic Survey and later president of the Massachusetts Institute of Technology, helped create the "map room" in the McKinley White House. Pritchett met with the president regularly and took detailed notes of a conversation on May 2, 1899, in which McKinley unburdened his frustrations with the press and politics in his quest to avoid war with Spain. Pritchett concluded that "there could be no question of his firm belief that, if left alone, he could have settled the matter without war." Pritchett added his own explanation of the emotional pressures brought to bear on the president:

> Perhaps few people realize the excited state of feeling in which men in public life lived during the first months of 1898. One who turns back to the files of the daily press of those days and reads the frantic appeals which were poured out, will understand in what a ferment the public mind must have been to accept such expressions of prejudice and passion.[96]

McKinley and his chief of staff, John Addison Porter, had created a comfortable, friendly environment for reporters in the White House, but with the deepening of the Cuban crisis, this no longer served to give an advantage to the president. McKinley and Porter relied heavily on George Cortelyou, Porter's staff assistant who became, in effect, the press secretary. Cortelyou was both gregarious and well informed. He cultivated good relations with the reporters and built up their confidence in the McKinley administration.[97] These innovative public-relations measures served the administration well for a several months, but events soon overwhelmed the McKinley White House. Lee's triumphal journey, Proctor's subdued, persuasive report, the government's preparations for war, and the revolt of the Republicans in Congress combined with the factual reporting, rumors, and contrived stories to put enormous pressure on the reluctant McKinley. Elihu Root, a prominent New York attorney and a leading Republican, expressed the fear that the president might destroy the party itself because of his resistance to the public sentiment for war. McKinley's hopes for a diplomatic solution finally collapsed, and he acceded to his fellow Republicans by turning the question over to Congress for what was a foregone decision. His war message, such as it was, disappointed many among his most loyal followers. The outcome, however, seemed inevitable. By March 1898 the more circumspect newspapers had adopted the same approach as, if not the excessive style of, the yellow press.

CONCLUSIONS

In one sense the press played a different role in 1897 and 1898 than it had in the Chilean and British war scares earlier in the decade. The McKinley administration cited the public uproar incited by the jingo press in its diplomacy with Spain to work for a peaceful solution. This tactic worked at first. Dupuy de Lôme and his colleagues in Madrid were disturbed by the warlike mood reflected in the newspapers. The pressure applied to the Spanish government seemed to have an impact. Spain moderated the reconcentration policy and moved in the general direction signaled by Washington, but these changes came too late. Worn down by press and public agitation, McKinley followed the example of the Harrison administration in the 1892 crisis with Chile: he sent Spain an ultimatum on March 29. Spain did not comply with all its demands, however, and at the same time, Redfield Proctor, Fitzhugh Lee, and other public figures were offering information and opinions that seemed to justify war preparations. These comments received extensive exposure in the press, thus heavily weighting the information system toward an aggressive position vis-à-vis Spain, which made diplomacy very difficult if not impossible.

Historians Richard Hamilton and W. Joseph Campbell argue persuasively that it was not the press alone that caused the McKinley administration to decide for war, but this interpretation is posited on the assumption that the press can by singled out and/or removed as a contributory causative factor in the outcome of this crisis. By viewing the press and the telegraph network as parts of a large information system, however, it becomes evident that the daily newspaper had become an integral part of politics, elections, and policy formulation. Senators, congressmen, ambassadors, secretaries of state, presidents, and prime ministers read newspapers as a matter of routine. The public in general and voters in particular also received much of their news from this form of mass communication. President McKinley and his Republican supporters understood this point and eventually acted on it. John Maxwell Hamilton and his research team have concluded that newspapers set the agenda for the national debate and that the Cuban issue was at the top of the list of issues that concerned citizens.[98] Offner's conclusion by implication recognizes the role of this information system in spreading the news to the public and thereby motivating politicians to act in an attempt to gain reelection. From McKinley's perspective, "it was better to blame Spain for inhumane conditions that required intervention than to act because of domestic political pressures. But in the end, it was these pressures that brought about military intervention. Republican legislators made war on Spain not to obtain control of Cuba but to retain control of Washington."[99] These "domestic political pressures" were brought to bear by political figures such as Proctor and Lee, but it was the mass-circulation press tied together by telegraph wires that gave these pressures an impact that was multiplied by the near simultaneous appearance of front-page stories from New York to San Francisco. Although not a singular causative factor, the press as a function of the new information system within the framework of national politics played a major role.

The imbalance in the information system so prominent in 1891–1892 and 1895–1896 had reappeared. With its stream of atrocity stories, condemnations of Weyler, extensive coverage of the ill-considered statements of Dupuy de Lôme, and the explosion of the *Maine*, newspapers gave the impression of emerging popular support for war. The information system carried the statements of a diverse array of politicians and newspaper editors who were motivated by different causes but who came to advocate the same policy by March and April 1898. William Randolph Hearst used sensationalism to boost the circulation of the *New York Journal*, while Whitelaw Reid took on the role of the rational commentator to boost the Republican Party. Adolph Ochs emphasized objective journalism in the pages of the *New York Times*. In the last months all three favored preparation for war. The skeptical *New Orleans Picayune*, which for months had refused to join the press chorus, came to accept war as necessary in early 1898. Republican senator Redfield Proctor presented a compelling perspective on conditions in Cuba.

Fitzhugh Lee was a prominent Democrat who promoted an anti-Spanish position and advocated the invasion of Cuba. All of these commentators were pointed in the same direction—war with Spain. McKinley's hopes for a peaceful settlement disappeared beneath a torrent that flowed through the information system and pushed the president to a reluctant, essentially political decision for belligerency.

On April 27, in the wake of the McKinley administration's declaration of war, the once-reluctant *New York Sun* editorialized a view that had been widely accepted for over a month: "We are all jingoes now." Only a handful of newspapers that specialized in business news, including the *Wall Street Journal*, maintained an antiwar position, and even that was sometimes "lukewarm."[100] The yellow press did not start the war, but Hearst of the *Journal*, Pulitzer of the *World*, Ochs of the *New York Times*, Reid of the *New York Tribune*, Joseph Medill of the *Chicago Tribune*, and most editors and newspapers from coast to coast, from large cities and small, formed a broad consensus. The nation's press, diversified in ownership, political preference, and editorial style, nevertheless was tied together by this commitment and the telegraph wires that conveyed news and opinions.

CHAPTER ELEVEN

Information Flow, the U.S. Press, and the War with Spain

AN OBSCURE CROSSROADS in the tropical wilderness in eastern Cuba became the point of intersection in the careers of the leading journalist of the United States and a rising political star, who, within three years, would become that nation's president. Richard Harding Davis, the experienced but still youthful reporter was on the front line of the first important land conflict between U.S. forces and the Spanish army. His personal relationship with Theodore Roosevelt became much closer on June 24, 1898, when the amateur officer and chief organizer of the Rough Riders had his baptism under fire and, in Davis's judgment, emerged as a leader of courage and ability. Davis wrote his account of the battle of Las Guásimas soon after the firing stopped that morning, and his story appeared in the pages of the *New York Herald* two days later, transmitted by telegraph cable from the Caribbean to New York City. Roosevelt, however, was not the only person to receive laudatory coverage. In the next few weeks the victory over the Spanish on San Juan Heights and the sinking of the Spanish squadron off Santiago brought a flood of names and courageous deeds into the information system in the United States. In the confusion of the armed conflict, there were several officers, both regular and volunteer, who held firm under heavy fire. Roosevelt's rise to media preeminence was not as quick as is often assumed. Davis, however, happened to be standing close to Roosevelt at Las Guásimas. The words of the nation's leading war correspondent drew attention to Roosevelt in newspapers across the country at a key moment. Over a period of months Davis gave the publicity-seeking Roosevelt exposure he would eventually exploit to the fullest.[1]

Davis's account and the stories written by other correspondents moved from southeastern Cuba to New York by means of the submarine cables that connected Cuba, Jamaica, and Haiti to the United States. There was no convenient cable station near the front, so reporters had to resort to various methods to get their hurriedly composed stories to the nearest station. The U.S. military established a telegraph office at Playa del Este in the Guantanamo Bay area. It was the closest, but because of the volume of prioritized military messages, it was available only after a long delay. The United States and Haiti Telegraph and Cable Company's station at Môle St. Nicholas was the next closest. Opened to cable traffic on December 1, 1897, this telegraph

Theodore Roosevelt and Richard Harding Davis formed a mutually beneficial collaboration in Cuba in 1898 (Prints and Photographs Division, Library of Congress, Washington, D.C.).

office was the newest in the region.[2] Although it was just across the Windward Passage from Cuba, reporters found the telegraph agents there to be less concerned about rapid transmission and competition for scoops than their counterparts in Jamaica. The cooperative telegraphers of the West India and Panama offices on the British Empire's island south of Cuba were preferred by the hard-driving reporters. A glance at the map indicates that Puerto Antonio on Jamaica's north coast was closer to eastern Cuba, but, according to Arthur Brisbane (managing editor of the *New York Journal*), the telegraphers at the Kingston office (on the south coast) responded with the alacrity that the reporters wanted. The records of West India and Panama indicate a substantial increase in cable business during this period. The typically narrow profit margin of this struggling company expanded as a result of the flood of telegrams to newspaper editorial offices in the United States.[3] In the wake of the intensity and confusion of June and July 1898, the Jamaican stations and also the Haitian office received large if irregular flocks of impatient reporters with high hopes of securing elusive scoops.

The reporters covered a war that was of brief duration (little more than three months) and that inflicted relatively few casualties (less than four hundred U.S. servicemen died in combat) in comparison with the Civil War of 1861–1865 and World Wars I and II. However, the combatants in the Spanish-American War faced serious problems. The following summary presents a brief overview. By 1898 both the Cubans and the Spanish were showing signs of material and psychological exhaustion. The intervention of the United States created difficulties for the Spanish. Although Cuba had been the center of attention for three years, the United States struck first in Manila Bay, the main harbor in the Philippines—a Spanish colony in the western Pacific. Admiral George Dewey launched a devastating attack on the decrepit Spanish fleet and gave the United States a decisive victory in the first days of the war. Nevertheless, the United States faced major challenges in Cuba. The War Department had had no experience with large-scale combat since the Civil War. The United States army had been preoccupied with small-unit confrontations against Native Americans in the trans–Mississippi West for three decades. In April 1898 the United States assumed the task of fighting an experienced professional army. Washington also had to organize an amphibious landing in Cuba. Before land combat could take place, U.S. warships had to clear Caribbean and Gulf waters of any elements of the Spanish Navy that could threaten the heavily laden and slow-moving troop transports that would carry the invasion force from Tampa to a beachhead in Cuba. The first task, therefore, was for the U.S. Navy to locate and destroy or at least contain the Spanish vessels. It launched a search for the squadron commanded by Admiral Pascual Cervera, a competent and courageous leader who had at his disposal antiquated and poorly maintained warships. The U.S. Navy suspected Cervera's force was marginal, but the U.S. public and press harped on the threat not only to the invasion vessels but also to East Coast cities. It came as a relief, therefore, when the navy finally verified the presence of the Spanish fleet in Santiago Bay in southeastern Cuba.

The focus of attention turned to the Cuban port of Santiago and its environs. The U.S. Navy blockaded the mouth of Santiago Bay, and the U.S. Army made an uncontested landing nearby thanks to the support of the Cuban insurgents. The U.S. Army—consisting of regular forces and some volunteers such as the Rough Riders led by Colonel Leonard Wood and Lieutenant Colonel Theodore Roosevelt—moved inland to take the high ground east of the city of Santiago, where they could threaten Cervera's squadron in the harbor below. The march inland was not easy, however. U.S. forces encountered heavy resistance at Las Guásimas, a point along the road to Santiago. This firefight tested both regulars and volunteers, including the highly publicized Rough Riders. The U.S. military prevailed at Las Guásimas and moved on toward San Juan Heights (consisting of both San Juan Hill and Kettle Hill). It remained for General William Shafter's forces to take San Juan Heights. They did so on July 1 in an operation that included the famous charge up Kettle Hill led by Lieutenant Colonel Roosevelt, among others. Cervera's ships were no longer secure in Santiago harbor. On July 3 he decided to attempt to break the blockade of Santiago Bay, a fateful decision. The U.S. Navy destroyed his entire squadron. With Cervera's defeat and Santiago under the guns of Shafter's army, Spain's capacity to make war came to an end.[4]

CHALLENGES OF COVERAGE: COSTLY CABLES AND DISPATCH BOATS

Richard Harding Davis was one of eighty-nine correspondents officially accredited to cover the campaign in Cuba by General William Shafter.[5] Most of these reporters worked for newspapers based in New York City and other large metropolitan areas where the press for several years been caught up in an intense competition to gain readers, circulation, and advertising revenue. The United States' war with Spain became the great story of the day. In this highly competitive environment, editors poured their resources into war coverage. Large type, hand-drawn illustrations, expanded telegraph and telephone facilities, multiple editions on the same day, additional editorial staff, and much-sought-after expensive reporters like Davis added greatly to their costs. Arthur Brisbane, managing editor of William Randolph Hearst's *New York Journal*, had the largest budget. Brisbane estimated that his newspaper spent $3,000 (approximately $76,000 in 2010 figures) per day above their typical operating costs. Extended over a period of several months, these costs were "enough to eliminate all the profits of the most profitable newspaper in America" according to Brisbane.[6] The editors of the ten metropolitan newspapers emphasized in this study did not have the budgets to compete with Hearst, Pulitzer, or James Gordon Bennett Jr. (publisher of the *New York Herald*), but they did pay the more affluent papers and the Associated Press for coverage from Cuba, and some managed to send a reporter to the island.

Among the largest costs were those stemming from the use of the telegraph. The most obvious was the expense of the telegraphed message itself. In spite of cost reductions in the early 1890s, the international cable companies provided a service that, at fifty to eighty cents a word, remained expensive by the standards of smaller, less profitable newspapers. This level of expense meant that, for example, the Associated Press 675-word story on Lieutenant Richard Hobson's daring exploit that appeared in the *New York Times* on June 5 cost approximately $300 to $380 simply to have the words transmitted (some messages were coded and combined words, however, so the number of words actually transmitted was less than the number of the story that appeared in print). This amount was close to the total annual income for most U.S. citizens (the typical urban worker earned about $490 a year).[7]

Once the reporter had assembled his notes and written the story—whether one hundred words or a thousand—the next step was to place the text (usually in code) in the hands of a telegrapher. Getting this script from the battlefield to a telegraph office depended on dispatch boats, an irregular, privately owned and operated fleet of small ships, many of which were only marginally suited for the large swells and rough seas between Cuba and its neighboring islands, Jamaica, and Hispañola. The editors in the United States were determined to have their war stories quickly, and so they rented these boats for that purpose. Ship owners extracted huge rentals from the newspapers, between $5,000 and $8,000 a month. The editors also had to pay for war insurance, as much as $2,200 a month. Hearst's *Journal* incurred expenses of $50,000 per week, while Pulitzer's *World* spent $30,000 over the same period. The Associated Press spent $284,210 from March to October 1898 for its Cuban coverage, including $124,554 on cables and $141,402 on dispatch boats and the salaries and expenses of reporters.[8]

In the early days of the war, reporters assembled in Tampa and Key West, awaiting action. Bottlenecks in the organization of the invasion army and the threat of the Spanish fleet meant that the first conflicts were likely to occur at sea. Perhaps 150 correspondents (many unaccredited) assembled at Key West, alert to the activities of the navy. But news in May was scarce. Reporters became restless and overeager. Ralph Paine of the *New York World* was aboard that newspaper's dispatch boat, *Three Friends*, and witnessed a small clash involving the U.S. Navy's *Hornet* and *Dupont* and Spanish shore batteries near Havana. This "battle" was hardly decisive, but it was the only event of consequence at the time. The *Three Friends* found itself in a race with the Associated Press's *Dauntless* to reach Key West with the first version of the story. The *Three Friends* won the boat race, but the AP temporarily regained the lead on shore as a result of the athleticism of a long-legged reporter who leapt onto the pier as the *Dauntless* approached. That same agile reporter reached the only hackney coach in the vicinity. Unfortunately for the AP, this unnamed reporter became entangled in fisticuffs at the entrance to the telegraph office, and the "scoop" apparently went to Ernest McCready of the *New York Herald*, who had been aboard the *Three Friends* with Paine.[9]

Once Shafter's forces reached Cuban soil at Daiquirí, land combat became a reality, and reporters were on the scene. Their accounts of the fighting became the staple of front pages across the nation. The dispatch boats were the vital links with the telegraph stations. The reporters faced not only the ordeal of witnessing and describing the battles but also the challenge of then carrying their text by horseback or muleback to the coast and then hurrying on board one of the small dispatch boats to dash, insofar as their craft could, to the telegraph station. George Kennan observes that fifteen to twenty of these dispatch boats operated routinely between eastern Cuba and Môle St. Nicholas, Haiti, and Port Antonio and Kingston, Jamaica. At the peak of their operations, newspapers and press associations employed as many as twenty-five boats. The datelines of stories in late June and the first few days of July provide a trenchant record of the movements on land and sea of these reporters. For example, the lead story on the battle of Las Guásimas in the *Atlanta Constitution* on June 27 carried the dateline "Juragua Cuba, June 25, 4 p.m. via Kingston, Jamaica, June 26, 10 a.m." The combat occurred on the morning of June 24, so the reporter (probably

Robert Cramer) composed his account, made his way to the coastal town of Juragua the next afternoon and reached Kingston by dispatch boat the following morning.[10] The Associated Press apparently used the same dispatch boat to transport its account of the battle from Juragua to Kingston.[11]

The most difficult duty for reporters may have been on board dispatch boats waiting with Admiral Sampson's fleet for the outcome of the standoff with Cervera in Santiago Bay. As George Kennan observes:

> All of the newspaper d[i]spatch boats were small, many of them had very limited coal-carrying capacity, and some were nothing but sea-going tugs, with hardly any comforts or conveniences, and with no suitable accommodations for passengers. The correspondents who used these boats were, therefore, compelled to live a rough-and-tumble life, sometimes sleeping in their clothes on benches or on the floor in a small, stuffy cabin, and always suffering the hardships and privations necessarily involved in along cruise on a small vessel in a tropical climate and on a turbulent sea.[12]

When Cervera finally exited Santiago Bay and Sampson's fleet responded with a devastating rebuttal, the long, agonizing wait was over for correspondents. These small craft had coal and water capacity for about ten days, which would likely have been the point at which the reporters would become physically and psychologically exhausted.[13] The costs of covering this battle were large both in terms of time and expense. The *New York Herald* expended $6,400 to get the story from the Cuban coast to the New York office.[14] The Associated Press paid $8,000 to cover this event.[15] In spite of these costs—both fiscal and physical—the reporters seemed to relish their experience in retrospect. John Spears recalled that "every man thought himself fortunate in being there" to cover a major episode in the war.[16]

MILITARY CENSORSHIP

The U.S. military attempted to control the flow of information from the war zone around south Florida, Cuba, and adjacent Caribbean islands, but its efforts had limited success at first. Personal connections between officers and reporters were the cause of these "leaks." Freewheeling reporters often exploited opportunities presented by cooperative military and naval officers about to embark on special assignments. Such was the case when the army organized a mission to supply ammunition and food to the Cuban insurgents on a lumbering old side-wheeler, the *Gussie*. Captain J. H. Dorst commanded the operation in what was supposed to be secrecy, since the *Gussie* faced the risky task of unloading its cargo on Cuba's north coast, not far from Havana. However, Dorst mentioned his plans—in confidence—to reporter Poultney Bigelow of *Harper's Weekly*. Apparently other reporters gained access to this information. On May 10 the *Atlanta Constitution*, the *Washington Post*, and the *San Francisco Examiner* carried details on the plans for the *Gussie*'s departure on that same day, thereby giving interested Spanish observers several days warning, since it would take the slow-moving *Gussie* a few days to reach Cuba.[17] The *Chicago Tribune* carried its version of the *Gussie*'s departure the next day. "The transport *Gussie*," wrote W. J. Rouse, the *Tribune*'s correspondent, "sailed for Cuba amidst a din of cheers, bearing the first regular troops of the United States to Cuban soil."[18] Dorst allowed several reporters on board the *Gussie*. When the ship finally approached the Cuban coast at Mariel, its passengers found the shore lined with Spanish troops, who promptly opened fire. There followed several frustrating days. The *Gussie* moved along the coast, actually managed to land three horses, and, with the help from U.S. Navy escort ships, survived fairly heavy exchanges with Spanish artillery. But the supplies never reached land. The *Gussie* returned to Key West on May 15 with its mission unaccomplished.[19]

Another breach of security occurred in mid-May. Lieutenant R. M. Brainard commanded the U.S. Navy's *Uncas* on a supposedly secret voyage to Havana to arrange for a prisoner exchange in order to obtain the release of two advance agents, or spies, who had been working for the United States in Cuba. Brainard knew and respected Sylvester Scovel, the *New York World*'s energetic reporter. Brainard allowed Scovel aboard the *Uncas* on this "secret" mission, and, once Brainard returned to Key West, Scovel dashed to the telegraph office to file his "exclusive." Brainard's career was in jeopardy, so Scovel convinced navy officials that he had "stowed away" on the *Uncas* unknown to the commander, thereby protecting the lieutenant. But this second violation of naval security in a few days brought tighter restrictions on reporters.[20]

The U.S. Army controlled the telegraph offices in Key West, Tampa, and New York City that were the conduits for messages about the war. In New York, Lieutenant Grant Squires assumed repsponsibility for the censorship duties in the days following the *Gussie* incident, and Lieutenant John Miley (General Shafter's aide-de-camp) established censorship in the Tampa office.

Miley enforced restrictions by requiring press messages to be examined by military personnel before they could be transmitted to their respective editorial offices. Any correspondent who tried to get around military censorship would be punished by being denied the credentials required to accompany the army in the invasion of Cuba. This threat had a chilling effect on the reporters gathered in Florida. The *Atlanta Constitution* complained that the censorship of troop preparations for the invasion was so severe that "correspondents are practically limited now to the veriest camp gossip." The *Constitution* claimed that these orders came directly from Brigadier General Adolphus Greely of the Signal Corps.[21]

The military's institutional arrangements for censorship were actually made before the *Gussie* incident. Admiral Sampson took over the cable office in Key West on April 23, and Greely placed censors in the six international telegraph offices in New York on April 25.[22] Control of the telegraph system passed into the hands of Greely's Signal Corps over the next few weeks. Greely was especially concerned about the possible press and public reactions to censorship duties and assigned Captain James Allen to the Key West office because of his confidence in Allen's judgment. Censorship was not limited to the press; messages by businesses, private individuals, and foreign governments were also subject to it. The telegraph companies involved readily agreed to U.S.-government supervision of their facilities and the messages they conveyed. The British-owned West India and Panama Telegraph Company noted the introduction of military censorship in Key West almost as a matter of routine. West India and Panama headquarters in London made no complaints or plans to protest.[23] Such was not the case with the reporters in Florida, however. A large part of the Signal Corps' work in this area focused on the press. As early as April 24 a reporter for Joseph Pulitzer's *New York World* complained that U.S. censorship was "as strict as any Spain ever enforced."[24]

The censors' intimidation of reporters intensified because of the events of mid-May and then grew even more pronounced because of an exposé by Poultney Bigelow published in the pages of *Harper's Weekly*. Bigelow's article appeared on June 4 and resulted in the cancelation of his credentials to serve as a correspondent in the invasion of Cuba; this punishment was apparently the most severe of those directed against a reporter by the military during the war. Bigelow's revelations included the poor living conditions of the army units stationed in Tampa. He specified that the heavy shirts and trousers worn by the soldiers were unsuitable for the tropics. He also pointed out the absence of fresh fruits and vegetables and the excess of "greasy pork and beans" in the soldiers' diet. He complained about the ineptitude of many officers, officers who had recently risen in rank—many to general—without the necessary experience or competence. His most devastating criticism came in a comparison with Germany's military: "The war is thirty days in swing. It took less time than that for Germany to put half a million well-trained men on the French frontier in 1870. To-day we look in vain for a single regiment fit to take the field." The reporter was aware that his essay was a threat to his future as a war correspondent, but he justified touching on such sensitive issues: "Nobody dares complain for fear of appearing unpatriotic. Still it will do us no harm to hear a little of the truth."[25] This rationalization, however, did not persuade the U.S. Army.

After Bigelow had exited the ranks of war correspondents and Scovel and others had borne the brunt of official scoldings, the government's message to the press became clear—no criticism of conditions in the camps and no premature revelations concerning operations. General Greely claimed that he had no intention of violating freedom of the press and, instead, wanted the cooperation and good will of journalists. His postwar reports and reminiscences emphasized these points, as did a letter he wrote to Captain Jasper Ewing Brady, who had assumed the role of military censor in Tampa, dated June 7, 1898. In his letter, Greely explained the many duties of the censor to Brady, noting that the telegraph operators in the employ of Western Union were under Brady's jurisdiction and that he (Brady) could refuse to transmit messages that compromised security. However, Greely asserted that this power should be used with care: "I propose to act only in extraordinary cases or where the public interest . . . demands such interference. It is for you to abstain from such complications so far as possible."[26]

Lieutenant Grant Squires, the censor in New York, and Howard Giddings, a captain in the Volunteer Signal Corps, represented the censors as having a more activist role in harnessing the press. Squires thought his "most important duty was to edit or to hold back from publication the press dispatches from the seat of war," and he was disturbed by the "sensational, if not always truthful narratives of the newspaper correspondents, written in their perilous positions."[27] Giddings expressed similar misgivings about the character of the wartime reporters, who, "with imperative orders from their papers, and unlimited money at their disposal, were determined to learn all there was to know and publish it. Some of them are suspected of being so unscrupulous as to prefer news which was intended to be kept secret, and if censorship fell with much vigor on the press at any time, these gentlemen are to blame for it." General Greely expressed his most serious concerns to President McKinley about false stories in the press regarding the presence of the Spanish fleet in Cuban waters during the tense time when Cervera's whereabouts were unknown to the U.S. Navy. The head of the Signal Corps, however, refrained from advocating censorship of such reports.[28]

While Greely's dictates about the careful use of censorship appear to have been ignored to an extent by Squires and Giddings, this seeming disregard may have as much to do with chronology as selective memory. The imposition of censorship in May and early June seemed especially onerous for reporters and may account for the misbehavior that disturbed Squires and Giddings. In that part of the war, reporters had difficulty in finding news worthy material. By mid-June, however, the active phase of the conflict was commencing with the imminent invasion of Cuba, giving reporters much to cover. Soon thereafter, reporters seemed to come to terms with the expectations of the Signal Corps.

HEROIC DEEDS

By June reporters understood that emphasizing the theme of the heroic in their stories was one of the most surefire ways to succeed in their quest for the front page. The description of heroic action would easily pass through the censor (provided the story gave no information about future operations) and would have the additional strength of appealing to the public's preference for tales of daring deeds. Unfortunately for U.S. soldiers and journalists alike, the bumbling management of the invasion preparations in Tampa, the awkward and unopposed landing at Daiquirí, and the ability of the experienced and well-armed Spanish troops to inflict casualties all combined to make heroic action a rare occurrence in the first days after the invasion.[29]

Once the U.S. Navy was confident of the location of Cervera's squadron in Santiago harbor, Admiral Sampson devised a plan to block the mouth of the bay and thereby trap the Spanish. Volunteers were to sail a large collier (or coal-carrying vessel), the *Merrimac*, into the mouth of the bay, where it would be exposed to the heavy fire of Spanish guns, and maneuver the ship to block the entrance. The final step was to scuttle the ship, the enemy gunfire aiding in its sinking, in effect sealing the Spanish warships in Santiago harbor. Assistant naval constructor Richard Hobson volunteered to lead this hazardous operation and on June 1 carried out the plan almost to perfection. The only part that went wrong was that the *Merrimac* sank in a position that did not fully block navigation through the channel.[30]

This last point proved to be an inconsequential detail for the newspapers and public in the United States. Reporters did not witness the scuttling of the *Merrimac,* and so their accounts were based on information supplied by sailors aboard the mission's escort ships and the Navy Department in Washington. Telegraphed reports datelined from Môle St. Nicholas or simply "Mole" appeared beneath headlines such as the *Chicago Tribune*'s "Deed of Courage at Santiago Bay" and "Heroes of the Merrimac."[31] The *New York Times* headlined, "Lieut. Hobson's Daring Exploit," and carried the AP's overstatement: "A single deed of magnificent American daring has lost the Kingdom of Spain the powerful fleet of Admiral Cervera."[32]

The laurels of heroic status came to rest on Hobson's brow in less than seventy-two hours, the cable lines from Haiti and Jamaica signaling various accounts of the young officer's action. Historian and journalist Charles Brown concludes that with this flood of excessive praise "the American people realized that they had a new national hero—one to compare to Dewey." Hearst's *New York Journal* was perhaps the most expansive, placing Hobson alongside John Paul Jones, James Lawrence, Matthew Perry, and William Barker Cushing.[33] The *San Francisco Chronicle* in the meantime exclaimed in a headline, "No Honor Too Great for Lieutenant Hobson and His Men," while in a banner headline, the exuberant *San Francisco Examiner* proclaimed, "Hobson and His Men Win Undying Glory."[34] The *Atlanta Constitution* bore the expense of a reporter on the scene and implied that this reporter interviewed eyewitnesses aboard the *Merrimac*; its headline for June 6 read, "Young Lieutenant Was Cool and Calm as He Arranged for His Desperate Deed, and It was with a Steady Hand He Carried It Out without Fear."[35]

Hobson and his crew survived the episode but were unable to avoid capture by the Spanish Navy. Even the misfortune of their capture appeared in ebullient prose on the front page of the *New York Times:* "Spanish Capture the Heroes, but Cervera Makes a Chivalrous Offer of Exchange."[36] (Eventually the United States and Spain arranged a prisoner exchange. Hobson and his crew returned to Sampson's fleet in good shape.[37])

The excitement so evident in the reportage on the events of June 1–2 continued for a least a week in most large newspapers, but the flaws in and, in fact, the ultimate failure of Hobson's mission began to creep into some reports. The *New York World*'s Alexander Kenealy was among the first to suggest that the *Merrimac*'s position in the straits did not in fact block the harbor.[38] By the second week in June the voracious press was searching for a new story of derring-do, one that would break the monotony of Sampson's blockade and Cervera's inaction.

The doldrums of the Santiago blockade gave way to a flurry of journalistic intensity in the coverage of the activities of navy lieutenant Victor Blue. The question that plagued Admiral Sampson and officials in Washington was whether all of Cervera's force was in the harbor. Only two or three Spanish warships in Caribbean waters could threaten the passage of the troop transports from Tampa to the landing site in southeastern Cuba. Blue's task was to penetrate Spanish lines with the help of Cuban insurgents, find a point from which to observe the Spanish squadron at anchor in Santiago Bay, and then carefully compose a report on the number and identities of these vessels. If the entire squadron was in the harbor, the way was clear for Shafter's invasion force to leave Tampa. In spite of the physical difficulty and personal risk, Blue accomplished his mission. The *New York Times*'s headline stated that Blue's report removed the doubt: "All of Cervera's Ships Located in Santiago Harbor."[39]

Blue's adventure moved him to the forefront in the press's quest for heroes. The *New York Times* saw him as "one of the naval heroes of the present war." The protagonist of a "bold exploit," Blue successfully completed "a dangerous trip of seventy-two miles to make observations."[40] The *San Francisco Examiner* placed a portrait of Blue on page 1 with a caption describing him as "a man of great courage."[41] In a shorter, less effusive article, the *San Francisco Chronicle* related the same story.[42] Although the coverage of Blue in the press generally did not equal that of Hobson, the *New York Times* described him as a modest "southern gentleman" who inspired confidence in his fellow officers and clearly fit the heroic mold.[43] The *Washington Post* praised his "brave exploit."[44] In another contrast to Hobson, Blue completed his mission. His report opened the way for the invasion of Cuba, but the course of fame was fickle in the U.S. press in the summer of 1898.[45]

ONE AMONG MANY HEROES

The images of Hobson and Blue in the telegraphed stories from Haiti and Jamaica qualified as dramatic and heroic, but they could not compete with the singular presence of Theodore Roosevelt, who claimed a large place in history as the leader of the charge up the Kettle Hill segment of San Juan Ridge on the eastern side of the city of Santiago. Roosevelt achieved heroic status and then capitalized on it, as his meteoric rise to the governorship of New York later in 1898, to the vice presidency of the United States in 1901, and, after the assassination of President William McKinley, to the position of chief executive of the nation attests. Roosevelt's rapid rise to power was affected by his special rapport with the press. During the war, Roosevelt formed a close and mutually rewarding relationship with Richard Harding Davis. Both men were aware of the influence of the newspapers that fed off the telegraphed stories from the Cuban front, and both men exploited their opportunities to the fullest.

The key event was the battle for Las Guásimas, which took place a week before the events on San Juan Ridge. The geographical setting of this clash between the U.S. invasion force and the Spanish defenders was along the Camino Real that led to Santiago. This road was in bad shape—rutted, pitted, and overgrown—after three years of neglect. It ran upward across ridges and hills and was flanked by long stretches of heavy tropical undergrowth, a likely setting for an ambush. The Spanish forces, however, had planned to fall back toward Santiago, apparently preferring to concentrate on the outskirts of the city. But the appearance of the U.S. Army changed their plans. The battle of Las Guásimas took place on June 24, the day the Spanish had intended to pull back. This battle might well have never taken place because, from the U.S. side, General Shafter had ordered the advance unit under ex-Confederate officer General Joseph Wheeler to avoid engagement with the enemy. But "Fighting Joe" was eager for combat.

The forward unit of Wheeler's force was under the command of General Samuel Young. Made up of army veterans of the campaigns against the Native Americans in the 1880s and early 1890s, Young's 464 men handled themselves well. They and the Spanish exchanged fire, and then Young ordered his men forward. After some confusion in the dense vegetation, they pushed the Spanish from their defensive positions.[46]

Colonel Leonard Wood commanded the approximately five hundred men of the Rough Riders located behind Young's regular army unit. The fighting soon included the Rough Riders. Since Wood's men were not regular army personnel, there was some concern about how they would perform under fire. But Wood and his second-in-command, Lieutenant Colonel Roosevelt, rallied their men, and, in spite of the difficulty in seeing what the other U.S. units along the battle line were doing, the Rough Riders charged ahead through the undergrowth and, along with Young's regulars, participated in driving the Spanish from the battlefield. After about an hour and a half, the U.S. forces knew that they had achieved victory, but it came with a cost. On the U.S. side, sixteen were killed and fifty-two wounded. The Rough Riders suffered over half these casualties: eight were killed and thirty-one wounded.[47]

The Rough Riders had comported themselves well, but they were the rear half of an advancing army and, in strictly tactical terms, they were important but not decisive to the battle's outcome. Military historian Ivan Musicant gives them credit for a "good part of that victory," but General Young's men were unquestionably at the forefront. Roosevelt was a stalwart among the untested volunteers, but he was subordinate to Wood, who was equally steady in the thick of the fight.[48] The question arises, how did Roosevelt's role in this battle reach the status of singularly outstanding, ahead of the contributions of Young, Wood, and other officers and men who performed well on June 24? While this question leads toward the slippery terrain where public perception, image creation, and journalistic objectivity intersect, it is appropriate to note that historians, media analysts, and several contemporary observers reached similar conclusions. The key to understanding how Roosevelt emerged as the greatest hero from the pack of potential heroes lies in his cultivated relationship with Richard Harding Davis and the consequent advantages that the lieutenant colonel of the Rough Riders secured with the press in general.

Davis was already a celebrity and a celebrated journalist in the United States. His reportage on the Cuban insurrection, his travel account *Three Gringoes in Venezuela and Central America*, his popular novel *Soldiers of Fortune*, and his rapport with the cultural elites of New York and Washington made him the most prestigious reporter in Cuba in 1898, overshadowing Sylvester Scovel, James Creelman, and Stephen Crane. Roosevelt was fully aware of Davis's preeminence and made an effort to stay in close proximity. In the initial phase of the march to Las Guásimas, both Roosevelt and Davis had the good fortune of riding instead of walking. Roosevelt had his own mount, and Davis, because of an attack of sciatica, received the benefit of an army mule. The two rode together, with Davis, the veteran of tropical travels, pointing out native flora to the interested Roosevelt.[49]

Once the fighting started Davis was the only able-bodied reporter with the Rough Riders, and, for most of the battle, he stood beside or near Roosevelt.[50] In later accounts, Roosevelt credited Davis as being the first in their combat zone to locate the Spanish.[51] The two shared a battlefield experience that fortified their relationship, but they also shared large parallel personal ambitions. As Davis's biographer Arthur Lubow explains, "The partnership of Roosevelt and Davis, forged at this nameless crossroads, was the high point of Davis's journalistic career, and the start of Roosevelt's final ascent to the White House. Each man instinctively recognized the importance of the other to his success."[52]

Las Guásimas was a critical point in the Roosevelt-Davis collaboration and also in the U.S. press coverage of combat in Cuba. Early accounts from the scene of battle indicated that U.S. troops had blundered into a bloody ambush. Later in the day a more positive version reached the reporters on the Cuban coast, and, on the next day (June 25), the *Washington Post*, the *Atlanta Constitution*, the *New York Tribune*, the *Chicago Tribune*, the *San Francisco Chronicle*, and the *San Francisco Examiner* carried brief, sketchy front-page accounts of a U.S. victory.[53] On the twenty-sixth Davis's story appeared in the *New York Herald*, and similar versions occupied prominent places in newspapers from coast to coast. Davis saw the battle firsthand and shared Roosevelt's perspective on the conflict, and his version of events carried much weight in the U.S. press. News of this engagement spread throughout the information system on June 26 and over the next few days. The Spanish opened fire from concealed positions along the roadway, and U.S. troops responded. Within an hour and a half they drove the Spanish from the area. On June 26 and 27 newspapers in New York, Boston, Philadelphia, Washington, Atlanta, Chicago, and San Francisco carried telegraphed accounts that contained the same information but expressed in different words. The *San Francisco Examiner* was the most flamboyant, featuring a banner headline across the front page flanked by a portrait of Roosevelt. The *Philadelphia Inquirer* used a two-column headline and large type. Other papers had multicolumn coverage accompanied by large maps or illustrations. The accounts presented Young, Wood, and Roosevelt as leading their respective units into battle and delivering a decisive defeat to the Spanish. However, Roosevelt, even though he was the lowest-ranking officer of the three, gained additional coverage in the *New York Times*, the *Chicago Tribune*, and the *San Francisco Examiner* with a colorful but militarily insignificant detail that had him "narrowly escaping a bullet himself which struck a tree alongside his head."[54]

By June 26–27 U.S. forces were in command of Las Guásimas, and Roosevelt was gaining command of the press coverage. The *Chicago Tribune* set the standard with the subtitled headline "Roosevelt's Narrow Escape" atop a five-paragraph section on the battle.[55] The *New York Tribune* employed a similar format with the headline reading "Roosevelt Leading the Attack."[56] Although this coverage did not carry his byline, Davis had set the journalistic ranking of the main characters, locating Roosevelt at the top. The next day the *New York Times* and the *Washington Post* featured an Associated Press interview with Roosevelt and one of his officers (unnamed in the dispatch) in which the second in command of the Rough Riders was the central figure. According to the unnamed officer, "Lieut. Col. Roosevelt, in front of his men, snatched a rifle and ammunition from a wounded soldier, and, cheering and yelling with his men, led the advance."[57]

Over a two-day period, Roosevelt moved from the flank to the center of newspaper coverage of the battle of Las Guásimas, the main story of the war effort at that time. Historians of this battle rate the work of General Young and Colonel Wood as being of at least equal and most probably of greater importance, but these two officers had not cultivated the good will and concentrated attention of Davis and did not achieve the public-relations advantage of an interview with the Associated Press.[58] The point is not to diminish Roosevelt's actions on the battlefield but to place them in their proper context, which requires taking into account his talent for winning the campaign for press attention before, during, and after the battle.

The Rough Riders' charge up San Juan Hill on July 1 carried Roosevelt to a higher level of national recognition in the press, but the piecemeal reports of the Spanish defeat on July 2 from among which the War Department in Washington and newspaper editors from the capital to San Francisco were sorting out the details did not immediately acknowledge Roosevelt's legendary charge. The next day a new story broke, flooding the cables and editorial offices for several days thereafter—the army's capture of the hills east of Santiago and the navy's victory over Cervera's force in the waters south and west of Santiago. Roosevelt's triumph was one of many newsworthy accomplishments. For example, the *Chronicle* and *Examiner* of San Francisco both gave significant attention to Roosevelt and his action under fire: "Lieutenant Colonel Roosevelt led waving his sword. Out into the open men went, and up the hill. Death to every man seemed certain."[59] The San Francisco papers used reports directly from unnamed correspondents in Cuba, but the *Boston Globe* and the *Philadelphia Inquirer* relied on the Associated Press and releases from the War Department. Their coverage gave scant attention to Roosevelt and the Rough Riders.[60]

The *New York Times* accorded Roosevelt some attention, but it was balanced by similar coverage of the exploits of other officers. The *Times*'s editorial observed that the volunteer Rough Riders held their own with the regular army's Tenth Cavalry: "The Roughriders with Roosevelt at their head kept pace with the regulars."[61] A few days later the *New York Times* reported that Roosevelt was nominated for promotion from lieutenant colonel to colonel and that several other officers were nominated for promotion as well.[62]

THE DECISIVE BATTLE

The final blow to the Spanish Empire in the Western Hemisphere came on July 3 when Sampson's fleet met Cervera's hopeless charge and sank or disabled every enemy vessel. The "scoop" on this seminal engagement went not to any of the eager reporters but to Lieutenant Colonel James Allen of the Signal Corps, who telegraphed the War Room in the White House from his base at Playa del Este that Sampson had gained the decisive victory. Allen's telegram appeared on the front pages of major newspapers across the United States: "Siboney office confirms statement that all the Spanish fleet except one warship is destroyed and is burning on the beach." Allen's second telegram was definitive: "The destruction of Cervera's fleet is confirmed."[63]

News of Sampson's triumph broke the mood of growing anxiety in the White House and throughout much of the nation that had arisen because of General Shafter's surprisingly pessimistic appraisal of his army's situation outside Santiago. U.S. newspapers carried accounts of Shafter's uncertainty during the daylight hours of July 3, but Allen's notification and confirmation of the destruction of Cervera's ships spread across the nation's telegraphic network the same evening. The *Atlanta Constitution*'s headline summarized the mood swing: "From the Deepest Gloom to Most Exultant and Hilarious Rejoicing." The *Constitution*'s Washington correspondent stressed the public response to the fast-breaking news: "A glorious fourth, after all. . . . The streets are filled with people rejoicing over the glorious news that has just come of the destruction of Cervera's fleet. The air is filled with the shouts of the patriots and Sampson's name is on every lip."[64]

Most of the bylines on July 4 registered Washington as the origin of the dispatches because the news of the victory came through Allen's Signal Corps and Sampson's official notification of the president and the Navy Department. A handful of reporters actually witnessed the battle, but they would take a few days to write and send their stories to editorial offices in the United States. The dispatch boat *Hercules*, hired by the *Chicago Record*, followed the course of the battle; from his position aboard the vessel, reporter Henry Barrett Chamberlain was able to witness and describe the sinking of the Spanish ships. George E. Graham of the Associated Press was on board Commodore Schley's *Brooklyn* and the *New York Sun*'s T. M. Dieuaide was on the new battleship *Texas*, while "Chappie" Goode, also of the Associated Press, was with Admiral Sampson on the flagship *New York*. Over the next few days their eyewitness reports appeared on the front pages of newspapers across the country.[65]

Charles Brown notes with good cause that the dateline assumed an especially prominent place in the reporting of the Spanish-American War and, in the case of the sinking of Cervera's fleet, this typically minor and sometimes omitted part of the journalistic record acquired new significance. For example, the unidentified AP correspondent who examined the wreckage of the Spanish fleet on the afternoon of the battle datelined his story so that the site of his observations and the movement of the account were clear to the reader: "Ten miles West of the Entrance to the Harbor of Santiago de Cuba, July 3, 4 P.M. via Kingston, Jamaica, July 4, 4:45 A.M."[66] A second AP account appeared in the *New York Tribune* also on July 5: "Ten miles West of the Entrance of Harbor of Santiago de Cuba, Sunday, July 3, 4 P.M. by the Associated Press Dispatch Boat Wanda, via Kingston, Jamaica, July 4."[67] Hearst's newspapers carried the following dateline on July 5: "On Board the '*Examiner-Journal*' Dispatch Boat Simpson, Off Santiago harbor (via Kingston, July 4)."[68] The *New York Herald* and the *Washington Post* shared an account of Sampson's victory with the following dateline: "Off Santiago de Cuba, Saturday, by Dispatch Boat Golden Rod to Port Antonia [*sic*] Jamaica."[69]

One of the longest datelines accompanied the story filed by one of the largest personalities involved in war coverage, William Randolph Hearst. The owner-editor of the *New York Journal* went to Cuba aboard his own yacht, the *Vamoose*, to do his part as a frontline correspondent. He missed the battle itself, but his dateline the day after was a story within itself: "On the quarterdeck of the Spanish cruiser Vizcaya, seen by many New Yorkers during her visit to their harbor and now lying beached on the southern coast of Cuba with gaping shot wounds in her port side, her bow blown out by the explosion of some forward magazine or torpedo, blackened by fire, beaten by waves and as nearly a collection of scrap iron as a once splendid battleship can be, July 4. By the dispatch boat Sylvia by way of Port Antonio, July 5."[70]

DAVIS, ROOSEVELT, AND THE CAMPAIGN FOR PUBLIC ACCLAIM

The destruction of the Spanish fleet pushed aside Roosevelt's charge up San Juan Ridge in the press coverage. After July 3 the newspapers and wire services presented the army's victory east of Santiago as basically a preliminary engagement that forced Cervera's fatal exit. By the end of the first week in July the newspaper press was giving more attention to Roosevelt's actions at Las Guásimas than to his risky dash to the top of Kettle Hill. How did the battle for San Juan Ridge come to eclipse Las Guásimas and the U.S. Navy's triumph off Santiago in the development of the Roosevelt legend? The answer lies mainly in the publicity skills of Davis and Roosevelt. For several months the two worked (Roosevelt was a well-known, published author before the war) to embellish the public image of Roosevelt's accomplishments in combat. They concentrated on the dramatic, photogenic, climactic charge up San Juan Ridge rather the scramble to escape the ambush amid the tropical vegetation of Las Guásimas. Davis took up the challenge of contrasting Roosevelt with Brigadier General Hamilton Hawkins, the white-haired veteran of the Civil War who led regular army soldiers to the top of the elevation that actually bore the name San Juan Hill and that was larger than Kettle Hill (Roosevelt's point of victory). According to Davis's biographer Arthur Lubow, in Davis's newspaper account that was published on July 14 "Hawkins has a featured role, and Roosevelt is the star." For Davis, "General Hawkins, with hair as white as snow and yet far in advance of men thirty years his junior, was so noble a sight that one felt inclined to pray for his safety: on the other hand Roosevelt, mounted high on horseback and charging the rifle pits at a gallop and quite alone, made you feel that would like to cheer." Why the observer would not fear for the mounted, nearsighted Roosevelt and could not rouse a cheer for the dashing Hawkins touches on the predispositions that the observer brought to the battle.[71]

Davis was persistent and prolific. Roosevelt's heroic image received wide circulation in three print media formats of the era: the newspaper, the magazine, and the book. He wrote a bylined article for the *New York Herald* and a series of articles for *Scribner's* magazine that were later published as a book: *The Cuban and Porto Rico Campaigns*. Davis continued to expand on Roosevelt's battlefield accomplishments, putting an ever greater emphasis on San Juan Ridge. Roosevelt enhanced his heroic image with the publication of *The Rough Riders* in 1899 in which he gave approximately equal space to Las Guásimas and San Juan Ridge but presented the latter as the culmination of the land war in Cuba.[72]

Military historians looking back on the event have identified several officers who also played important roles. General Samuel Sumner, who took the initiative on ordering the charge, commanded Roosevelt's group along with three regiments of regular cavalry, all of which advanced alongside the Rough Riders up Kettle Hill. One of these three was the African American Ninth Cavalry, which received little mention in the newspaper accounts. Too their left, General Jacob Kent directed five infantry regiments up San Juan Hill, one of which was led by General Hamilton Hawkins. A total of about 6,600 U.S. soldiers took part in the assault. In the elation of victory even Roosevelt admitted that dozens if not hundreds who made it to the top of Kettle Hill in the early waves thought they were leading the way: "As for the individual men, each of whom honestly thought he was the first on the summit, their name was legion."[73]

Roosevelt understood the full context of his moment of glory and also grasped his debt to Davis. The Davis-Roosevelt collaboration bore fruit several times in the print media, and for many years the two revisited those events in articles, books, speeches, and in their personal correspondence.[74]

The Roosevelt-Davis collaboration pricked the sensitivity and perhaps the jealousy of contemporaries. Charles Johnson Post, a member of the New York infantry volunteers, wrote a memoir of the campaign in Cuba that cast aspersions on Davis—"a brilliant and superficial writer to whom glamour was always gold." Davis's attraction to glamor weighed heavily in his account of Roosevelt and the Rough Riders, according to Post: "In addition to having Teddy as its second in command, the regiment had its own press agent in Richard Harding Davis, to whom the human beings not listed in the Social Register were merely varied forms of pollution."[75] Although not quite so acerbic, British journalist John Black Atkins claimed the newspaper reading public in the United States, fed by Davis and like-minded reporters, "had singled out the amateurs (the Rough Riders and other volunteers) for special applause and taken them familiarly to their hearts as 'Teddy's Terrors.'"[76]

Ironically, Davis himself criticized those reporters who praised the volunteers at the expense of regular army men. These correspondents: "did a very great injury to the men of the regular army, in persuading the public at home that the volunteer is an effective fighting machine, instead of making it clear that he is an 'amateur' and, as such, is a menaces and a danger to the safety of the country."[77] Post and Atkins would have agreed but would have placed Davis in this group of irresponsible correspondents.

THE TRANSFORMATION OF THE CUBAN IMAGE

The victories on San Juan Heights and off Santiago Bay brought an end to the fighting in Cuba. The hagiographic coverage in the U.S. press was accompanied by the erasure of the image of the courageous Cuban insurgents that had been so prominent in the same newspapers from 1895 to 1897. In its place a new version of the Cuban soldier began to appear. The disciples of José Martí and Antonio Maceo did not fare well in this phase of U.S. press commentary. Journalists from the New York and Chicago papers and the wire services sought stories about the accomplishments of their compatriots in uniform, an understandable preference in terms of the interests of the newspaper reading public but one that did a disservice to Cubans who had been fighting the Spanish not just for three months but for three years (many for longer because of service in the 1868–1878 war). The prejudice against Cubans of African lineage was particularly prominent. There was a marked contrast between the negative coverage of the Afro-Cuban soldiers in 1898 and the laudatory commentary on the triumphs and tragic death of Maceo—the "Bronze Titan"—of 1896 and 1897. U.S. reporters and editors had somehow managed to move beyond the typical racial prejudice of the era in their depictions of Maceo, but stereotyping and indifference resurfaced in 1898. Journalists generally accepted the U.S. Army's prejudiced accounts of the role of Afro-Cubans in the war effort. Historian Philip Foner suggests that this refusal to give credit to these insurgents was due in part to the numbers of white southerners in the U.S. military: "The contemptuous term 'nigger' became the one most commonly used by American officers and soldiers—and newspaper correspondents—to describe the Cuban fighters." This prejudice was not limited to southerners, however, because much of the distorted reporting in the summer of 1898 appeared in the dispatches by reporters from cities such as New York, Cleveland, Detroit, and San Francisco.[78]

Cubans of Hispanic descent also found themselves subject to the harsh judgments of reporters and editors. The *New York Times* offered a negative assessment of all of the island's participants in the war. This gross misrepresentation appeared in a simple sentence that reflected the opinion of U.S. soldiers and reporters: "The Cubans did little or no fighting." They avoided combat and, according to this account, appeared only to forage for loot.[79] These slanted conclusions were belied in accounts of the bloody offensives against the Spanish in 1895–1897, a large story covered in some detail in the *Times* and many other U.S. newspapers. The noble crusaders against

Spanish imperialism inspired by José Martí and the Cuban junta became the uncooperative, irresponsible troublemakers of 1898. The most conspicuous example was Calixto García, a veteran of some of the most difficult fighting of 1895–1897. Historian Louis Pérez succinctly describes General Shafter's decision to exclude García and other Cuban officers from the Spanish Army's formal surrender in Santiago on July 17, 1898: "Cubans not only seemed to have vanished from the war (in U.S. military reports and press coverage): they also were even excluded from the peace."[80] U.S. officials dominated the ceremonies and raised their national flag to symbolize Spain's final defeat in the Americas. The U.S. press covered the story with patriotic zeal, convinced that the island of Cuba was an appropriate Caribbean theater for the enactment of rituals of national unity and expressions of national pride.[81]

The low opinion of Cuban soldiers carried over to expressions of doubt about the ability of Cubans to create a stable government. The *New York Tribune* argued that the McKinley administration's commitment to Cuban independence was misplaced. In the *Tribune*'s editorial judgment, the island was not prepared for self-government.[82] The *New York World* used the vocabulary of racism to support its case against an independent Cuba.[83] *New York Times* reporter Stanhope Sams worked with General Shafter's staff officers and relied on the U.S. military's point of view. His comments about the Cuban population were damning. The press and politicians in the United States had created a false image of Cuba in the nation's information system. The idealized heroes of the Cuban insurgency were, on this "closer" view, unprepared to form a nation: "There are no Cuban people. There are not freemen here to whom we could deliver this marvelous island. We have fought for a spectral republic. . . . If we are to save Cuba, we must hold it. If we leave it to the Cubans, we will give it over to a reign of terror—to the machete and the torch, to insurrection and assassination."[84]

The U.S. public's interest in the war led to the publication of several books. Many of these volumes—lengthy tomes with flags and other patriotic symbols emblazoned on stout covers—were part of the national celebration. Two ex-Confederate officers made their contribution. Ex-consul Fitzhugh Lee and General Joseph "Fighting Joe" Wheeler authored a retrospective entitled *Cuba's Struggle against Spain* in which they avoided derogatory language but expressed thinly concealed skepticism regarding Cubans of African descent in two paragraphs that contrast with a dozen pages of promotional commentary on the material resources of the island. Lee and Wheeler sought to reassure their readers: "The black and colored Cubans are not numerous enough or strong enough morally or physically to cause them to preponderate to such an extent as to Africanize the island. There is no danger that the island will ever pass from the control of the white race." In short, Cuba was to become a fertile field for U.S. business in spite of the island's population.[85]

Two other bulky volumes provided a remarkable contrast that encapsulates the sudden reversal of the image of Cuba in the print media of the United States. Quesada and Northrop's *War in Cuba*, intended to inspire the support of the government in Washington and the U.S. public in general, was a 544-page tome that extolled the "daring deeds of Cuban heroes and patriots" like Maximo Gomez and Antonio Maceo and that featured pleasant portraits of prewar life in cities, villages, and the countryside of the island.[86] Three years later Congressman J. Rankin Young and journalist J. Hampton Moore sent to press *Reminiscences and Thrilling Stories of the War by Returning Heroes*, a 766-page mélange of first-person accounts provided by "the heroes themselves; stories which once woven into the text books of the schools of the nation, will obtain for the brave contemporaries of our own times, places in history with those of our forefathers." The title of chapter 5 of this book, "Eyewitness Accounts of the First Battles on Cuban Soil," makes it sound as if there had been no fighting in Cuba before May 1898. Theodore Roosevelt, Leonard Wood, "Fighting Joe" Wheeler, and Richard Hobson shared pages with ordinary soldiers and sailors, but there was virtually no mention of the Cuban insurgents.[87]

The version of the Cuban war presented by Quesada and Northrop is barely visible in Young and Moore's pages. It is almost as if the two books describe two separate historical events. Quesada and Estrada Palma must have experienced a sense of distortion and disorientation similar to that endured by José Andrade, Ezequiel Rojas, and William Scruggs in the outcome of the Venezuelan boundary dispute. The leaders in Washington turned their attention to transatlantic rivalries with London and Madrid and placed the interests of Venezuelans and Cubans far down on their list of priorities. The press of the United States followed this line and rejoiced as the triumphs of the U.S. Navy and U.S. Army in 1898 held the attention of readers and sold newspapers and books.

THE NATURE AND CONTENT OF THE INFORMATION FLOW

The Spanish-American War saw an impressive use of modern communications technology that brought reports from the front lines in Cuba to the front pages in the United States within a day or two. War correspondents acquired reputations for courage that earned them a status close to that of the soldiers and sailors who were under fire. The hurried composition of battle-scene reports and the dash to the telegraph office that often involved horseback rides and dispatch boats constituted notable accomplishments in journalism history. The telegraphers, telegraph lines, and submarine cables also worked well during the war. The wires of West India and Panama, Cuba Submarine, United States and Haiti, and International Ocean of Western Union functioned with little or no interruptions except for those imposed by military censors and war-related cable cutting. The stresses of war tested the reporters and cable companies, and they responded well.

These technical accomplishments made possible the movement of information from the handwritten notes at the scene of battle through thousands of miles of copper wire and onto the front pages of newspapers across the United States and eventually Europe and Latin America. The impact of the information system, however, must also be measured by the content of the messages conveyed. Viewed from this perspective, the U.S. newspapers had a mixed record. Some individuals received highly favorable coverage, and some fared poorly. Two who were acutely aware of what news over the wires could do for their careers were Theodore Roosevelt and William Shafter. Roosevelt's close relationship with Richard Harding Davis and other reporters helped to bring him favorable coverage in stories from Las Guásimas and Kettle Hill. Shafter, by contrast, underwent some of the sharpest criticism in the Cuban campaign to a large extent because of his dismal assessment of his army's prospects immediately after Roosevelt's moment of glory in the victorious advance up San Juan Heights. Shafter's excessive weight and illness in the tropical heat added to reporters' skepticism about his capacity to command. Shafter's frustration reached a breaking point in his confrontation with Sylvester Scovel at the formal Spanish surrender of Cuba in Santiago on July 17. The pugnacious Scovel had been a critic of Shafter and had often defied the military's guidelines for the press. When Scovel approached Shafter in front of a large crowd, the general's anger got the better of him. According to historian Joyce Milton, Shafter exclaimed, "You son of a bitch . . . you and all of your tribe are goddamned nuisances." Scovel admonished the general to avoid the use of such language, whereupon Shafter apparently punched the reporter in the face. The general's reputation took another severe blow. Roosevelt, by contrast, enjoyed his exchanges of compliments with Davis over the next few years. Both men used their Cuban adventures to build their respective reputations.[88]

The center of attention was Cuba, but the White House was nevertheless often in the news. The McKinley administration understood the importance of the press. The president's staff clipped articles and editorials from newspapers.[89] From these sources, McKinley and his advisors followed the coverage of the war in several newspapers. Chief of staff John Addison Porter and staffer George Cortelyou (in effect the administration's press secretary) supplied journalists with information in typed press releases and personal briefings.[90] In spite of these cordial relations, the anxieties of wartime leadership led McKinley to exert presidential power and restrict the press. The president used censorship to block stories about troop movements—especially the perilous journey of the invasion force from Tampa to the landing site in southeastern Cuba.[91] The administration also employed this authority to punish the press when it criticized the armed forces, as exemplified in the revocation of Poultney Bigelow's credentials.

Nevertheless, the White House maintained a good reputation. One of the country's most tenacious reporters, Ida Tarbell, wrote an article for the July 1898 issue of *McClure's* in which she noted that McKinley accepted the presence of journalists as a part of White House routine: "Representatives of half a dozen or more papers are always to be found, and during Cabinet meetings and at moments of grave importance the number increases many fold."[92] Cortelyou worked closely with the press. He was sensitive to the mood and the well-being of the president and the motives and mind-set of the reporters.[93] The president even devoted time to welcoming Sylvester Scovel to the White House only weeks after the assertive reporter's confrontation with General Shafter in Santiago.[94] In spite of its use of censorship, the McKinley White House emerged from the war on good terms with the press.

The press itself took some telling blows from commentators. Exaggerated accounts of battles and invented stories about conflicts that never took place headed the list of journalistic wrongdoing. To many media critics and historians, the worst aspects of the "yellow press" came to the fore during the Spanish-American War. The most notorious and overrated example of bad behavior on the part of the yellow press was Hearst's boastful telegram to illustrator Frederick Remington that implied the editor of the *New York Journal* had exercised the power to bring about war between the United States and Spain.[95] In terms of day-to-day journalism, poor journalism seemed to be a function of the newspapers' voracious appetite for fast-breaking stories. The newspaper reading public and the editors who supplied them with news had come to expect the telegraph to furnish a flow of information only one or two days old. The correspondents in Cuba had only a few hours or, in some cases, perhaps a few minutes to decide what to write, and then they had to take their story by dispatch boat to Jamaica or Haiti. The pressure on the reporters was intense, and some of them jumped aboard the dispatch boats before the significance of the event was clear. The premature stories about the "invasion" of Cuba by the ill-starred *Gussie* and the overeager quest for heroes in the quiet days of May and June fit the pattern whereby the press manufactured stories to fill the wires and ultimately the front pages.

In their postwar retrospectives, most journalists chose to emphasize more favorable aspects of their work in Cuba. Robert Cramer of the *Atlanta Constitution* defended himself and his colleagues as devotees of the journalistic quest for the truth who got caught up in difficult circumstances. Cramer stressed the physical hardships involved in reporting the war.[96] By the time combat was under way, it does seem likely that the members of the press had neither the time nor the need to concoct their stories, but many of them did have a preference for the heroic deed or, at least, for dramatic action that could be portrayed as heroic.

The direction of the information flow shifted significantly when the U.S. government abandoned diplomacy and decided on belligerency as its official policy. Generals and admirals replaced diplomats as sources of government messages. The press, with a few exceptions, contributed to "the martial spirit" with patriotic prose. Probably the most perceptive and adroit participants in the flood of patriotic militarism were Theodore Roosevelt and his supporter Richard Harding Davis. They managed to build up Roosevelt as the preeminent war hero of 1898, outdoing the bylined, telegraphic inundation celebrating Admiral Sampson after his defeat of Cervera's squadron off Santiago. The information system had become an electric and print cheerleader for the war effort susceptible to the manipulations of the clever and ambitious.

CHAPTER TWELVE

The Panama Conspiracy

THE PRESS COVERAGE OF the Cuban revolution for independence and the Spanish-American War established that reporters from the United States were capable of probing into the internal affairs of a Hispanic political entity (Spain's colony of Cuba) and tracking the course of naval and military conflict in the West Indies, albeit often with a sensationalist and patriotic slant. The Cuban controversy and the Spanish-American War were front-page news, and the circum-Caribbean—especially the long-discussed Panama Canal—likewise held the public interest. It is somewhat surprising, therefore, that a plot to break the province of Panama away from Colombia to enable the United States to build the canal escaped the attention of scores of journalists who had considerable experience in the area. The press did encounter rumors about a revolutionary conspiracy but could uncover no more information than a few vague stories that were enticing yet unsubstantiated. Secrecy was essential for the conspirators because their goal was to enlist the clandestine support of the administration of President Theodore Roosevelt in a plan to sever Panama from Colombia.

Room 1162 of New York City's Waldorf-Astoria hotel was the location for the planning of this political revolution that would impact the isthmus of Panama, the governments of Colombia and the United States, and eventually patterns of commerce and investment in the Americas, Europe, and East Asia. The occupant of the room and host for this meeting was Philippe

Bunau-Varilla, the French engineer who had worked on the ill-fated canal project in the 1880s and recently had turned his attention to political lobbying and the revival of the canal enterprise. Bunau-Varilla greeted his guest, Dr. Manuel Amador Guerrero, at about 10:30 a.m. The two had known each other since the uprising led by Pedro Prestán in 1885. Amador was a seventy-year-old Panamanian physician and political activist. The purpose of the meeting on the morning of September 24, 1903, was to plan the revolt—but with some major differences—that Prestán had been unable to carry out. Instead of seeking to instigate an amorphous movement to overthrow established authority on the isthmus, the two conspirators of 1903 had a more specific agenda—the independence of the province of Panama from the nation of Colombia so that a new government could be created that would proceed with the signing of a treaty with the United States to grant to the rising power of the Western Hemisphere control of a strip of land that reached from the Caribbean to the Pacific to build an interoceanic canal. Unlike Prestán, Bunau-Varilla and Amador did not want to arouse the peasants and workers of Panama. Their goal was a quick and relatively quiet coup that would leave power in the hands of the local Panamanian elite of lawyers, businessmen, and politicians.[1]

Beyond this revolutionary goal, the two men did not have much in common. Bunau-Varilla had the connections with the power structure in the United States that Amador lacked. Only three weeks before, Amador had been ejected from the office of New York lawyer William Nelson Cromwell, the head of the Panama Railroad Company and, in Amador's early vision of the revolt, a worthy coconspirator. Cromwell refused to join the conspiracy not out of any concern for legal proprieties but largely because he feared that the failure and public exposure of the revolt would cost the Panama Railroad Company its concession from the Colombian government. His hopes for a powerful ally in New York crushed, Amador turned to Bunau-Varilla in near desperation. Bunau-Varilla already had established contacts with Mark Hanna, the influential Senator from Ohio, and had worked with President William McKinley. After McKinley's death, he gained the confidence of President Theodore Roosevelt and key members of the State Department. Bunau-Varilla sensed that the assertive chief of state and his advisors would cooperate in the accomplishment of Panamanian independence. The confident, energetic French engineer and the confused, disappointed Panamanian doctor devised a plan that relied on the connivance of the president of the United States.[2]

Both Amador and Bunau-Varilla had witnessed the U.S armed intervention on the isthmus in 1885 and were aware that warships dispatched by cabled messages could reach Panamanian waters in a few days. In their conversation of September 24 and in later meetings both men made it clear they understood that the confidential telegraphic communications that tied Panama City on the Pacific to Colón on the Caribbean and from there to New York and Washington would be necessary to their endeavor. Amador had arrived in New York with an extensive code for the sending and receiving of telegraphed messages with his coconspirators in Panama. Bunau-Varilla also used this code.[3] The two men conceived and executed the revolution that led to the overthrow of Colombian authority on the isthmus and the protection of the newly formed government from Colombian forces by the U.S. Navy.[4] The uncertainties of separatist revolts in relatively isolated areas such as Panama, the potential complications involved in the movement of naval vessels, the months and even years needed to establish a new government, and the lengthy time usually required in the diplomatic recognition of such a new government posed formidable challenges.[5] In addition, Bunau-Varilla wanted treaty negotiations with the United States to get under way promptly so as to open the way for the building of the canal through Panama. The negotiation of a similar treaty between the United States and Colombia had taken nearly two years.[6]

Bunau-Varilla, Amador, and their associates could not have moved quickly without submarine cables. The pages that follow here document the role of telegraphy in these exploits. West India and Panama linked Panama City to Colón and Colón to Jamaica and then, through Cuba Submarine and International Ocean, to Western Union in the United States and its numerous connections to Europe. Scrymser's Central and South American had an alternate route from Panama City north through Mexican Telegraph with its connections to Western Union. By the early 1900s, these lines, as with most submarine cables, were fairly reliable. Diplomats, journalists, businessmen and, in this case, revolutionary conspirators could use them with some confidence that their messages would reach the intended recipients.[7]

West India and Panama, however, experienced the disruptive effects of a natural disaster at the eastern end of the Caribbean in early May 1902. The eruption of the Mount Pelée volcano on the island of Martinique destroyed the port of St. Pierre and also West India and Panama's cable ship, the *Grappler,* which happened to be in the harbor at the time. The *Grappler* sank, and all fifty-six crew members perished. The company's onshore staff also lost six of its employees. West India and Panama restored cable service to the nearby islands of St. Vincent, St. Lucia, and Grenada within a year, but Martinique had no cable service until 1907. Mount Pelée's eruption was a brutal reminder of the power of nature in the Caribbean, but in spite of this disaster and the annual threats brought by the hurricane season, telegraphic services in the circum-Caribbean continued to provide regular and generally reliable service.[8]

PANAMA OR NICARAGUA?

The natural cataclysm of May 1902 was contemporaneous with an international crisis that favored Bunau-Varilla and Amador. The government of Venezuela was in debt to European bankers and had little hope of meeting its financial obligations. Venezuelan president Cipriano Castro adopted a defiant approach that aroused the ire of bankers, business leaders, and politicians in Britain, Germany, and Italy. After months of pointed warnings and threats of intervention, the British and Germans decided to blockade Venezuela's main ports. From December 1901 through February 1902 the joint German-British blockade remained in place. The German Navy was particularly aggressive, sinking some of the poorly armed vessels of the Venezuelan Navy and destroying a fortress near the mouth of Lake Maracaibo. Newspapers in the United States splashed these German actions across their front pages and, in a belligerent tone, portrayed Germany (Britain had backed away from such aggression) as a bully with the intention of grabbing territory in the Caribbean—a blatant violation of the Monroe Doctrine. This crescendo of protest assumed some of the characteristics of a war scare. President Roosevelt, who had taken

French engineer Philippe Bunau-Varilla was adept at public relations, lobbying, revolutionary plotting, and the manipulation of information through the international cable system (Philippe Bunau-Varilla, *Panama: The Creation, Destruction, and Resurrection* [London: Constable, 1913], opposite title page).

a noncommittal position during the early phase of the dispute, became concerned about Germany's intentions. At the height of the anti-German outbursts in the U.S. press, Germany's ambassador to Washington, Speck von Sternberg, attempted to calm the diplomatic waters, commenting that "we have no more intention of violating the Monroe Doctrine than we do of colonizing the Moon." As historian Nancy Mitchell has shown, German goals were modest and depended on British support.[9]

Nevertheless, Germany's aggression had captured the attention of the press and politicians. President Theodore Roosevelt was suspicious of the German presence in the Caribbean. He found support from Admiral George Dewey, the hero of the battle of Manila Bay in 1898 and head of the General Board (the strategic planning body of the U.S. Navy); Secretary of State John Hay, the nation's foreign policy chief; and Mark Hanna, a powerful senator. To Roosevelt and his advisors in 1903, the signs with respect to German aims in the Caribbean were ominous, which further pressured the United States to move ahead on the transisthmian canal.[10]

With the security of the United States tied to the canal project, the Panama vs. Nicaragua question became an issue of broad national concern. When Bunau-Varilla initiated his public-relations and lobbying effort in the United States in 1901, however, he faced an uphill struggle against the widely accepted notion that a canal through Panama had no future, given the debacle of the 1880s. Senator John Tyler Morgan of Alabama championed the Nicaragua route, and many legislators and newspaper editors agreed with him.[11] Richard Harding Davis visited Panama in 1895 and recorded his impression of the abandoned French project in his popular travel account *Three Gringos in Venezuela and Central America.* Davis had followed the news of the collapse of this engineering effort and its attendant financial scandals, so he anticipated evidence of a large-scale disaster in a tropical wasteland. His first impression fit his preconceptions: "The ruins of Panama fill one only with melancholy and disgust, and the relics of this gigantic swindle can only inspire you with contempt for yourself and your fellow-men."[12]

Davis was attuned to the public mood in the United States, and his initial response reflected this mood that seemed to take hold in the mid-1890s. Yet portions of Davis's evaluation of the canal project would have given Bunau-Varilla some encouragement. After all, Davis was a journalist. He took note of perceptions that were not consistent with his preconceptions, and he found much to challenge the negative image of what remained of the French undertaking. He was surprised to see that the French company took care to protect its locomotives and dredging equipment so that the project could be restarted. Davis claimed no special knowledge of engineering in tropical lowlands, but he expressed his admiration for the canal excavation work because "there is so much of it and that it is so well done. It looks as though only a little more energy and a greater amount of honesty would be necessary to open the canal to traffic."[13]

But Davis confined his praise to a few paragraphs and, like most of those interested in a transisthmian canal, assumed that the United States would turn to Nicaragua. Bunau-Varilla, summoning his expertise as an engineer and his talent as a publicist (he had a firm command of English), thus decided to embark on a speaking tour of major cities in 1901. He presented his case to business and civic leaders such as Chicago's Marshall Field and Cyrus McCormick, and the latter secured a lecture date for the French engineer at Princeton University. Bunau-Varilla also produced a pamphlet with the title *Panama or Nicaragua?* and mailed copies of it to every member of the U.S. Congress and all the nation's governors, the editors of four thousand newspapers, and businessmen engaged in industry, merchandising, and shipping. He emphasized the engineering advantages of the Panama route, including a transit time of twenty-one to twenty-four hours contrasted with an estimated sixty-four hours for the longer Nicaraguan route, but the point that gained the greatest public attention was the earthquake issue. The projected path through southern Nicaragua ran close to Mount Momotombo, an active volcano. This issue suddenly vaulted to the forefront of the debate in May 1902 with the disastrous eruption of Mount Pelée. About 1,500 miles east of Nicaragua and with no geological connections to Momotombo, Mount Pelée suggested to the excitable press and public a striking parallel. Bunau-Varilla took advantage of this false but nevertheless popular analogy. He even found and subsequently publicized a Nicaraguan stamp that prominently displayed Mount Momotombo. The engineer-turned-propagandist found a clever spin on the news by which to gain a public-relations advantage.[14]

THE FIRST CONSPIRACY CONFOUNDED

With newspapers, politicians, and the public shifting toward Panama, President Roosevelt encountered no more than the expected opposition from Senator Morgan and his loyalists when he decided on the Panama route. Roosevelt was sensitive to the public debate, but his decision had more to do with engineering than with political persuasion.[15] The first method adopted for securing the right to build the canal was to approach Colombia and propose a treaty. Secretary of State Hay thus initiated the difficult process of negotiating with Colombia. The subject of much controversy in both Washington and Bogotá, this treaty tried the patience of all concerned. Hay persevered and, with the help of Colombia's representative in Washington, Tomas Herrán, signed the document in January 1903. The U.S. Senate ratified it on March 17. The politics of treaty ratification were more fractious in Colombia. That nation had recently endured a destructive civil war, and it saw the economic potential of the canal as a means of dealing with its financial burdens. Colombian president José Manuel Marroquín wanted to revise the financial arrangements with the New Panama Canal Company (Bunau-Varilla's firm) and to increase the payments to be made by the U.S. government to give his administration larger sources of revenue. The Colombian Senate agreed with its president and so rejected the treaty on August 12, 1903. This turn of events undercut the arduous negotiations pushed by Roosevelt. The president was frustrated and angry, but he also was determined to proceed in spite of Bogotá's rejection of the treaty.[16]

Bunau-Varilla's company held the assets of the French canal enterprise of the 1880s, including the partially completed excavations, the railroad and its equipment, and the concessions granted by the Colombian government. Bunau-Varilla placed their value at $40 million and understood that these assets would have to be purchased by the United States in order for the building of the canal to proceed. However, because the company's concessions with Colombia would lapse in 1904, these assets would go to the Colombian government unless quick action were taken. So the French engineer and business promoter became a propagandist and conspirator. He composed an essay that appeared in the Paris newspaper *Le Matin* on September 2, 1903, in which argued openly for a revolution in Panama to create an independent nation. This new nation could then sign a treaty granting the United States control of a fifty-mile sliver of land across Panama that would include the assets of New Panama Canal. Bunau-Varilla also concluded that the United States would have the right to intervene through its naval and military forces to protect transit across the isthmus and also to protect the newly independent state of Panama. The author sent an English version of this incitement to revolt to several political and business leaders in the United States, including President Roosevelt.[17]

This revolutionary tract outlined a plot that coincided with rumors that had already begun to circulate in various publications. The *New York Times* carried an account of discontent on the isthmus on August 31 under the headline, "Rejection of Canal Treaty Likely to Lead to Revolution."[18] From Bogotá on September 14 the *Times* reported that the Colombian government "appears to realize the immanent peril of the Isthmus of Panama declaring its independence."[19] The *Washington Post* carried a similar story on its front page that emphasized the revolutionary inclination of the Colombian armed forces stationed in Panama.[20]

That the Colombian government was aware of revolutionary ferment on the isthmus did not mean that it would therefore be able to suppress the movement. Many officials in the Panamanian provincial government and in local law enforcement were sympathetic to the canal project and to independence. In addition, the conspirators in Panama did take care to conceal the specifics of their plans. Led by Manuel Amador with the support of several local leaders, the movement was limited to the upper echelons of Panamanian society. This elite was accustomed to excluding ordinary citizens—especially anyone who might be harboring plans similar to those of Pedro Prestán eighteen years earlier. The main problem for the revolutionary elite was figuring out how to establish connections with the Roosevelt administration in Washington in order to secure the recognition of the new government and, even before that, how to arrange for naval/military protection in order to block Colombian efforts to put down the rebellion. Secrecy was essential to this part of the revolutionary scheme. Amador was aware that numerous efforts at revolt in the nineteenth century had failed because the national government was able to intervene and to reassert its authority.[21]

Amador served as the revolutionary agent who went to the United States to make the necessary connections. The telegraph was crucial during this visit because Amador had to keep his fellow conspirators on the isthmus apprised of his work. To ensure secure exchanges between Panama City and the New York/Washington orbit, Amador's group established a telegraphic code to disguise their messages. Although their creating such a code might suggest dramatic intrigue, it should be noted that most private companies and government agencies (including the U.S. State Department) used their own special codes to prevent the leakage of information to the press or to commercial or political rivals.[22] Amador's code contained over eighty words that stood for phrases or sentences likely to be used in the group's planning. For example, "Canada" meant "Do not be worried by the delay. All is well." "Boy" carried the meaning "Nothing has occurred which necessitates modification."[23]

Amador sailed for New York on August 26. During the five-day voyage, he conversed with J. Gabriel Duque, the Cuban-born owner of the *Panama Star and Herald*, the main English-language newspaper on the isthmus. Duque had heard of the conspiracy, and apparently Amador spoke rather freely with him. Once on shore, Duque moved more rapidly than the elderly Amador. He first went to the offices of William Nelson Cromwell on Wall Street. Cromwell was a man of singular importance for the conspirators because he was head of the Panama Railroad, chief counsel for New Panama Canal, and known to favor the revolution. Two days later Duque was in Washington, where he had an agreeable conference with Secretary of State Hay, who allegedly said that the United States would protect the new government after the revolution but would not enter directly into the conspiracy. Next, Duque made a major blunder, at least from the plotters' perspective. He visited Colombia's representative Tomás Herrán and told him the whole story or, at least, enough to sound alarm bells. Herrán cabled Bogotá with news of the conspiracy and then notified the Colombian consul in New York of the prospective role of the Panama Railroad in the conspiracy, threatening the company with the loss of its concession from the Colombian government should it proceed. In a matter of just a few days Duque had managed to discredit Amador and undercut the revolutionary plot.[24]

The unwitting Amador met with Cromwell twice. The first meeting, on September 2, went well, but the second a few days later was a debacle. Cromwell had heard of Herrán's cables to Bogotá and took seriously the threatened cancelation of his company's concession. He literally pushed the bewildered Amador out of his New York office and severed all connections with the Panamanian conspirators. Rejected and confused, Amador telegraphed Panama City using only one word, a word that was not part of the prearranged code: "Disappointed."[25] The revolution was on the verge of failure.

BUNAU-VARILLA TAKES CHARGE

Distraught but not defeated, Amador found a new path on the road to revolution by renewing his acquaintance with his associate from the tumultuous events of 1885, Philippe Bunau-Varilla. The Frenchman had connections in New York and Washington that Amador lacked, while the Panamanian doctor had ties to the conspirators on the isthmus—something that Bunau-Varilla needed. The meeting in Room 1162 of New York's Waldorf-Astoria on September 24 propelled the revolution on to the fast track.[26] The two men were far from complete agreement on the conduct of the revolution in their first discussion, but by their second session on October 15 Bunau-Varilla had devised a plan that seemed viable. Amador was to return to Panama and initiate the revolt in early November. Bunau-Varilla assured the Panamanian that once the coup d'etat had been effected, the United States would defend the revolutionaries from the Colombian armed forces. Bunau-Varilla and Amador were to use coded telegrams as a means of keeping the clandestine revolution on its intended course.[27]

Bunau-Varilla presented his plans with confidence because before his second session with Amador he had arranged meetings with high-ranking U.S. officials, including President Roosevelt. On September 29 he met with Professor John Bassett Moore of Columbia University, who was a top-level legal advisor to the Roosevelt administration. Moore assured him of the president's preference for Panama over Nicaragua.[28] During the next two weeks Bunau-Varilla talked with Francis Loomis, the assistant secretary of state (the two had met a few years earlier in Paris). Loomis introduced him to Secretary of State John Hay. On October 10 Bunau-Varilla met with Roosevelt. The exact content of their discussion remains subject to some dispute, but it is evident that Bunau-Varilla predicted a revolution. Roosevelt was not disturbed to hear about the political unrest on the isthmus and continued the conversation with interest. Some months after the independence of Panama, Loomis wrote a letter in which he asserted that Roosevelt gave no indication of support for the revolution. Even Bunau-Varilla in his memoir indicated that while Roosevelt's manner seemed sympathetic to the cause, "No word had been pronounced, no concealed meaning had been attached to any sentence which would constitute a tie between us. His liberty was as complete as my own."[29] Apparently Loomis, Hay, and Roosevelt wanted the United States to be able to deny accusations of fomenting the dismemberment of Colombia.[30]

In spite of the silence of U.S. officials, Bunau-Varilla carried from his meetings with Loomis, Hay, and Roosevelt an understanding that the United States would quickly come to the aid of the new government of Panama. At the same time he grasped the importance of official deniability for the U.S. government.

Buoyed by his conversations in Washington, Bunau-Varilla returned to the task of encouraging Amador to implement the revolution. The two met in the Waldorf-Astoria on October 15 and again two days later. The Frenchman seized the initiative. He told Amador that the movement had no need for mercenary gunboats to safeguard Panama's shores from Colombian interference because the U.S. Navy would take on that job. The revolutionary budget of $6 million that Amador had requested was much more than what was needed. Bunau-Varilla committed $100,000 of his personal funds to support the revolution, much of which would be used to bribe the Colombian military and civilian officials in Panama. Amador agreed with these plans. In their second meeting Bunau-Varilla strongly insisted on his role in the final stage of the plan—the negotiation of the treaty between the new government of Panama and the United States. Bunau-Varilla had little confidence in the Panamanians to arrange a treaty that would satisfy his expectations. The officials of the new government would most probably attempt to secure some provisions for their own purposes. Bunau-Varilla, who had been waiting for years to exploit this opportunity, made clear to Amador that he—Bunau-Varilla—and only he was to negotiate and sign this treaty as the representative of Panama. Amador objected but to no avail. A few days later Bunau-Varilla went so far as to dictate to Amador the content of a telegram that the leaders of the new government were to transmit from Panama to Bunau-Varilla in the United States that would officially designate him to carry out these duties, thereby formalizing his authority. Bunau-Varilla insisted that his experience in dealing with U.S. officials made him uniquely qualified for this role.[31]

On October 16, Bunau-Varilla made a quick trip to Washington in between in his last two meetings with Amador to confer again with Hay. Although the secretary of state was careful to avoid any direct commitment for U.S. naval or military support, he provided the revolution's leading conspirator with another kind of support. He presented Bunau-Varilla with a copy of Richard Harding Davis's popular fictional story of revolution in Honduras conceived and directed by the eponymous Captain Macklin. Davis had viewed the intervention of "civilized" outsiders in Latin American politics as a means of dealing with the region's political ills at least since the mid-1890s publication of *Soldiers of Fortune*. Davis's Captain Macklin had as a coconspirator an experienced French army officer. Bunau-Varilla recalled Hay's comment: "The young, ambitious American and the old French officer, who as head of the [revolutionary] army displays in all his acts the generous disinteredness of his race, [are] both charming types of searchers for the Ideal." Presumably the "Ideal" was the spread throughout Latin America of what Davis considered civilized society and organized government. Bunau-Varilla left the meeting in a "spiritual daze," clutching *Captain Macklin* and its symbolic content. Hay had provided him with a work of popular fiction that constituted "the subtle symbol, the password," advocating a revolution in Panama that would have the clandestine but unmistakable endorsement of the U.S. government. Bunau-Varilla concluded that "it only remained for me to act."[32]

With Amador on the way to Panama, Bunau-Varilla made the financial arrangements. Again the submarine cable expedited things. Bunau-Varilla borrowed $100,000 against his personal securities in Credit Lyonaise. This French investment firm, according to Bunau-Varilla, was "accustomed to make cable transactions for me during my sojourns in America." By October 26, the funds had been moved from France to the Heidelbach Ickelheimer Bank in New York. Bunau-Varilla's access to these assets meant that he could disperse funds quickly if the need arose, and it soon did.[33]

REVOLUTION BY ELECTRONIC COMMUNICATION

When Amador left New York, it seemed that the independence of Panama was inevitable. The Panamanian cabal would spearhead the coup, bribing the local officials as necessary. The U.S. Navy would turn out to protect the new nation from Colombian intervention. The revolutionaries would have their independence, and Bunau-Varilla would set up the treaty and then sell New Panama Canal's concession to the U.S. government. The Roosevelt administration would have both its deniability and its canal zone. All did not go as planned, however. Events of October 29 abruptly destroyed Roosevelt's deniability, and the telegraph was the culprit in the disruption of the plot. On that day Amador, using his code name "Smith," sent Bunau-Varilla the disturbing news that a large contingent of Colombian troops were en route to Colón on a gunboat, threatening the coup. The coded cable read "Fate News Bad Powerful Tiger," which decoded as "For Bunau-Varilla, Colombian troops arriving [on the] Atlantic [coast] within five days. More than 200." Then Amador added an uncoded plea for help: "Urge Vapor Colon." This message was a call for the dispatch of a U.S. warship to Colón.[34]

Obviously Roosevelt, Hay, and Loomis wanted to avoid engaging in any action that would draw the United States directly into the revolution, but this telegram relayed a turn of events that meant that the U.S. Navy would have to move its warships into Panamanian waters before Amador and his revolutionaries initiated the insurgency. Bunau-Varilla hurriedly contacted Loomis, who, on first blush, was noncommittal. Bunau-Varilla attempted to justify his plea for U.S. intervention by recalling the disastrous burning of Colón in 1885 (although neither Amador nor any of his followers had any intentions of emulating Prestán). Early in the following day (October 30), Bunau-Varilla paced about Lafayette Square near the White House and the State Department, pondering an unannounced visit to Hay's office, when he encountered Loomis. Much to Bunau-Varilla's relief, Loomis stated that a repetition of the "deplorable catastrophe of 1885" was intolerable. Bunau-Varilla took this voluntary declaration to mean that the U.S. Navy would soon appear off Colón and that the revolution should go ahead as scheduled.[35]

On his return to New York from Washington, Bunau-Varilla stopped off in Baltimore to send Amador the brief but decisive telegram that concerned the arrival of U.S. warships: "Thirty-six hours Atlantic, forty-eight Pacific." He dispatched the telegram late on October 31, and Amador received it the following morning in Panama.[36] The conspirators then took steps to carry out the revolution, emboldened by the news that they had the support of the United States government.[37]

The Roosevelt administration had joined the conspiracy by means of cable telegraphy. The navy had begun making preparations for its participation in defending an independent Panama as early as October 17. Roosevelt had ordered four ships into the vicinity, including the *Boston,* which he directed to proceed to a position off Nicaragua. He also had alerted the army's general staff to the possibility of troop assignments to Panama. But these orders stopped short of sending U.S. vessels directly to Colón or Panama City as part of the revolution.[38] The overt action in support of the revolution began with the October 30 enciphered telegraphic message to the USS *Nashville,* under the command of John Hubbard, from Charles Darling, the acting navy secretary: "Secret and confidential. Proceed at once to Colon. Telegraph in cipher the situation after consulting with the United States consul. Your destination is secret. Telegraph in cipher your departure from Kingston."[39]

Hubbard's vessel arrived in Colón on November 2 at about 5:30 p.m., a few hours later than Bunau-Varilla had predicted in his cable to Amador but a day *before* the outbreak of the revolution. Also on November 2, the U.S. Navy telegraphed orders to three other warships to "proceed with all possible dispatch to Panama." Two of these ships, the *Boston* from San Juan del Sur in Nicaragua and the *Marblehead* from Acapulco in Mexico, were to anchor off Panama City. The gunboat *Dixie* with four hundred Marines on board followed in the wake of the *Nashville* for Colón.[40] Telegraphed messages from Washington to San Juan del Sur, Acapulco, and Kingston brought the Roosevelt administration deeply into the plot for Panamanian independence. A decade later, Roosevelt stated in an interview that the revolution began after the revolutionaries "saw the smoke of an approaching steamer," which turned out to be the *Nashville*.[41]

Meanwhile, Bunau-Varilla had been waiting nervously for news. On November 1 he found reassurance in the *New York Times*. There he read a brief and apparently routine notice: "The American cruiser *Nashville* left this morning with sealed orders. Her destination was believed to be Colombia." The dateline was "Kingston, Jamaica 31st Oct." From his perspective, the departure of the *Nashville* on the day after his conversation with Loomis in Lafayette Park was more than mere coincidence. The Roosevelt administration was taking action to support the revolutionary plot. A reassured Bunau-Varilla used the cable code word "Boy" to let Amador know that "nothing had happened which requires modification." The revolution was to proceed.[42]

Also during this period of uncertainty, Roosevelt received a lengthy telegram that reinforced his Panama policy. From Paris on October 31, William Nelson Cromwell, speaking for New Panama Canal, abandoned his stance of formal ignorance of the movement for Panamanian independence: "I am authorized in the name of the president of the canal company with the unanimous approval of the board of directors at its meeting today . . . to give you, personally, and to the government of the United States, the assurance of their loyal support, which they firmly maintain, and to express to you their confidence in the success of your masterful policy." Cromwell also sent his law partner, William J. Curtis, from New York to Washington to wait to hear Roosevelt's response. The president soon confirmed receipt of the cable and his satisfaction with its content.[43]

Bunau-Varilla' reassuring November 2 message to Amador was according to plan, but the plot nearly fell apart largely because of the bungled delivery of a telegraphic message from Darling to Hubbard of the *Nashville*. The *Nashville* was in Colón harbor, and the telegram reached Hubbard by way of the office of U.S. consul Oscar Malmros, also stationed in Colón. Darling's cabled orders were clear:

> Secret and confidential. Maintain free and uninterrupted transit. If interruption threatened by armed forces, occupy the line of the railroad. Prevent landing of any armed force with hostile intent, either government or insurgent, either at Colon, Porto Bello, or other point. Send copy of instructions to the senior officer present at Panama upon arrival of Boston. Have sent copy of instructions and have telegraphed Dixie to proceed with all possible dispatch from Kingston to Colon. Government

> forces reported approaching Colon in vessels. Prevent their landing if in your judgment this would precipitate a conflict.[44]

The problem for Hubbard was that he did not read the cable until sometime between 10:30 and 11:00 a.m. on November 3. The Colombian troops, more than four hundred of them, were already on shore and outnumbered the U.S. force of forty-two *Nashville* crewmen in Colón. For whatever reason, the cable did not reach Hubbard promptly on November 2.[45] Whether the message was delayed at the cable company (probably West India and Panama, since the telegram went to the station in Colón) or held up owing to some inefficiency in Consul Malmros's office, or whether there was a problem in the delivery of the paper message, remains uncertain.

Hubbard assessed the situation and, late in the afternoon of November 3, sent a message to Washington:

> Receipt of your telegram of 2 November is acknowledged. Prior to receipt this morning about 400 men were landed here by the Government of Colombia from the (gunboat) Cartagena. No revolution has been declared on the isthmus and no disturbances. Railway company declined to transport these troops except by request of the governor of Panama. Request has not been made. It possible that movement may be made tonight at Panama [City] to declare independence in which event I will [message mutilated] here. Situation is most critical if revolutionary leaders act.[46]

Historian Richard Turk has located two versions of corrections of the mutilated portion of this telegram. In one Hubbard said he would protest against Colombia's sending of the troops and in the other he said the opposite—that he "cannot protest" against the actions of Colombian forces. The content of this portion of the telegram remains unclear, but there was no attempt to discipline or criticize Hubbard by the Roosevelt administration. Quite the opposite—President Roosevelt commended the commander for his part in the Panamanian episode.[47] More importantly, Hubbard's presence on the isthmus was a sufficient sign of U.S. willingness to intimidate the Colombian effort to block the insurgents. Also, the first telegram stated unequivocally that Hubbard was to block the Colombian force and thereby clear the way for the revolution. The United States had joined the conspiracy by electric communication in its intentions even if the execution of the official orders was incomplete at this point.

Commander Hubbard did carry out a decisive maneuver that removed the threat of the gunboat *Cartagena* from Colón harbor. Hubbard's *Nashville* cleared the decks for action, trained its guns on the *Cartagena*, and sailed menacingly toward the lightly armed Colombian vessel. The Colombian captain quickly grasped his disadvantaged position and departed the harbor. One of Colombia's naval assets was driven from the scene.[48]

Fortunately for Amador and his followers, further direct intervention by the U.S. Navy turned out not to be necessary. The conspirators cleverly separated the two Colombian generals on the isthmus from their soldiers by escorting these officers across the isthmus by rail, entertaining them with a grand reception in Panama City, and then placing them under arrest. The Panama Railroad readily participated in the subterfuge and then, with the help of provincial officials and General Esteban Huertas (a defector from the Colombian military), the conspirators managed to bribe the officers and key enlisted men in the Colombian army. Bunau-Varilla wired the necessary funds from his New York bank to the Brandon Bank in Panama. Colombian general Pampilio Gutierrez received $14,445, and the somewhat more patriotic Colonel Eliseo Torres eventually accepted $8,000. General Huertas, whose defection on November 3 was crucial to the movement, provide helpful insights into the use of telegraphy by both Bunau-Varilla and the U.S. government. Amador approached the general with an offer of "glory and riches" and then pointed out that one U.S. cruiser had already arrived in Colón and that others were on the way. Amador argued that it was futile to resist the power of the United States and nor did it make sense to reject the offer of financial rewards (supplied by Bunau-Varilla from New York). According to David McCullough, "Huertas is said to have remained 'impassive' for a moment, then put out his hand. 'I accept.'"[49]

Amador and his colleagues proclaimed Panamanian independence at about 6:00 p.m. on November 3. Cable communications alleviated the anxieties of Bunau-Varilla and the Roosevelt administration. At about 10:00 p.m. Amador cabled Bunau-Varilla in uncoded Spanish: "Proclama Independencia del Istmo sin sangre" ("Proclaim independence of the isthmus without bloodshed").[50] On the same evening Vice-Consul Felix Ehrman notified the State Department that the "uprising occurred tonight Six. No bloodshed. . . . Government will be organized tonight."[51] After Amador and Huertas had distributed their bribes to the military, the possibility of Colombian interposition melted away. On November 6 Secretary Hay cabled notice of Washington's de facto recognition of Panama as an independent nation to Bogotá.[52]

The revolution was a fait accompli. Ehrman cabled Hay on the same day: "The situation is good. Isthmus had obtained so far success. Colon [*sic*] and interior provinces have enthusiastically joined independence."[53]

THE TREATY

As complicated as the gaining of Panamanian independence had been, it was only the first step. Neither the Panamanian leadership nor Bunau-Varilla could pause to enjoy the fruits of their labor. Instead they immediately directed their attention to the new canal treaty with the United States, and, on this point, the conspirators parted company. Bunau-Varilla had made clear his intention to become the sole negotiator for the new Panamanian government in earlier discussions with Amador and had telegraphic confirmation of this arrangement. Amador and his Panamanian associates—especially Federico Boyd—did not regard these arrangements made during the hectic days leading up to the revolution as binding. For nearly a month Bunau-Varilla squabbled with the Panamanians regarding his qualifications to serve as the formal representative of the new nation.[54]

Bunau-Varilla had the upper hand in this struggle. He worked harmoniously and quickly with Hay to produce a new treaty. These two men signed what was arguably the most important document in the early history of the new nation on November 18, 1903, in Washington. The treaty granted highly favorable terms to the United States. According to the analysis of historian John Major, Hay practically dictated the key terms in the document. This treaty gave the United States much greater authority over the canal zone than the previous treaty with Colombia had—as much as if it were sovereign over the ten-mile wide strip of land that divided the new country.[55] Bunau-Varilla, in a real sense subservient to Hay and Roosevelt, had leverage over the Panamanians because of his connections in the United States. Washington had entered an asymmetrical relationship with the nascent and only partially organized government of Panama. Amador and Boyd hurried to the United States to assert the interests of their country only to be greeted by Bunau-Varilla at the Washington train station with news that the treaty already bore the signatures of Hay and himself. According to Bunau-Varilla's account, Amador "nearly swooned on the platform," David McCullough records that an angry Boyd reportedly slapped Bunau-Varilla in the face.[56]

The near faint and alleged slap did nothing to slow the inevitable movement toward ratification of the treaty by the U.S. Senate. However, Amador and Boyd did have influence in Panama, and Bunau-Varilla had to call on his reserves of energy and inventiveness. Again, Bunau-Varilla made good use of the telegraph. On Saturday November 21 he cabled the entire treaty to Panamanian minister of foreign relations F. V. de la Espriella. Under the circumstances, the minister hesitated. His first telegraphic response to Bunau-Varilla came two days later, and it was negative. The adroit Bunau-Varilla found a sensitive point to probe—the new government's fear of the recently rejected mother country. Colombian president Rafael Reyes was currently in Washington to undertake discussions with the Roosevelt administration. Although the two governments were now on unfriendly terms, Bunau-Varilla employed the telegraph to play on the fears of the Panamanians. In a 370-word cable to Minister de la Espriella on November 25, Bunau-Varilla asserted it was possible that if Panamanian officials did not sign the Hay-Bunau-Varilla treaty immediately on its arrival in Panama (by boat and due to arrive on December 1 or 2), the United States would negotiate a new treaty with Colombia and, in effect, undo Panama's independence. The tactic worked. The next day, de la Espriella and his associates cabled that the government would sign the treaty as soon as it arrived.[57] The Panamanian ratification became official on December 2, 1903. On February 23, 1904, the U.S. Senate, as expected, ratified the treaty.[58]

THE U.S. PRESS RESPONSE: THE LIMITS OF JINGOISM

The Panamanian revolt for independence and the complicity of the United States in that revolt dominated the front pages of newspapers across the country for more than a week. Yet the reaction in the press to Roosevelt's act of national assertiveness was considerably different from the jingoistic enthusiasm and approbation that filled the telegraph wires and news columns in the 1890s in response to the Chilean crisis, the Venezuelan boundary dispute with Britain, and the conflict with Spain. Leading editorial voices in the press were critical of the Roosevelt administration. The president took offense at the condemnations of what he thought was a necessary action.

One significant aspect of these contrasts between the 1890s and 1903 was the success of Amador, Bunau-Varilla, and the government in Washington in maintaining secrecy. The revolt of November 3 was front-page news on November 4, but in the weeks leading up to this event, rumors of plots and vague references to the well-known unrest on the isthmus did not furnish substantial information about what was to transpire in Panama. In the previous confrontations and conflicts, reporters had access to information from the presidents, Congress, the State Department, and the War

Department, as well as cables from Santiago, London, Havana, and Madrid. There was no 1903 version of Sylvester Scovel in Panama uncovering evidence of the Colombia's problems in governance on the isthmus that could parallel reportage on Spain's mismanagement of Cuba.[59]

The Panamanian revolt caught the nation's leading newspapers by surprise. After a two- or three-day interval (presumably to sift through reports from Panama, Bogotá, and Washington) most newspapers editorialized in favor of Roosevelt's participation in the revolt to protect the revolutionaries and the newly independent nation for the sake of opening the way for a U.S.-built canal. The *Washington Post* saw the United States' action as a justifiable removal of an inefficient, uncooperative government in the name of progress.[60] The *Chicago Tribune* praised the explanation provided by Illinois native Secretary of State John Hay, who, in the judgment of the paper, had used his rhetorical skills to convince most reasonable doubters to join the ranks of Roosevelt supporters.[61] The *New York Tribune* took a businessman's measure of events and concluded that an independent Panama in close alliance with (but not domination by) the United States would be a much more stable and therefore favorable environment for the large construction project and its heavy investment of U.S. resources than the unstable and unfriendly nation of Colombia.[62] The *San Francisco Chronicle*, the *Philadelphia Inquirer*, the *Boston Globe*, the *Atlanta Constitution*, and the *New Orleans Picayune* also rendered favorable opinions on events in Panama.[63]

The press evaluation, however, was not unanimous. The *Literary Digest* surveyed fifty-three newspapers that offered editorial judgments in November 1903 and found that a substantial number—seventeen—opposed the president's actions. The influential *Springfield Republican* chastised the Roosevelt administration, which "has virtually seized the isthmus by force, and now, with shocking hypocrisy, defends its wickedness as in line with the most honored precedents and highest morality." Also among the naysayers were Joseph Pulitzer and William Randolph Hearst. These champions of jingoism in 1898 had become critics of Roosevelt's adventurism in 1903. Hearst's *San Francisco Examiner* concentrated in its news coverage on demands for a congressional investigation of the role of the United States in the revolt. Editorial headlines summarized this view: "Induced Panama Revolution Is National Scandal" and "American Honor Dragged in the Mud in Panama."[64] Pulitzer's *World* faulted the president for adding to the list of wrong-headed U.S. aggressions dating from the war with Mexico of 1846–1848.[65]

The most telling criticism came from the *New York Times,* which launched an extended editorial and reportorial campaign focused on the dark side of events in Washington and Panama. The November 5 editorial scored the United States for its interference in Colombia's domestic politics. The following day the editorial headline read, "A National Disgrace," and it was directed specifically at President Roosevelt: "If Mr. Roosevelt pursues a course which will result in the dismemberment of Colombia and thereafter proceeds with the Panama Canal undertaking, . . . all the world will conclude that the instigation of the revolt, interference by us to prevent suppression of it, and the recognition of the independence of the new little republic were one and all measures of our national policy determined upon in advance by our Government and executed by our President to the end that in that way all obstacles to the construction of the Panama canal might be removed." The editorial concluded that Roosevelt had "put a stain upon the country's good name by such a policy of dishonorable intrigue and aggression."[66]

E. L. Godkin lashed out at Roosevelt in the pages of the *New York Evening Post* and the *Nation,* a prestigious journal of opinion. Godkin used sarcasm to belittle Roosevelt's insistence that he had done the cause of the civilization a service by establishing a "protectorate" over the infant nation of Panama.[67] J. P. Gordy in the *Forum* pointed out the abnormally short time it took for Washington to recognize the new republic—only ten days—while the diplomatic recognition of the nations of South America had taken seven years in the early nineteenth century.[68]

Roosevelt's sensitivity to complaints and compliments in the popular press came through in his private correspondence. The president's lengthy letter of appreciation to editor George Brinton McClellan Harvey of *Harper's Weekly* indicated that Roosevelt felt a need for press support.[69] And he attacked his critics. To British writer James Bryce, Roosevelt asserted his conviction that Godkin of the *Nation* lacked the moral fiber to be taken seriously as a political commentator: "Personally I regard him as a very insincere and very untruthful man, for I do not consider a lie any less a lie when due to spite, envy, and malevolence than when due to pecuniary self-interest."[70] The criticism in the *New York Times,* the *Nation,* the *New York Journal,* the *New York World,* and other press organs angered Roosevelt, who saw them as "a small body of shrill eunuchs" who did not grasp the larger purposes of his actions.[71] The contrast between his elevation to heroic status in the newspaper and periodical press of 1898 and the mixed response to his actions in 1903 was, for him, painfully obvious. Roosevelt had mastered the information system in 1898 and used it well in his rapid rise to the presidency, but the unfavorable reaction to the Panama episode threw him off balance, and he continued to defend his action in public forums even after he left the presidency.[72] Jingoism and an aggressive foreign policy had found some limits in the U.S. press. The use of manipulation and intimidation fell outside the boundaries of acceptability in the opinion of some prominent media voices.

CONCLUSIONS

Bunau-Varilla and Amador used the cable effectively to carry out their plan to engineer a revolution in Panama under the protection of the U.S. government. Amador's telegraphic warning on October 29 of the imminent arrival of the Colombian troops led to Bunau-Varilla's inducing the Roosevelt administration to break with its policy that combined a close watch on events and marginal support with deniability. Loomis, Hay, and Roosevelt avoided incrimination as far as they could. However, Bunau-Varilla's presentation of the information in the Amador telegram created a crisis in Washington, and the administration's subsequent telegrams committed the U.S. Navy to an active role in the conspiracy to oust the Colombian government from Panama. Amador continued to send cables to Bunau-Varilla until the revolution for independence was completed. Essential to this coup was the bribery of Colombian officials largely using funds transferred by cable from Paris to New York to Panama. Soon thereafter Bunau-Varilla used the telegraph to outmaneuver his former coconspirator, sidestepping Amador to secure the ratification of the new treaty to protect his own financial interests.

While Bunau-Varilla mastered the employment of cable communications at every crucial turn, the U.S. government was outwitted at key moments. Bunau-Varilla used the information system to orchestrate a successful public-relations campaign that promoted Panama as the first choice for a canal over Nicaragua and also managed to use the cable network to pull Loomis, Hay, and Roosevelt into active participation in the revolutionary conspiracy that culminated in the events of October 29–November 2. The U.S. Navy sent its orders to the *Nashville* and its other warships by cable, a command process that worked well and helped to ensure the viability of the insurgency. The navy fumbled its job, however, with the telegram of November 2 to Commander Hubbard in Colón, which was either sent too late or not read in time by Hubbard. Hubbard's forty-two-man contingent on shore faced a possible confrontation with over four hundred Colombian troops.

Matters became even more uncertain as a result of Hubbard's mangled telegram to Washington on the afternoon of November 3, but the navy escaped the potentially serious consequences of these miscommunications when the bribery offensive launched by Bunau-Varilla and Amador eliminated the opposition. Bunau-Varilla, a witness to armed intervention directed by telegraphy in 1885, employed the cable system to achieve his ends in 1903. The engineer, publicist, and conspirator understood the international communications system.

In contrast to Bunau-Varilla, the Roosevelt administration stumbled badly. Bunau-Varilla and Amador drew Washington into their scheme, and when this complicity became known to the U.S. press, Roosevelt and his top advisors faced a flurry of criticism from the *Times*, the *Journal*, and the *World* of New York as well as from the *Nation* that troubled Roosevelt in spite of support from other media organs. They found fault with the imperial maneuvers of the navy and the State Department in the creation of the new nation of Panama and the sudden and obviously premeditated annexation of the canal zone.

Beyond the immediate responses to Bunau-Varilla's handiwork and Roosevelt's heavy-handed intrusions, there was an alarm raised in certain quarters over the domineering presence of the United States in Panamanian independence, which sent loud and ominous signals throughout Latin America. The techniques of dismemberment used against Colombia could be applied to other nations in the hemisphere. The U.S. metropolitan press was swept up in the drama of the revolution, but two major newspapers grasped the potential for problems in the relations between the United States and its neighbors. The *New York Times* extended its criticism of Roosevelt's policies and reached a disturbing conclusion: "The South American republics see us investing them, pressing ever nearer and nearer, first Cuba, then Porto Rico, now the Isthmus of Panama. Nothing is lacking to establish and complete the presumption that we have entered upon a policy of conquest and control."[73] Joining the *New York Times* unexpectedly in this line of thought was the *Chicago Tribune*. A strongly Republican paper and a supporter of the Roosevelt-Hay policy, the *Tribune* nevertheless issued a similar editorial warning: with the Roosevelt administration "firmly planted at the gateway to South America, the inhabitants of that continent will be more suspicious of the United States than they are now. They will be all the more distrustful of this country after it shall have become the absolute master of the isthmus."[74] Bunau-Varilla and the Roosevelt government used the technology of telegraphy and controlled the flow of information to their advantage, just as the U.S. Navy did to secure its domination over the waters of the Caribbean in 1898, but the successful deployment of electrical and mechanical innovations in communications and naval weaponry did not carry with it ipso facto the admiration and trust of other nations in the aftermath of the Panamanian revolt of 1903. These events would produce unintended and largely unforeseen consequences.

CHAPTER THIRTEEN

Celebrations of Heroism and Power

NEWSPAPERS, CABLES, AND CORRESPONDENTS provided a steady stream of headlines and front-page stories that depicted the triumphs of the U.S. armed forces in the Caribbean and the Philippines. The public response to battlefield heroics was adulatory, and the market for publications was vigorous. In the afterglow of these victories several writers used the medium of the printed word to carry on an extended celebration. Journalists shared these stories with politicians, political commentators, and military and naval officers. Celebratory publications spanned in the information system, taking the form of everything from telegraphed reports and editorial cartoons in newspapers to large, profusely illustrated books. The criticism of the Roosevelt administration's conduct in Panama in 1903 by a portion of the press had dissipated, and the thrust of opinion in the popular media became triumphalist as the U.S. government succeeded in the construction of the canal where the French had failed. Two large, interconnected themes emerged in this explosion of patriotic prose: the heroic actions of men in uniform and the expansive power of the United States. Victory in combat translated as power in peacetime. Heroism easily merged with concerns about the longer view—the roles of the United States and its armed services in the postwar environment. Heroic men became symbols of a powerful nation capable of competing with the European nations and Japan in the imperial age. The public purchased these publications in great quantities. The information system reflected these values. Concerns about the political pressures created by yellow journalism waned, and the idea that careful diplomacy and calm policy making called for discretion in the use of telegrams came to carry less weight. The armed services took the moment to make the case for upgrading their material and human resources, and individuals—most notably Theodore Roosevelt—exploited their personal popularity.

The cultural values that supported triumphs on the battlefield and the creation of a U.S.-dominated empire rested on the ideal of vibrant, virile manhood. Kristin Hoganson's seminal work draws heavily from newspapers, magazines, and speeches that stressed the importance of an expansive foreign policy that advocated a martial approach—an emphasis on the capacity to use war or the threat of war—in international issues. These same masculine ideals served in defense of the commercial and political domination of Cuba, the outright annexation of Puerto Rico and the Philippines, and the taking of the canal zone in Panama. Hoganson emphasizes the connection between gender history and foreign policy. It was men who enacted these aggressive policies in the Caribbean and the Philippines and defended them in the print media and in public speeches in the United States. The jingoes saw war as a means of building national character and espoused empire as a means of extending national power. Strengthening "American manhood" was a matter of national security, and it also sold newspapers and books.[1]

CELEBRATIONS OF HEROIC DEEDS AND IMPERIAL POSSESSIONS

Newspapers from coast to coast abandoned objectivity to boast of the accomplishments of the U.S. Navy and the U.S. Army in the war with Spain. William Randolph Hearst, editor and owner of the *New York Journal*, led the nation in patriotic exposition in his New York newspaper and also on the other side of the continent in the *San Francisco Examiner*. He claimed that the defeat of the Spanish in Cuba meant that "July 1st will live in the world's history as the date of a splendid achievement."[2] The usually circumspect *New York Times* headlined the story of the Spanish surrender in Santiago with "Our Flag Flies Over Santiago," "A Thrilling Ceremony," and "Bands Played 'The Star-Spangled Banner.'"[3] The *Philadelphia Ledger* noted the growing number of warrior-celebrities: "We are making heroes so fast now that fame will soon have to lay aside her roll and start a directory." The *New York Sun* opined: "Our capture of Santiago [is] . . . an achievement of extraordinary brilliancy." And the *Pittsburgh Post* claimed that the Spanish defeat served as vengeance for the deaths of the *Virginius* crew and passengers in 1873.[4]

Books took a large share in the field of patriotic publications, and politicians joined with journalists to draw attention to the heroics of the returning veterans. J. Hampton Moore, a reporter for the *Philadelphia Ledger* for over a decade and more recently a leader in the Republican Party in the Philadelphia area, collaborated with James Rankin Young, a member of the House of Representatives from Pennsylvania, to produce a volume of personal accounts by the participants titled *Thrilling Stories of the War by Returned Heroes* that downplayed the contributions of Cubans while lauding the performance of the U.S. forces. The editors praised William McKinley's Republican leadership, and they also applied their hyperbole to global politics: "All Europe stood amazed at the sweeping victories wherever the American flag was planted. The wonderful success of Admiral Dewey at Manila stirred the interest if not the fear, of foreign potentates, as it had never been stirred before." General A. C. Parkerson put together a similar volume, *How Uncle Sam Fights*, which featured over forty photographs of U.S. warships, including the *Oregon*, *New York*, and *Texas*, plus views of the *Maine* before and after the explosion. Editor Parkerson, assisted by "military and naval experts," included a sixty-page section that concentrated on the fighting in Manila Bay and the defeat of Cervera near Santiago. The patriotic language praised U.S. officers and sailors. Theodore Roosevelt weighed in with a series of articles in *Scribner's Magazine* that appeared in book form under the title *The Rough Riders*. According to biographer H. W. Brands, Roosevelt produced "a tale of heart-swelling heroism."[5]

Accounts of heroic exploits were not limited to men in arms; journalists also risked life and limb to report from the fields of battle. Richard Harding Davis displayed coolness under fire at Las Guásimas and San Juan Heights.[6] Robert Cramer of the *Atlanta Constitution* described the challenges that he and other correspondents faced that reached a national readership in the *Literary Digest*. Cramer exalted the pure reportorial motives of his colleagues "to get at the truth" and noted the risks that they took: "They swam streams to get it. They starved by day and shivered by night. . . . They dodged bullets and slept with yellow fever germs. . . . They walked hundreds of miles through mud and ran small boats through raging seas. They endured more hardships in a single day than could with justice have been distributed through a lifetime."[7]

Journalists, soldiers, and sailors endured the dangers of battle for a variety of reasons, but one post hoc explanation of their derring-do came to focus on the imperial power suddenly acquired by the United States. The annexation of the Philippines turned into a controversial issue in the United States, and the decision to accept the overseas empire in the Far East that Spain ceded to it became formalized only after an extensive debate. Politicians and newspaper editors were generally agreed on the virtues of commercial expansion through trade and investment, but the annexation of territories such as the Philippines and Puerto Rico carried with them governmental responsibilities that could give rise to social and political problems (as the Filipino insurrection of 1899–1902 indicated). William McKinley, so reluctant to enter the fight against Spain, became a defender of U.S. annexation of the Philippines. In numerous public addresses he connected U.S. power overseas to the essence of national identity. Expansionism, in his view, was congruent with the national mission of extending the virtues of representative government and the free enterprise system to other places on the globe. This message resonated well with the public, and he won reelection to the presidency in 1900, having secured a place for assertive foreign policy within the popular culture of the nation.[8]

The celebration of expansion was a central theme in the pages of *Our New Possessions* by Trumbull White. The volume had an encyclopedic quality, listing the agricultural and mineral wealth of the colonies. But White indulged in geopolitical boastfulness: "The sun will never set on American Possessions." He used a kind of bravado to sweep aside the ambiguities of the nation's new imperial status: "For good or for ill, the United States has entered upon a colonial policy of expansion, a policy which forces us into the position of a world power, deep in the complications of international politics and the Eastern Question."[9]

Although Cuba was not a formal colonial "possession," White included it in his volume because, as the author admitted, the island would be dominated by U.S. commercial interests. The victories over Spain in the fight for Cuba made "a name for the United States such as it had not before among the other nations of the world." White reveled in a piece of imperial rumination: "I felt the strong impression that a new era was dawning for the unhappy island with the first landing of American forces on Cuban soil."[10]

Celebrations of the spoils of victory from the Spanish-American War were echoed in the excitement of building the Panama Canal. Theodore Roosevelt was once again a central figure in newspaper columns, magazine articles, and books. Stung by the public controversy around his administration's involvement in the Panamanian revolt and its rapid takeover of the canal zone, Roosevelt responded to criticisms (many of them well placed) of the faltering construction work in 1905 and 1906. The president made the building of the canal a priority, and his personal involvement as well as his selection of leadership brought positive results. Improved sanitation defeated mosquitoes and yellow fever, and innovative management and engineering made the canal opening in 1914 a national triumph. Roosevelt took full advantage of the public-relations opportunities during his presidential term. He made several speeches, met frequently with selected journalists, and, in November 1906, visited Panama to see and be seen at the construction site. Roosevelt described this work as a campaign of military dimensions accomplished by heroic men who exemplified the martial spirit that had carried the nation to victory in 1898. Publications flowed from the presses, with "more than twenty-five full length books and innumerable articles" extolling these accomplishments. Perhaps the most impressive of these publications was Willis Abbot's *Panama and the Canal in Pictures and Prose.*[11]

Abbott's pride in these possessions was shared by James Creelman, a veteran reporter for the Hearst papers who was wounded at the battle of El Caney. Creelman recovered from his wound and subsequently took an assignment in the Philippines, where his enthusiasm for scoops was undiminished. By his own account, Creelman was a case study in what Hoganson has identified as the assertion of "American manhood" in the 1890s. The reporter, of course, shared the masculine pride in military victories and new colonies, and his memoir makes reference to a phenomenon that Hoganson considers to be one of the root causes of the assertions of imperial manhood: the appearance of the "new woman," a female who entered politics or the professions with the capacity to compete with her male counterpart.[12]

Creelman described a race between himself and the unnamed wife of a competing journalist who stepped in for her injured husband in an effort to be the first to send a telegraphed message from Manila to the United States of the first formal pronouncement by the new U.S. administration of the Philippines regarding plans for the islands. Creelman and "the bright-witted young wife" of his fellow journalist took their copies of the pronouncement and dashed through the city streets, first to the censor and then to the telegraph office in what was, in part, a horse race. Creelman claimed that spectators lined the streets to see the race: "Invalids leaned out of the hospital windows, natives stood still and stared, store keepers cheered in their doorways, a horde of yelping dogs raced after us in the trailing dust," and war hero General Henry Lawton "sat in front of his headquarters and laughed when my hat blew off."[13] How Creelman observed all this during his frantic race defies explanation, but his account of his motives amounted to a revealing statement an aggressive male journalist competing with a "new-style woman" for a scoop: "Chivalry had vanished; courtesy was forgotten. It was a struggle for news, fierce and sexless—the old-style man against the new-style woman. To surrender the road to my rival meant a defeat that could not be explained by cable. The modern newspaper and its thirsty presses take no account for the amenities of life. It has one supreme law—send the news and send it first."[14]

Creelman's naïve, almost adolescent description of his own behavior is more important than the outcome. Nevertheless, it is worth noting that Hearst's man in Manila won by selecting a more sympathetic censor and a more cooperative telegrapher. Beyond voicing attempts at masculine humor and local color, the text resonates with male braggadocio and patriotism.

ALFRED THAYER MAHAN ON THE PROJECTION OF NATIONAL POWER

Well-known figures in the armed forces wrote numerous books and articles that detailed their postwar reflections for popular consumption. Two men with ties to the Confederacy, Fitzhugh Lee and Joseph Wheeler, collaborated on an account of the land war in Cuba.[15] Alfred Thayer Mahan, however, dominated this genre. His international reputation was established by the classic study *The Influence of Sea Power upon History* in 1890, and he continued to turn out books and articles at a steady pace through the decade. He owed his acclaim to a work of history, but the Spanish-American War imparted a profundity to his comments on contemporary events. Mahan produced three widely read books: *The Interest of America in Sea Power* (1897), *The Lessons of the War with Spain* (1899), and *The Problem of Asia and Its Effect upon International Policies* (1900).[16]

To argue that Mahan's perspective falls into the same category as the ebullient Creelman would be off the mark, but it is clear that the events of the last part of the decade stimulated the naval historian and strategist to reassess the place of the United States in world affairs. It was also be amiss, however, to ignore Mahan's desire for publicity. He wanted his writing to reach not only military and political leaders but also the mass market. He implored Little, Brown, and Company to advertise the forthcoming *Interest of American in Sea Power* in venues "where popular feeling may be strong."[17]

He also complained to his publisher when sales for one of his volumes did not reach expectations.[18]

While he benefited from the sale of his publications and probably enjoyed the satisfaction of popular acclaim, Mahan's quest for publicity was motivated by professional concerns. The admiral was troubled by the misinformation that often appeared in newspaper accounts and the cumulative impact of these skewed and sometimes sensational stories on public opinion, which, in turn, had an impact on the deliberations of the nation's leaders. He understood the connections between news from foreign lands, the U.S. press and politics, and the formulation of national policy. The discussion of strategic issues in a representative democracy with an open press challenged the serious analyst.[19] Therefore, Mahan deliberately sought to influence the public. He frequently spoke on history as well as current events, and his name appeared in the *New York Times*, which highlighted his comments.[20]

Mahan's interest in the projection of U.S. national might through a strong navy predated the Spanish-American War. *The Interest of America in Sea Power* was a compilation of magazine articles he wrote from 1890 to 1897 that explored several issues in this area. The volume's date of publication was timely—December 8, 1897—little more than two months before the explosion of the *Maine*. Mahan had already developed a broad perspective on the importance of sea power in U.S. strategy. He advocated the annexation of Hawaii, control of the anticipated transisthmian canal, and U.S. predominance in the Caribbean and Gulf of Mexico, all of which required a large fleet of warships. He supported the Monroe Doctrine, but was troubled by its vagueness. Mahan saw the British as allies in the new international order and urged reconciliation between Washington and London in the aftermath of the Venezuelan boundary dispute. He saw Germany in a different light—a potential rival in Latin America, where German settlers and merchants had created civilian bases for military and imperial adventures that might push the United States to defend the ambiguous implications of the Monroe Doctrine.[21]

The victory over Spain, the annexation of Hawaii, the Philippines, and Puerto Rico, and the temporary responsibility of governing Cuba (1898–1903), along with the long-term strategic and economic hegemony over this island, gave Mahan a new set of conditions to examine. Defense of these far-flung possessions weighed heavily on the U.S. Navy. Mahan, ever aware of press misdirections and popular misunderstandings, wrote to clarify two key points. The first pertained to the concept of defense as understood in traditional strategic discourse and the more specialized understanding of this concept in the context of the nation's new status as a world power. Mahan agreed that in the political sense the United States would continue to deploy its navy only if "we are forced into war." In the strictly military sense of the term, however, which is what Mahan wanted the public and politicians to understand, a navy limited to a defensive posture would most likely be defeated by the mobile sea power of a rival. In Mahanian doctrine, the navy must take the initiative, concentrate its warships, and seek out and defeat the enemy at sea. Island possessions such as Hawaii and the Philippines as well as the mainland United States would best be defended by planned offensive action: "A navy for defense only, in the *political* sense, means a navy that will only be used in case we are forced into war; a navy for defense only, in the *military* sense, means a navy that can only await attack and defend its own, leaving the enemy at ease as regards his own interests, and at liberty to choose his own time and manner of fighting. . . . Among all masters of military art—including therein naval art—it is a thoroughly accepted principle that mere defensive war means military ruin, and therefore national disaster."[22]

Mahan's prose was typically academic and formal. However, in his approach to planning for the defense of the new possessions within the context of U.S. politics, an underlying excitement sometimes broke through. In articles published in *Leslie's Weekly* (and quoted in the *New York Times*) and the *Independent*, Mahan adopted a more direct, forceful manner. He argued that the vagueness of the Monroe Doctrine could be overcome by invoking it to protect a transisthmian canal: "Therefore, if we intend to make good the affirmation of the [Monroe Doctrine], we must be prepared to resist, forcefully if need be, any attempt to obtain adjacent territory of ports which may serve as stations for a navy hostile to ourselves."[23] Mahan insisted that a large, well-prepared navy was the best means for the peaceful enforcement of the Monroe Doctrine:

> Such a lamentable alternative as war is best avoided by evident readiness for it; by due preparation, timely made, to support our interests by arms. Now the one preparation for war, in a maritime region like the Caribbean, is a navy large enough to be effective. Ports are quite secondary. They are necessary to a navy which needs a local base of operations; but

> they are useless without it, especially in the Caribbean, because conditions there are so backward as to give no local resources.[24]

The annexation of the Philippines presented new challenges to U.S. strategy, which now had to take into account the Pacific and East Asia. James Creelman's frantic carriage race through the streets of Manila contrasted with Mahan's serious and at times ominous considerations of this new role. In *The Problem of Asia*, Mahan engaged in geopolitical analysis by concentrating on Russia's need for a warm water port on Asia's Pacific coast and the consequent potential complications for Japan, China, the European imperial powers, and the recently arrived colonial resident, United States. Even in a volume devoted to such global issues, Mahan took care to reiterate the essential importance of the Caribbean, the future canal, and the Monroe Doctrine.[25] Mahan's discussions of the geography of Asia and Europe, rivalries in the Pacific and the Caribbean, and the advantages of cooperation with Great Britain made his case for the formulation of policies in Washington based on a grasp of world geopolitics and were an expression of his hopes for influencing public opinion through the print media.[26]

THEODORE ROOSEVELT AND NATIONAL POWER

Mahan's ideas on national power had much in common with those of Theodore Roosevelt. As a rising star in the political firmament, Roosevelt parlayed his heroic status after the Spanish-American War into a victorious campaign as the Republican Party's nominee for governor of New York in the fall of 1898.[27] Early in his postwar ascension in politics, Roosevelt took advantage of his access to the press to publicize his views on the role of the United States in global affairs. The occasion was the New York State Republican Party's nominating convention for the elections of November 1898. The location was Carnegie Hall in New York City. This large arena was filled to capacity, with five thousand to six thousand enthusiastic party members inside and perhaps four thousand on the streets outside. "Well dressed and orderly," this assemblage waited eagerly for the appearance of "Teddy." The *New York Times* provided its city, state, and national readership with a detailed account of the proceedings under the column headline "Cheer for Roosevelt."[28]

With a flair for capturing the attention of a mass audience, Roosevelt employed a direct, dramatic speaking style for the receptive group assembled in Carnegie Hall. Although he spoke (and wrote) in energetic prose that contrasted with the often stodgy, academic writing of Mahan, the two men agreed that the new international status of the United States must be upheld by powerful armed forces. As a former assistant secretary of the navy (who ordered the preparation of warships in the Pacific prior to the battle of Manila Bay) and the hero of the army's campaign in Cuba, Roosevelt could not resist integrating a broad statement on the nation's foreign policy into his opening speech in the campaign for the governorship of New York. He saw military preparedness as central to a unified national purpose: "Greatness means strife for Nation and man alike. A soft, easy life is not worth living, if it impairs the fiber of brain and heart and muscle. . . . Our flag is a proud flag, and it stands for liberty and civilization." He implored his audience in Carnegie Hall as well as the newspaper readers to support the McKinley administration in the election of 1900 because its defeat would mean a retreat by the United States from its recently won position as a world power.[29]

Echoing Mahan, Roosevelt argued for the buildup of the navy and army: "The surest way to avert war . . . is to be prepared to do well if forced to go into war. If we don't prepare for war in advance, then other powers will have a just contempt for us."[30] He then turned to the recent past to fortify his argument. The 1891–1892 crisis with Chile during the presidency of Republican Benjamin Harrison came to a peaceful conclusion because Washington was able to use the threat of war to intimidate Chile: "Nothing but our having the ships and being ready to use them made Chile keep the peace, and as a mere matter of expense the war which the navy thus averted would have cost many times over more than the whole cost of the navy for the past fifteen years."[31]

Roosevelt then cited the more recent example—the Spanish-American War—but he juxtaposed the victories at sea with the panic that swept the East Coast cities during the agonizing weeks when the whereabouts of Admiral Cervera's fleet was unknown. Then he drove home his conclusion that the United States needed an even larger navy not only to secure its far-flung possessions and commercial ties but also to defend its own coastline. At this point Roosevelt again followed Mahan in his use of the world "defend:"

> When I say defend, I wish it understood that I use the word in its proper sense, for the only defensive that is worth anything is the offensive. A peaceable man must not brawl, but when forced to fight, if he is worth his salt, he will defend himself by hitting and not parrying. We want a navy that will defend us from the enemy by punishing that enemy; a

> navy which can strike as Hull and Perry struck in the days of our grandfathers; as Farragut and Porter struck in the civil war, and as Dewey and Sampson struck in the war that is just over.[32]

The colonel of the Rough Riders also called for a stronger army, one to take on the duties of the new empire: "We need a far larger regular army than we now have, and we need to have it trained on a larger scale than it has been trained for the last quarter century." This imposing force was not required to maintain domestic tranquility, "but we do need it to protect our interests abroad."[33]

The Republican partisans roared their approval. As if to provide patriotic punctuation to the rousing speech, according to the *New York Times*, "At the conclusion . . . everybody rose and sang 'The Star Spangled Banner.'"[34] Roosevelt's case for expanding the armed might of the nation to match its imperial grasp found resounding endorsement in Carnegie Hall and also in the pages of the *New York Times*.

Roosevelt's aggressive nationalism, the prideful accounts of journalists, the thick celebratory books that spewed from printing presses tended to drown out the recognition of potential problems in the nation's new role in international affairs. For example, George Kennan, a journalist whose coverage generally supported the war effort, focused his attention on the medical care for ill and wounded soldiers in 1898 and found many inadequacies the food supply, logistical planning, and command decision making that were sufficiently widespread as to call into question the fitness of the U.S. Army to meet its new international responsibilities.[35] The press controversy that arose in August regarding the scandalous conditions in army camps in Cuba and the United States concentrated on the immediate need to save the lives and maintain the health of the soldiers, but Kennan's extensive critique pointed out systemic weaknesses that would persist after the alleviation of conditions in the camps. Admiral Mahan, a proponent of an even larger navy, warned that the Spanish fleet was hardly an up-to-date armed force; their boats were not comparable to the warships of a major world power and so going up against them did not constitute a legitimate test of the fitness of the U.S. Navy. And Mahan faulted Commander Winfield Scott Schley for his poor decisions in the crucial hours before the battle of Santiago. This significant tactical error by a high-ranking officer did not reflect well on the navy in Mahan's view.[36] Yet these and other telling criticisms did not negate the patriotic excitement following victory in battle and the nationalistic celebrations of imperial expansion. A generation later Walter Millis termed this mix of pride, propaganda, and patriotism "the martial spirit," a phrase that captured this amalgam of militarism and popular culture.[37]

The doubts Kennan and Mahan voiced in public forums were indicative of larger problems that confronted the leadership of the armed forces. The absence of a centralized command structure that could combine the operations of both the army and the navy was a glaring weakness, as evidenced in the case of the invasion of Cuba. The demands for more warships, a larger army, and additional practice maneuvers were responses to obvious shortcomings, but there were many practical but less publicized needs. The addition of transport and supply ships, the establishment of coaling stations in strategic areas (especially Latin America and the western Pacific), and the challenges of governing the new colonies (especially the Philippines) imposed heavy burdens. The creation of bureaucratic channels to coordinate the work of the two armed services and to bring together military officers and civilian officials required attention to detail and behind-the-scenes negotiations that were often slow moving and unproductive. In short, the excitement associated with victory and heroism in the Spanish-American War tended to obscure the less dramatic aspects of maintaining an empire and playing a role in global politics.[38]

COMMUNICATIONS AND NATIONAL POWER

Theodore Roosevelt's meteoric rise in politics was indicative of the political potential of the new information system. The nationwide coverage of his exploits in Cuba secured his political future in a matter of a few weeks if not days. Only three days after the engagement on San Juan Heights—on July 4—and again on July 6 and about a week later (probably July 12), Henry Cabot Lodge wrote to Roosevelt about his growing stature: "You have won yourself a high place already as one of the popular heroes of the war."[39]

While Lodge emphasized the role of cabled information in Roosevelt's rising status, the same system also garnered public attention for itself and its role in the assertion of national power by the government in Washington. In the early 1890s news coverage of cable telegraphy diminished as its operations became routinized and more reliable. The Spanish-American War, however, brought a new wave of public attention to the communications system. Newspapers across the United States explained the electronic nerve system that connected the White House, the War Department, and the U.S. forces around Santiago. The United States' rise to the position of world-class power was accompanied by numerous references to the telegraphic lines that enhanced and publicized this exercise of national might. News dispatches from Washington, Playa del Este (Guantanamo Bay), and Siboney Beach made clear that Lieutenant Colonel James Allen, the Signal Corps expert on the scene, had established the electronic ties between General Shafter, headquartered in Cuba, and General Adolphus Greely of the Signal Corps and President McKinley in the nation's capital. The press and the public depended on those slender copper wires.[40]

The timing for the establishment of these electric connections was fortuitous. They became the nerve centers for relaying the news of the July 1 victory over the Spanish. On Saturday, July 2, the press broadcast accounts of the U.S. triumph but in subdued and occasionally ominous terms. Datelined Washington, the story was based on General Shafter's terse telegram to the War Department: "Had a very heavy engagement to-day (July 1) which lasted from 8 AM until sundown. We have carried their outworks, and now in possession of them." The headlines employed phrases such as "Our Army Is Victorious" in the *New York Times* on July 2, but by the next day uncertainty had taken hold. The extra edition of the *Boston Globe* on the evening of the third emblazoned the headline, "In Doubt: Anxiety Everywhere at Lack of News from the Front," on its front page.[41]

The information system burst forth with the news of Sampson's July 3 defeat of Cervera on the following day. The *San Francisco Examiner* used exalted patriotic prose: "Here is a Story of glorious victory" in which "deeds of personal heroism marked the battle."[42] Doubts and anxieties about Shafter turned into a celebration of military superiority: "Shafter Demands Surrender," read the *New York Times* headline.[43] The *Atlanta Constitution*, the *Philadelphia Inquirer*, the *Boston Globe*, and the *New York Times* reprinted Shafter's telegram to the War Department: "Early this morning I sent a demand for the immediate surrender of Santiago, threatening to bombard the city. I believe the place will be surrendered."[44]

Alongside Shafter's telegram on page 1, all of the metropolitan newspapers published a succinct cable from Lieutenant Colonel Allen: "The destruction of Cervera's fleet is confirmed."[45] Sampson's telegraphic report sent from the Siboney station on July 3 was characterized by an appreciation for public relations that Commander Walker had spoken to in the context of the Panamanian expedition of 1885: "The fleet under my command offers the nation a Fourth of July present(—) the destruction of the whole of Cervera's fleet. Not one escaped."[46]

The work of the Signal Corps garnered favorable press notices. The *Literary Digest* featured contemporary coverage on submarine cables during the war, and the *Electrical World*, a publication for professional engineers, had two lengthy essays on the same subject.[47] The *New York Times* ran seven articles on cables in April and May 1898 and published two editorials that expressed approval of cable cutting in times of war.[48] Professor George Wilson of Brown University delivered a lecture at the Naval War College on the importance of submarine cables in international affairs that was punctuated by the observation that during the fighting in Cuba, exchanges of messages between Washington and U.S. forces near Santiago transpired in as little as twelve minutes.[49]

Lieutenant Colonel Allen and the Signal Corps held a prominent place in the news coverage of these events. The War Department, the White House, and the army and navy units on the Santiago front had become a part of the daily conversation of U.S. citizens.

COMMUNICATIONS AND THE CULTURE OF EMPIRE

The information system made up of telegraph wires and printing presses was the main conduit in the spread of the ideas and values that underlay the new imperial status of the United States. One of the leading proponents of U.S. expansionism was Henry Cabot Lodge, the Massachusetts senator. Lodge's 1899 book *The War with Spain* contained an account of the spread of the public's acceptance of the justification of the war. Although Lodge did not mention newspapers and the telegraph by name, their functioning was strongly implied in his conclusion that in the time after the destruction of the *Maine* on February 15 and before the decision for war on April 11, the information system had made possible what approached a national consensus: "Diplomats might plan, and twist, and devise, and exchange notes, and deal with all the forms so futile at a great crisis, but the American people had made up their minds that the only real and possible solution was the end of Spanish rule in Cuba." Given the tense circumstances, this conclusion led to war.[50]

The cable network continued to supply the press and publishing houses with information and opinions that extolled the victory over Spain and, in the aftermath, endorsed the concept of an overseas empire for the United States. The formal annexation of the Philippines sparked controversy, but there was only scattered opposition in the press to the pursuance of commercial expansion and the construction of more powerful armed forces. The decision to take the Philippines became an extreme position—the establishment of a formal empire—within the larger discussion of the new role of the United States as a world power. Two well-known newspaper magnates weighed in on this debate. Whitelaw Reid, owner of the *New York Tribune*, was an active Republican who had been Benjamin Harrison's vice-presidential running mate in the unsuccessful campaign of 1892 and a member of the U.S. commission that negotiated the Treaty of Paris ending the war with Spain. Reid and the *Tribune* were to some extent typical of most of the ten big-city newspapers central to this study. The *Tribune* was supportive of McKinley and Roosevelt and the policies that placed the United States on the path to a place among the world's powers.[51]

More of a maverick was Henry Watterson, editor of the *Louisville Courier-Journal* and a partisan Democrat who, before 1898, opposed the idea of expansion and the growing war sentiment in the United States. But Watterson underwent a transformation in the first six months of 1898 and by war's end was an advocate for an activist foreign policy, commercial expansion, and the creation of a formal empire. Like Republicans Reid, McKinley, and Roosevelt, Watterson found expansionism to be consistent with the ideals of the founding fathers. His editorials in the *Courier-Journal* stressed that a new era was dawning in world history and that the United States had to adjust to it. Although his position put Watterson at odds with some Democrats and Democratic-leaning newspapers, the Kentucky editor insisted that his purpose was to pull his party to a position that made it possible to claim the expansion issue as part of its own political arsenal.[52]

Watterson, Reid, and other newspaper voices produced information and shared opinions that reached millions of readers, but probably the most widely read prose on the war was a relatively short (about 1,400 words) essay by Elbert Hubbard, a journalist and promotional writer. Hubbard's "A Message to Garcia" was a didactic tale of personal initiative and endurance that was printed in the mass-circulation *The Patriotic Reader for Seventh and Eighth Grades and Junior High School* and, in other formats, adopted as required reading by the U.S. Navy, the U.S. Marines, and the U.S. Military Academy at West Point. The central character was Lieutenant Andrew S. Rowan of the U.S. Army, who received orders directly from President McKinley on April 9, 1898, to go to Cuba, to find General Calixto García and to make arrangements for military cooperation with the general's rebel army. According to Hubbard's version, Rowan did just that *on his own*. He met García and returned to the United States with the necessary information.[53] As is often the case with morality tales composed for educational purposes, Hubbard's story contained several factual flaws. Rowan relied on help from U.S., British, and especially Cuban forces throughout his trip. And there was the irony of the United States seeking and receiving help from García in April 1898 and then snubbing and insulting the same general at the Spanish surrender in Santiago in July of that year.[54] The main point of relevance to this study is that Rowan exemplified the characteristics of initiative, courage, and dedication to duty as part of the larger effort that brought the United States victory in Cuba, Puerto Rico, and the Philippines and entry onto the world stage as an international power. Like McKinley, Watterson, Reid, Lodge, Mahan, and Roosevelt, Rowan, as well as Hubbard through his didacticism and distortion, contributed to the popularization of martial triumph and imperial power.

McKinley personally ordered Rowan to carry the message to García, testimony that the transmission of messages was vital to the successful prosecution of the war. Rowan, no matter which version of his exploit is consulted, did his duty the hard way: he traveled by boat, on horseback, and on foot, taking several weeks to complete his task. The telegraph cable, by contrast, delivered messages to their destination in a matter of hours or even minutes, if the proper connections were available. Captain George O. Squier, a veteran of the Army Signal Corps, invoked his war experience in advocating the upgrading of the nation's military communications system. In Squier's view the outcome of the war was "largely a story of 'coal and cables.'" By that he meant that the U.S. Navy was able to corner and defeat Admiral Pascual Cervera's fleet in Santiago Bay because the Spanish were unable to notify their Atlantic fleet of the presence of a nearby supply of coal, information that would have allowed these vessels to return to Spain. If Cervera had followed this course, the threat of the Spanish fleet in the Atlantic would have served as a deterrent to the U.S. invasion of eastern Cuba and, according to Squier, the war would have been prolonged, leading to a much more costly outcome for both sides. Squier sought to amend Mahanian doctrine by making the point that submarine cables were crucial to naval operations: "It may be said, therefore, *that the very foundation of successful naval strategy is efficient and exclusively controlled communications, and the lack of them more serious than inferior ships.*"[55]

Squier cited the extensive submarine-cable system controlled by the British as the preeminent example for the United States to follow in developing a coordinated defense for its overseas territories and trade routes.

Although most of these cables were owned and operated by the Pender family of private companies, the British government had privileged access to them and in time of war, the British armed services would take over the cable offices. Squier pointed out that the armed services already operated cable and landlines in Cuba, Puerto Rico, the Philippines, and Alaska, and that this system could be expanded and linked together. He projected the cost of this communications network to be approximately $25 million, one-eighth the estimated cost of a canal across the Central American isthmus: "For the expense of three or four first-class battleships, the United States can provide herself with the most powerful means known for extending and preserving her commercial influence and for the speedy pacification and civilization of the people who have recently come under her control, and can secure a strategic advantage—military, naval, and political—which is necessary for her position as a world power."[56]

Squier's essay was not merely an exercise in abstract strategizing; it appeared in the pages of the *National Geographic Magazine*, a glossy-paged publication that reached a mass audience.[57] Nor was Squier alone in the promotion of improvements in military and naval communications technology. The head of the Signal Corps, General Adolphus Greely, published "The Signal Corps in War-Time" in *Century* magazine in 1903, and journalist Henry MacFarland wrote an article on this topic for the *Review of Reviews* in 1899.[58] These publications on military communications reached the public and the nation's highest legislative body at a propitious time. The combination of the strategic concerns highlighted by Squier and the ongoing war in the Philippines (1899–1902) led to a major expansion of the Signal Corps from 10 officers and 50 enlisted men at the time of the outbreak of the war with Spain in 1898 to 35 officers and 860 enlisted men by 1903.[59] The call for a submarine cable in the Pacific Ocean did not fare so well, in spite of the endorsements of McKinley and Roosevelt. Congress refused to act and left that field to the private sector, which was dominated by an agglomeration of British and U.S. firms. Nevertheless, the operations in the Spanish-American War and the discussion of strategy in its aftermath elevated the public perception of electric communications and turned such public perception into an important factor in the context of international power politics, leading the world's major nations, including the United States, to expend additional resources developing this ever-changing field of technology.[60] To boost their plans for expansion, Captain Squier, General Greely, and their supporters brought public attention to their cause.

POPULAR CULTURE AND IMPERIAL POWER

Squier and Greely wanted a secure cable network to convey intelligence reports and command decisions, but the popular press used the cables for entirely different purposes. The new information system was open to inputs that moved out into public view via the broad freedom of expression available across the United States. The acceleration of messages by the 1890s both fed on and found reinforcement in the aggressive nationalism of the United States. This section examines the cultural assumptions embedded in this electronically evoked nationalism.

The larger strategic discussions that reinforced imperialism mingled with journalistic portraits of heroism and battlefield victories intended to displace criticisms and doubts. The newspaper press tied together by the electric telegraph joined with magazine and book publishers to convey these messages of patriotism and power to a receptive public. The mood of the nation had been primed to receive these messages. Civic leaders, newspaper editors, politicians, educators, and reformers had orchestrated rehearsals for this patriotic concert over the course of the previous decade. In 1888 the "schoolhouse flag" movement came into being, its announced purpose being to raise the nation's flag over every school in the land. This effort, intended to heal the social and political wounds of the Civil War, coincided with the assertion of national power in the 1890s. The confrontation with Chile of 1891–1892, the run-in with the British over the Venezuelan boundary in 1895–1896, and the Spanish-American War of 1898 marked an upward trajectory for manifestations of national pride. The nation's telegraph network supplied stories for newspapers from New York to California about "schoolhouse flags" that were preliminaries for the Columbian Public School Celebration of 1892, an event that combined the commemoration of the first voyage of Christopher Columbus to the Americas with nationalism in the classrooms across the United States. Responding to a wave of newspaper editorials, political speeches, and public meetings, President Benjamin Harrison proclaimed October 21, 1892, to be a day of national celebration on which the Stars and Stripes were to fly over every schoolhouse in the land. Six years later the nation celebrated its martial victories and a new empire.[61]

The metropolitan press played a large role in rationalizing aggressiveness and in formulating popular images that symbolized this surge of nationalism. One of the most repeated justifications for the Cleveland-Olney stand against the British in 1895–1896 was that put forth by the Andrade-Scruggs duo—the Monroe Doctrine. The reformulated doctrine projected the United States as the hegemonic power in the Western Hemisphere. This version of the Monroe Doctrine was well received by most newspaper editors and politicians. The overwhelmingly favorable responses to the coverage of this issue by editors and politicians indicated that the new, expansive version of the Monroe Doctrine was widely popular.[62]

The Monroe Doctrine carried with it the pedigree of a long-standing historical status, but other justifications for it grew out of disdain, mistrust, and even fear of foreign nations. The confrontations with Chile inspired a mixture of belligerent boastfulness about defeating the highly regarded Chilean Navy with derogatory comments about the "bad qualities" of that nation's population. The U.S. press heaped mockery and contempt on the British and their empire in 1895, while it listed warships, accompanied by illustrations, that exemplified the might of the largest navy in the world. By 1898, three years of verbal assaults on Spain and its imperial policies in Cuba had come to fully focus on the person of Valeriano Weyler. The U.S. press saw Germany as a rival in the Western Hemisphere even though Berlin had limited goals in this part of the world. The *New York Times*, other big-city newspapers, and *Harper's Weekly* publicized the activities of Germany's new war vessels in the Caribbean in the 1890s. Their perceptions of this threat were laden with warnings of how Kaiser Wilhelm's aggressive nature could lead to German colonies being planted in the backyard of the United States.[63]

These pugnacious responses to the threats (imagined and real) from Chile, Britain, Spain, and Germany were contemporaneous with expressions of disdain and sometimes contempt for several other Latin American nations and peoples. The Venezuelans took the initiative in promoting the Monroe Doctrine in defense of their boundary but found themselves shunted aside in the press coverage of the final settlement between Washington and London. The U.S. newspapers exhibited a blatant inconsistency in their coverage of the Cuban revolution. In the 1895–1897 period, Antonio Maceo, an Afro-Cuban, appeared in the pages of these newspapers as "the Bronze titan," a hero of the rising of the Cuban masses against Spanish tyranny. After U.S. forces invaded the island these same Cuban patriots became a disheveled, untrustworthy lot whose machetes, clubs, and antiquated firearms had somehow become the implements of a ragtag army that did not deserve a place at Spain's formal surrender. The Cuban as troublesome riffraff became a typical stereotype in the U.S. press in the late 1890s. News coverage, editorials, and editorial cartoons were based on racist assumptions and distorted images.

These images strengthened feelings of national, racial, and cultural superiority in the United States and rationalized projections of its strategic and commercial hegemony in the Americas. Much press attention concentrated on the Teller Amendment, which disavowed any imperial motives on the part of Uncle Sam in the Spanish-American War by prohibiting the annexation of Cuba as a colony. The leaders in Washington, in accord with their chivalric self-concept and the ideals of noblesse oblige, would grant to the Cuban people their independence (this reasoning did not apply to the Philippines, which was annexed as a colony). The formal status of Cuban independence did not prevent the informal expansion of U.S. commercial interests on the island, however. Encouraged by the four-year military government in Cuba, U.S. entrepreneurs and investors poured their resources into the island, rebuilding the railroads, sugar mills, port facilities, and urban infrastructure. Cuba, in effect, became an economic appendage of the United States. This rapid expansion of U.S. interests in Cuba followed the pattern described by journalist-novelist Richard Harding Davis in his popular work of fiction *Soldiers of Fortune*. Davis's protagonist, mining engineer Robert Clay, symbolized the technical expertise and political savvy that would secure domination over the lands of the circum-Caribbean and much of South America. William Eleroy Curtis in his rather stuffy writing style promoted what Davis described in his novel. The opportunities outlined by Curtis appealed to the U.S. business community but was hardly the stuff of sensational derring-do. The mechanics of informal imperialism outlined by these two advocates rarely made the front pages of the metropolitan press. It took a fictional dramatization in *Soldiers of Fortune* to grab public attention.

Neither the engineering of copper mines nor the building of railroad trestles across mountain gorges, difficult and dangerous though they were, could hold public or press attention on the same level as the projection of U.S. strategic and military might across the oceans could. Roosevelt's grand strategy, in accord with Mahan's vision, based expansion on the readiness to engage in combat. The first step in armed preparedness was a strong navy. Roosevelt, Mahan, and their like-minded expansionists such as Henry Cabot Lodge never pulled back from their position that the nation's character and credibility were determined by its willingness to fight.

A stern, clean-cut "Uncle Sam" prepares to punish Chile while President Benjamin Harrison handles communications (*Washington Post*, January 19, 1892, 1).

"Disfigured," disheveled British lion displaying bandaged wounds from 1776 and 1812 while clutching the Venezuelan rabbit and confronting the aroused U.S. eagle bearing the national shield and the revered Monroe Doctrine (*Chicago Tribune*, December 20, 1895, 1).

This 1895 cartoon, titled "Uncle Samuel Getting the Old Gun Ready," gives the icon an aggressive, sinister appearance (*Chicago Tribune*, December 24, 1895, 1 [from the *New York Journal*]).

Uncle Sam and his dynamite gun "Monroe Doctrine" aimed at British interloper (*San Francisco Chronicle*, December 15, 1895, 16).

Spain's minister to the United States Enrique Dupuy de Lôme punishes himself in this mocking image (*Chicago Tribune,* February 11, 1898, 7).

The figure of U.S. consul Fitzhugh Lee assumes gigantic proportions as he departs Cuba to receive a hero's welcome in the United States (*Chicago Tribune*, April 10, 1898, 3).

EASTER GREETINGS TO THE PEARL OF THE ANTILLES.

Anti-Spanish sensationalism is contrasted with a softly feminine version of the arrival of Freedom and the achievement of Cuba Libre (*Chicago Tribune*, April 9, 1898, 1).

Highly idealized illustration of the U.S. landing at Daiquirí or Siboney (James Rankin Young, *Thrilling Stories of the War and Returning Heroes* [Chicago: Providence Publishing Company, 1898], 96).

Uncle Sam's wide-eyed, boastful expression captures the nation's celebratory patriotism (*Philadelphia Inquirer*, July 6, 1898, 1).

The world watches the U.S. victory off Santiago (*Boston Globe*, July 8, 1898, 4).

IMPERIAL IMAGERY

Roosevelt's predisposition to fight for the sake of national pride and expansion was reflected in the pen-and-ink illustrations that became increasingly common in newspapers across the country. As early as the 1891–1892 Chilean crisis, newspapers such as the *Washington Post* and *Chicago Tribune* featured a determined Uncle Sam with whip in hand, ready to take on the "troublesome" Chileans.[64] These illustrations, often called political cartoons, proved popular with readers and flooded newspaper pages during the Venezuelan boundary controversy. They were larger and more dramatic than previous illustrations and marked a turning point in the use of visual hyperbole. Although the artists used a variety of images to convey their messages, two types of illustrations are most relevant to this study. One featured depictions of gestures or actions that demonstrated the aggressive attitude of the United States, often represented by the character Uncle Sam. The second type of illustration was a realistic display of weapons such as artillery or naval warships. In these images the message was clear: the United States had the weapons and the will to use them.

The aggressive-gesture cartoon had its own range of assertiveness, spanning the spectrum from a picture of President Cleveland indicating to Prime Minister Salisbury that he should drop to his knees to honor the Monroe Doctrine in the December 18, 1895, issue of the *Washington Post* to a long-armed Uncle Sam shoving his fist in the face of the personification of Britain, John Bull, who was scrambling on his hands and knees to defend his claims on a large-scale map of Venezuela and the Caribbean in the *San Francisco Chronicle* of December 23, 1895.[65] Perhaps the culmination of these depictions of U.S.-British confrontation came in the *Chicago Tribune* of December 20, 1895, which contained an illustration of the U.S. eagle, drawn to it full height, screaming at a scruffy, disheveled British lion with appropriately labeled band aids covering the wounds of 1776 and 1812.[66]

While images of assertiveness appeared in masculine form, the symbolic representations of Cuba often took on a feminine physique. The *Chicago Tribune*'s illustration of Cuba Libre was a demure, long-haired maiden who waits on bended knee to receive the newly born embodiment of Freedom, also conceptualized in feminine form.[67] As Louis Pérez, Michael Hunt, Kristin Hoganson, and Amy Kaplan have concluded, the feminization of the Cuban cause fit the chivalric metaphor that was dominant in romance novels of the day, which were widely admired in U.S. popular culture. Pérez and

Hoganson provide revealing selections of political cartoons that exemplify the various visual metaphors prevalent in the press.[68]

Pen-and-ink illustrations of hegemonic gestures reappeared during the confrontations between the United States and Spain. The *Philadelphia Inquirer* depicted an inept, miniaturized Enrique Dupuy de Lôme being inelegantly dropped down the diplomatic stairs, while the *Chicago Tribune* pictured an amused Uncle Sam watching the allegedly masochistic Spanish minister switch on an electric-powered kicking machine that applied a series of blows to his rump.[69] Enjoying the defeat of Spain in July 1898, the *Inquirer* cartoonist portrayed Uncle Sam with his "tally sheet" listing the Spanish warships sunk by the U.S. Navy.[70] The *Boston Globe* carried a cartoon of Uncle Sam celebrating victory with a fireworks display that includes the message "Remember the *Maine*" in electric lights.[71]

Illustrations of weaponry during the Venezuelan crisis tended to concentrate on the navy, which had recently undergone expansion. The new, sleek battle cruisers bristling with high-powered guns and cutting through the water at high speed appeared in many illustrations. The remodeled protected cruiser *Chicago* was something of an exception with its three old-fashioned masts on display, but this ship was only one of over a dozen presented in detailed sketches in newspapers such as the *Chicago Tribune*. Also in the series was the state-of-the-art "first-class steel-armed cruiser *New York*," which had gun turrets, three funnel smokestacks, and a modern bridge.[72] The *San Francisco Chronicle* published a large cartoon that summarized the potential conflict with Britain. A cocky Uncle Sam leans against a large naval weapon pointed directly at the head of John Bull at close range and comments, "Our national dynamite gun—it's a long time since it's been tested but it will no doubt prove as efficient as of old." The gun carries the label "Monroe Doctrine" in large print.[73]

Illustrations of naval weaponry became more varied and more numerous in 1898. Perhaps the best known are the renderings of the ill-fated *Maine* that occupied the front pages of Pulitzer's *World*, Hearst's *Journal*, and many other newspapers in February. In the *Journal*'s illustration, which was headlined "Destruction of the *Maine* Was the Work of an Enemy," a mine sits beneath the *Maine*, with submerged cables leading from the explosive to the shore. The *World* featured the moment of the explosion, depicting human bodies and metal fragments hurtling through the air.[74] Several metropolitan newspapers carried illustrations of the *Maine* before the explosion, reminding the public of its sleek lines, modern construction, and seeming invulnerability.[75] This appearance of the *Maine* intact produced a jarring contrast with the *World*'s image of its fiery destruction.

Naval vessels figured prominently in newspaper illustrations throughout the weeks of waiting for the location of Cervera's fleet to be discovered. The depiction of the five battle cruisers of the Flying Squadron on maneuvers outside the port of Hampton Roads unrealistically places these ships in a close-packed battle line in order to project an aura of naval power that appealed to newspaper readers during this uncertain period.[76] Hobson's courageous but vain attempt to block the exit of the Spanish force from Santiago harbor received extensive attention. The *San Francisco Examiner* included an illustration of Hobson's feat imposed on a map containing details of the shoreline.[77] The celebrations of victory over the Spanish were epitomized in the *Atlanta Constitution*'s bracketing the news lead with pictures of the cruisers *New York* and *Brooklyn* steaming forward into the face of the reader.[78]

Illustrations ranged from humorous pen-and-ink cartoons to artistic renderings with a touch of aesthetic quality. The *Minneapolis Journal* published a compendium of illustrations by Charles Bartholomew that spanned 1898 and included patriotic representations of martial might often centered on the symbolic figure of the lean, long-legged Uncle Sam. Bartholomew even parodied himself and other editorial artists with an illustration that mocked the overuse of the image of Uncle Sam.[79] At the other end of the visual arts spectrum, Frederick Remington created portraits of training exercises, scenes of combat, and the aftermath of battle for magazines such as *Harper's Weekly* and *Collier's Weekly* as well as the *New York Journal*.[80]

Magazines and books added to the imagery of the United States as a military and naval power. *Harper's Weekly*, *Collier's Weekly*, and *Leslie's Weekly* extolled the accomplishments of the armed forces, emphasizing rifles, bayonets, artillery, cavalry, and warships.[81] Books added new dimensions to the publication of visual material. Their slick pages made use of photographs and applied color in both artists' paintings and touched-up photos. Two volumes are notable for the incorporation of these media. Parkerson's *How Uncle Sam Fights* included seventy-six illustrations, apparently all of them photographs. James Rankin Young and J. Hampton Moore's *Reminiscences and Thrilling Stories of the War by Returned Heroes* included seventy-two glossy-paged illustrations, seven of them in color.

Illustrated books on the Panama Canal were also popular. Photographs of huge Bucyrus steam shovels, railroad cars moving tons of mud and rocks, and the massive walls and locks of the canal itself juxtaposed with scenes of dense jungle vegetation, rocky mountains, and the ruins of the halted French project spoke volumes on the accomplishments of U.S. engineers, the management team, and, not incidentally, Roosevelt and his administration. Author Joseph Pennell's *Pictures of the Panama Canal* enjoyed brisk sales and went through six editions in 1912–1913.[82] Abbot's *Panama and the Canal in Picture and Prose* included over five hundred black-and-white photographs and several painted color illustrations. Page 125 is a large reproduction labeled simply "The Man and the Machine," in which Theodore Roosevelt perches himself in a commanding position in the "monster steam shovel," the Bucyrus. On page 130 appears a fuzzy photograph of Roosevelt standing beside President Amador Guerrero of Panama, an unintentional reminder of the shadowy roles played by these two men in the plot to dismember Colombia and secure the canal zone for the United States.[83]

CONCLUSIONS

Through words and images the new technologies of cable communications and the mass press of the United States in the 1890s and early 1900s disseminated a set of cultural and political assumptions that rested on some of humanity's most basic and primitive values. To express this phenomenon in Hoganson's terms, "American manhood" carried the day. Nearly instantaneous messages described diplomatic confrontations, military and naval combat, and engineering feats. The emotional excitement was centered on organized aggression intended to expand the authority, territory, and prosperity of the nation that had recently reunited in the years following the Civil War. The electric information system made possible a nationwide outpouring of press opinion and public sentiment that would challenge cautious diplomats, heads of state, and corporate managers for decades to come. Yet the electric information system was not especially accessible to the members of another growing movement—those involved in the anti-imperialist struggle in the United States and Latin America. These criticisms were directed mainly against the United States and its propensity for military intervention and desire to expand corporate control of resources and investments in the region.[84] The editors, reporters, and managers who operated this information system chose to accentuate the popular tales of heroism and the exertion of national power and to largely ignore the protests against imperialism until well into the twentieth century.

Conclusion

The Ominous Triumph of Popular Culture

THE CREATION OF THE NEW information system at first seemed to offer many advantages to powerful institutions and few opportunities for ordinary citizens. Profit-making companies operated cable lines, and publishers printed newspapers. Governments and private corporations—especially newspapers—had the financial wherewithal to send messages. Ordinary citizens did not. The system seemed destined to reinforce smooth functioning of the existing political and social power structures. Such an early assessment of the new information system, however, would have proven incorrect. Governments and the press were soon pitted against each other, and the outcome of that struggle was unanticipated. And the meek and the lowly—especially the mass of the newspaper-reading U.S. public—acquired an influence in the flow of information and the formulation of policy that surprised and troubled government leaders.

PRIVATE-SECTOR PROFITABILITY

The growing political and cultural influence of the new information system was reflected in the profitability of its component parts. The telegraph companies and newspapers were private enterprises that enjoyed healthy profit margins during the last years of the century, including the lengthy depression of the early 1890s. Viewed from this angle, the information flow was supported by market forces that reflected the interests of the general public in the United States. The public purchased the newspapers that in turn relied on the international cable companies for news reports. These cable companies had depended on mining and shipping companies in the 1870s and 1880s, but the reduction of cable rates in the 1890s coupled with political and diplomatic crises and the Spanish-American War meant a large increase in telegrams for news reportage. For the international cable companies, these profits marked an improvement over the struggles of their early years. Eight of the cable companies that operated in the Caribbean and South America were headquartered in Britain, and their financial records were good to excellent with only a couple of exceptions (West Indian and Panama and West Coast of America). A small corporation with only £220,000 of initial capital investment, Cuban Submarine averaged £36,000 in annual profits, a 16 percent return on invested capital, over the period from 1890 to 1899. Brazilian Submarine, a more sizeable firm, had an annual average of £152,000 in profits during that decade, a 10 percent return, earning a very healthy £180,000 for 1899, a 12 percent return.[1] Even the phlegmatic West India and Panama experienced a boost in revenue as a result of the Spanish-American War. Its profits rose from about £30,000 per year from 1890 to 1897, approximately a 2.3 percent return, to £45,000 in 1898, a 3.6 percent return.[2] The full records of James Scrymser's companies are not available, but it is clear that his Central and South America telegraph company averaged a net profit of $480,000 from 1881 to 1891, earning a 6.1 percent return.[3] The larger transatlantic companies that connected the United States and Europe were also on a prosperous footing by the 1880s, including Western Union, which purchased two of these cables in 1881.[4] A powerful national monopoly, Western Union was one of the most prosperous corporations in the United States. Its gross revenues nearly doubled from 1880 to 1900, from $12.7 million to $24.8 million, and its profits were steady in spite of the depression of the early 1890s, reaching $5.8 million in 1889 and $6.2 million in 1900.[5]

One of the best customers for the cable companies was the Associated Press, a news-gathering cooperative supported by newspapers throughout the United States. Metropolitan newspapers received most of their coverage of international and national events from the AP, special news services, or their own foreign correspondents, all of whom depended on the electric information system. To pay for these expensive sources and the installation of new printing technology, newspapers came to rely on advertising as their main source of revenue. Income from advertising amounted to 44 percent of total revenue in 1879 and rose to 54.5 percent in 1899. The prosperity of the metropolitan press can be gauged from the increase in newspaper advertising revenue in this period: $39 million in 1880, $71 million in 1890, and $150 million in 1900. By the end of the century most big-city newspapers operated as profit-seeking and generally profit-making enterprises (with the prominent exception of Hearst's *New York Journal*). The huge expenditures for the coverage of the Spanish-American War were exceptional. The business manager, circulation manager, and advertising manager became important figures in the day-to-day routine of the large metropolitan newspapers.[6]

THE MISLEADING PROMISE OF THE NEW SYSTEM

This privately owned, profit-generating information system became an essential conduit for government messages. A central question in this study concerns who made the more effective use of this system—newspapers or governments? There are indications that in the 1870s and 1880s, governments made best use of this new tool. The British navy's gunboat *Niobe* responded to the emergency cable from Santiago, Cuba, in 1873 to put a stop to the murderous reign of General Juan Burriel in that Spanish colony. This telegram saved lives, and soon a series of telegrams between Madrid and Washington minimized international tensions and averted the war that may have resulted from public outrage in the United States. Secretary of State Hamilton Fish and Spanish diplomat José Polo de Barnabé calmed the turgid diplomatic waters through cable connections across the North Atlantic. Eight years later Chile and Argentina transcended their rivalry with the help of the good offices of U.S. diplomats. The negotiators used the recently completed transcontinental telegraph lines to conclude a treaty that fixed their boundary along the Andes. The U.S. ministers in Buenos Aires and Santiago received approbation from Secretary of State James G. Blaine in Washington for their work. The administration of James Garfield could take some credit for the "wire treaty" negotiated six thousand miles to the south.

In 1885 another type of crisis erupted in Panama, when itinerant laborer-turned-revolutionary Pedro Prestán led an outburst of unrest among the unemployed that resulted in the burning of the port city of Colón. The cable lines of West India and Panama, International Ocean, and Central and South American fed the news of the Panamanian insurrection to Western Union, and in a matter of ten days President Grover Cleveland dispatched marines and sailors to the isthmus and brought the revolution to an end.

Chile and the United States confronted each other in 1891–1892 in what amounted to the culmination of a quarter century of tensions and misunderstandings. Sparked by the deaths of two U.S. sailors on shore leave from the cruiser *Baltimore* in Valparaíso and magnified by the impatience of President Benjamin Harrison's administration with the pace of the Chilean judicial system, this crisis came to a head when Harrison demanded a decision on the "*Baltimore* affair." Harrison gave the Chileans barely two days to respond—a demand made possible by the new communications system. For several weeks prior to the issuing of Harrison's demand, newspapers in the United States had fanned the flames of martial aggression. This combination of patriotic extremism in the press and belligerence in the White House forced the Chilean government to capitulate to Harrison.

In these four cases, which seem to point to the rising trajectory of U.S. power in the Americas, four different presidential administrations used the international communications system to assert the authority of the United States. In 1873 Spain agreed to pay an indemnity for the deaths in Cuba. Chile and Argentina followed the friendly advice of U.S. diplomats to resolve their boundary disagreement in 1881. In 1885 Pedro Prestán fled Colón in the face of the U.S. intervention and, after his capture by Colombian armed forces, underwent an abrupt trial and immediate execution in Panama. Seven years later the Chilean government submitted to U.S. demands for an apology and financial compensation to the families of the two dead sailors. These cable-facilitated victories for the United States might lead to the conclusion that the new information system brought the greatest benefit to the most powerful nation involved in the crises.

FOREIGN INPUTS AND UNANTICIPATED RESULTS

Such a conclusion, however, is not warranted by the evidence in five other case studies. Key to understanding these events is the ease in accessibility to the information system. It is true that telegrams were expensive through the period under study, but there were ways of placing selected, strategically important messages into the system that were not prohibitively expensive. As the cable system spread, more stations were opened, including, by the 1880s, one in Santiago, Chile, and one in Caracas, Venezuela. And finally, the open communications system in the United States made it possible to insert printed material, such as pamphlets or newspaper articles, into the information flow. In the cases under examination, the protagonists were not agents of the U.S. government or any other powerful nation, but representatives of less influential countries and nascent political movements.

Chile's Marcial Martínez was one of the most adroit diplomats in the use of the information system. He chose the traditional print media to defend the interests of Chile in its dealings with the United States. His first task was to attack the argument that Chile, in league with the British, was violating the Monroe Doctrine through its control of the nitrate-rich territory recently won from Peru and Bolivia in the War of the Pacific. Chile faced humiliation and defeat as a result of the heavy-handed diplomacy of the United States. The corrupt practices of the Gilded Age politics of the United States spilled over into the diplomacy of Secretary of State James G. Blaine, who attempted to protect the claims of U.S. corporations in Peru by denying the nitrate properties to Chile. Martínez assembled and circulated a pamphlet that insisted that charges made by Peru and its advocates in the United States distorted the Monroe Doctrine for the sake of greedy speculators and their political cohorts, including Secretary Blaine. While this issue and Blaine's role in it were discussed in the U.S. Congress and press, Martínez contributed to the debate by circulating his pamphlet, thereby introducing the Chilean perspective into the information flow in the wire services and onto the pages of newspapers. Martínez exulted when Blaine's diplomatic and political maneuvers failed. Chile held on to the disputed territory.

Martínez used the information system to defend his country's interests. Mexican political rivals likewise employed the wires and press to air their own differences to an audience in the United States. The government of Porfirio Díaz courted U.S. corporations in order to entice investments into Mexico. News of political instability in Mexico carried the threat of stock liquidations and corporate withdrawals. Catarino Garza challenged Díaz by starting a revolution on the Mexico-Texas border. A clever propagandist, Garza gained access to the U.S. news media through local newspapers in Texas, an interview with the *New York Times*, and an attention-grabbing letter to President Harrison. Garza's tactics worked for a few weeks. His movement received prime—sometimes front-page—coverage. The Díaz administration fought back mainly through the media-savvy diplomat, Matías Romero, whose interviews, speeches, and articles rebutted Garza's propaganda by reassuring investors that Díaz remained in charge. Garza's revolt soon collapsed. U.S. newspapers abandoned the story, but this episode remains important for two reasons: Garza and Romero's use of the U.S. press and the spectacle of Mexicans engaging in a political confrontation via the U.S. information system. Propinquity and telegraphy had created a new level of political intimacy between the two countries.

In spite of their clever use of the print media, Garza and Romero did not capture the center stage in the newspaper press in 1891–1892, but two years later the unlikely duo of José de Andrade and William Scruggs created a propaganda campaign that touched a sensitive nerve in the U.S. body politic and soon became the subject of a major international controversy that stretched across four continents. Andrade, a Venezuelan diplomat, who, like Martínez, was conversant with the intricacies of U.S. politics, joined with diplomat-turned-publicist Scruggs to use the old-fashioned print culture to produce a pamphlet that defended Venezuela's territorial interests by portraying the British Empire as a violator of the Monroe Doctrine. Andrade and Scruggs, supported by Venezuelan foreign minister Ezequiel Rojas, sent *British Aggressions in Venezuela: The Monroe Doctrine on Trial* to politicians and newspaper editors across the United States and saw the gist of their story of a diplomatic disagreement portrayed as a morality play enter the electronic-communications network. Subsequent public pronouncements, military and naval preparations, and diplomatic exchanges became the stuff of front-page news for over a year. Prime Minister Salisbury and Ambassador Pauncefote were surprised and alarmed at the intensity of anti-British feelings expressed in U.S. newspapers and the U.S. Congress. Rojas, Andrade, and Scruggs entered an opening that Martínez, Garza, and Romero had probed. Their efforts in the short term were effective and at times bordered on the spectacular.

Andrade and Scruggs engineered a successful publicity campaign built on a pamphlet, but the Cuban junta employed a diversified media approach to elicit public support for the revolution in that Spanish colony. Their audiences were not only in New York, Chicago, and other big cities but also smaller cities and towns across the United States. The best-known media implants developed by Tomás Estrada Palma, Gonzalo de Quesada, and other activists appeared in the pages of Hearst's *New York Journal* and Pulitzer's *New York World*, and that duo also gained flattering biographical sketches in the champion of objectivity, the *New York Times*. The junta did not limit itself to print and the wire services. With ample funding, the junta organized public "sympathy meetings," carnivals, and theatrical productions in communities from Ohio, Indiana, and Illinois to Florida and Louisiana. Spanish general Valeriano Weyler unintentionally aided their cause with his censorship of U.S. reporters. The U.S. press was caught up in the excitement of the anticolonial struggle in Cuba. The melding of this propaganda campaign with press sensationalism produced a wave of publicity favorable to the junta.

The beleaguered and soon-to-be discredited Enrique Dupuy de Lôme, Spain's ambassador to Washington, fully grasped the punch of this campaign that was pushing public opinion and political leaders—including the reluctant President McKinley—in the direction of intervention. In the same letter that contained the infamous slur against McKinley, de Lôme called on the Spanish government to launch its own countervailing publicity campaign in the U.S. press. His recommendation went unheeded, and he left Washington in disgrace. The Cuban junta seemed to triumph at every turn in its exploitation of the information system.

Philippe Bunau-Varilla's efforts to persuade the U.S. government to build the transisthmian canal in Panama did not inspire a surge of support similar to that provoked by junta, but the wily Frenchman did use the print media to help convince the U.S. Senate. Bunau-Varilla's pamphlet *Panama or Nicaragua?* and his publicity campaign that featured the distribution of a Nicaraguan stamp displaying the image of a smoldering volcano aided the Roosevelt administration in building public and senatorial support for the location of the canal in Panama even if it meant the dismemberment of Colombia.

These five examples indicate that the spokesmen for Chile, Mexico, Venezuela, the Cuban revolutionary junta, and the French-owned Panama Canal Company were easily able to penetrate the U.S. information system. Their messages reached popular audiences in the case of Venezuela and the Cuban junta and smaller but politically important audiences in the case of Chile, Mexico, and the Panal Canal Company. These successes, however, should not obscure the long-term outcomes. In all five instances, Washington followed policies that were in its own national interest. Chile blocked U.S. meddling on the west coast of South America for a decade, but by the early 1890s the United States asserted itself in the *Baltimore* affair, much to Chile's dismay. The Garza-Romero exchange had little impact on U.S. policy toward Mexico, which meant that investments continued to flow southward from Wall Street. Bunau-Varilla's publicity campaign amounted to an invitation to assume a dominant role in the newly formed nation of Panama by building and operating the canal.

The protests against the emergent imperial power of the United States only fleetingly broke into the nation's information system. José Martí and Manuel Ugarte made compelling arguments against the assertions of U.S. might within the hemisphere but gained only limited exposure in the press. The U.S. annexation of the Philippines inspired a well-publicized debate in the U.S. Senate, and the subsequent Filipino insurrection brought a controversial military response from Washington. But the Roosevelt administration held firm in its expansionist policies. The anti-imperialists in the Philippines, Latin America, and the United States were frustrated. Only in the 1910s and 1920s did anti-imperialist protests break out of the narrow confines of specialized magazines and academic circles to gain the attention of the mainstream press.[7]

In contrast to the anti-imperialists, the Andrade-Scruggs duo and the Cuban junta had successes, but they proved to be short term. The impressive achievements of Andrade-Scruggs and the Cuban junta in the U.S. information system were ultimately undone. Venezuelans celebrated President Cleveland and Secretary Olney's rebuke of the British on the boundary issue only to find that, after months of negotiations between the two English-speaking powers, the settlement gave much more to the British than Caracas expected. The Cuban revolutionaries relished the public support expressed for their cause in the United States, but in the combat in eastern Cuba and the postwar peace settlement, U.S. officials neglected or shoved aside the Cuban patriots, taking a forceful role in the new nation's affairs. In short, the effective inputting of alternative agendas into the U.S. information system generally tended to arouse public interest in these issues, but this arousal, in turn, presented politicians and government officials with the opportunity to institute policies that were popular, patriotic, and congruent with the growing imperial agenda of the United States. These outcomes tended to be based on the more aggressive instincts of a recently reunited and reinvigorated nation rather than on the intentions of those external agents who had hoped to turn the United States in other directions.

CLANDESTINE CABLES

The same cable network that carried incitements from propagandists and dramatic news from battlefields also transmitted coded messages with carefully guarded purposes. In day-to-day operations, much diplomacy and foreign reporting was carried on through coded messages intended to provide security for diplomatic negotiations and newspaper scoops. Hamilton Fish, U.S. secretary of state, and José Polo de Barnabé, the Spanish minister plenipotentiary, used the secure, point-to-point transatlantic lines to resolve the differences between the United States and Spain during the *Virginius* crisis of 1873. Journalists Sylvester Scovel, George Kennan, and James Creelman reported via cable on the Spanish-American War.

The use of the telegraph to convey secret messages intended to disrupt government policy, however, was different from its use to resolve differences. Few examples of its potential for stealth can match the cleverness of Marcial Martínez during the difficulties between the United States and Chile in the aftermath of the War of the Pacific. The Chilean engineered a diplomatic defeat for the United States through the transmission of information by the international cables. One of Secretary of State Blaine's last acts was to dispatch his son Walker and William Trescot to Santiago to pressure the Chileans to abandon their claims to the nitrate territory. Martínez quickly ascertained that Blaine's successor, Frederick Frelinghuysen, decided to relent on this point. The Chilean diplomat cabled this information to José Manuel Balmaceda, the minister of foreign affairs, from Washington to Santiago via Paris on the lines of the British companies. Walker Blaine and Trescot did not know of this change in policy because the State Department used surface mail rather than submarine cables. Martínez enabled Balmaceda to surprise his visitors from the United States by reading to them their revised orders from Washington—a diplomatic defeat for the United States.

The Martínez-Balmaceda coup—impressive though it was in 1882—resulted in only a temporary advantage for Chile, but the Panamanian coup led by Philippe Bunau-Varilla brought long-term changes that ultimately affected the territorial integrity of the nation of Colombia and the strategic position of the United States. Not content with his propagandistic publications designed to push the choice of the Panama route for the U.S.-built canal, Bunau-Varilla entered into the revolutionary plot begun by Panamanian Manuel Amador to lead a secession movement from Colombia in order to finalize Washington's commitment to build the canal. Amador established a cable code, and he and Bunau-Varilla used this code to organize the elite-led revolt against the government in Bogotá. Critical to this plot was the tacit but not explicit or open support of the United States. In arranging this key element in the operation, Bunau-Varilla used secret cables and his connections in Washington to wrest control of the plot from Amador and the Panamanians. Bunau-Varilla celebrated Panamanian independence in New York as Amador and his followers carried out a risky but bloodless coup on the isthmus with the friendly presence of the U.S. Navy.

Bunau-Varilla was not done. The last step was the treaty between the United States and the new nation of Panama. Here Bunau-Varilla used the potential to communicate by cable with Panama to his advantage by sending no messages. As the self-proclaimed representative of Panama, he negotiated and signed the Panama Canal Treaty with Secretary of State John Hay. By avoiding contact with Amador on the isthmus, the Frenchman who had so cleverly used the information system in the independence coup simply went silent. Amador was stunned to find the treaty completed and signed when he reached Washington. Bunau-Varilla, a major stockholder in the Compagnie nouvelle de canal de Panama, received a huge payment when Washington purchased that company.

THE LIMITS OF GOVERNMENT-MANAGED INFORMATION FLOW

The Roosevelt administration emerged as the victor in the Panamanian episode, securing as it did control over the canal zone. By focusing on Panama as the concluding event, a premature analysis could argue that the U.S. government made the most effective use of the information system, but that would be based on faulty reasoning that ignores several important points. As explained in chapter 12, the Roosevelt administration blundered in the use of cables in the deployment of forces in and around Panama and faced serious criticism for its actions in the national and international press. The president took control of the canal zone because he flexed the military, naval, and financial power of the United States, but this move was not as popular in the United States as the entry into the Spanish-American war had been. In short, this assertion of national authority as portrayed in much of the U.S. press was unexpected and unwarranted—it was not in accord with editorial judgments or validated by public support.

Roosevelt's disappointment at the public response to his boldness in Panama was similar to what previous administrations had felt when confronted with the unexpected, seemingly unpredictable results of the movement of reports and opinions through the new information system. At this point, it should be recalled that the resolution of the *Virginius* crisis in 1873 and the U.S.-directed solution to the Chilean-Argentine boundary dispute in 1881 suggested that the cable could provide the means to resolve international disagreements. The Harrison government's use of the cable to intimidate Chile in 1891–1892 indicated that a powerful nation could use the cable to impose its will on less powerful countries and achieve popular approval at home. These successes, however, occurred before the information system was fully in place. By the mid-1880s and more emphatically in the 1890s, improvements in the international cable network, including more reliable and less expensive service along with innovations in newspaper printing and distribution and the expansion of advertising as a source of revenue, all gave the information system an internal dynamism—a life of its own. The popular newspaper press became the central public platform for the new information system. Presidents and politicians of both major parties often witnessed an information flow that seemed to run out of control. The political leadership not only saw the informational content of the system lurch in unexpected and often threatening directions but at times found that the system worked against the policies that they wanted to implement.

Commander John Grimes Walker of the U.S. Navy saw the possibility that the information system could produce results harmful to government policy goals. The 1885 intervention in Panama revealed both the organizational benefits of the new cable system and the possibility that unanticipated outcomes could have an adverse impact on public opinion. Walker and other naval officials used the telegraph and cable system to coordinate the intervention of over two thousand marines and armed sailors in a period of ten days, an operation that quickly restored order. Officials in Washington were even able to give detailed orders to Commander Bowman McCalla in Panama City (much to McCalla's frustration). Walker's efforts to use the telegraph to keep "the people . . . in accord with the [Navy] Department" worked well.[8] The navy sent an embedded reporter with the intervention force: Irving King of the *New York Tribune*. Press coverage, led by the *Tribune*, was favorable and the brief mission concluded successfully with a quick withdrawal.

It was King, however, who made the observation that the Panamanian population and other local residents soon turned hostile toward U.S. naval personnel. The *Tribune*'s man in Panama concluded ominously that "all the natives carry revolvers, and many of them would count it a holy and patriotic thing to kill one of the Americans."[9] The intervention ended soon thereafter and newspaper coverage directly from Panama became scanty and irregular. It is a speculative point but one worth mentioning that if the U.S. press had given more attention to evidence of anti-U.S. feelings in Panama, Commander Walker's concerns about public support could have become a larger issue subject to press criticism and partisan debate.

The Cleveland administration's brief intervention in Panama was a selective, targeted undertaking that seemed to fit the mood of the public, and President Benjamin Harrison used a similar approach is the field of diplomacy six years later. Harrison trapped the Chilean government in a corner and forced a quick capitulation to U.S. demands for the settlement of the *Baltimore* affair. This episode, however, was particularly notable because outpourings of patriotic support came to dominate the press coverage of the crisis. Bombastic calls for war and public displays of militarism appeared in newspapers from coast to coast far, exceeding similar outcries in 1885. The pace of news coverage accelerated, adding to the excitement. Congressman James Blount, chairman of the House Committee on Foreign Affairs, expressed the concerns of many political leaders when he complained of the rapidity of the information flow through the cables and onto the front pages of newspapers. He saw potential difficulties "in the confusion provoked by instantaneous discussion."[10] Experienced editor and press critic E. L. Godkin saw imbalances and inaccuracies in the patriotic coverage of the crisis and tended to blame the youthful immaturity of many of the reporters.[11] A London *Times* correspondent commented on the distortions and ineptitude of the U.S. press in "the accounts of battles which had never been fought . . . and interviews and events which had never occurred. All of these frauds upon the public must not be overlooked, and attention should be drawn to them to prevent further mischief."[12] This exaggerated coverage happened to reinforce Harrison's tough approach to Chile, but the government in Washington did not control or even modulate this patriotic explosion. The information system circulated this exaggerated, irresponsible journalism, and later administrations found it difficult if not impossible to pursue their goals in environments shaped by such spontaneous outbursts.

The Venezuela boundary dispute brought the intensity of public incitement to a higher, more dangerous level. At first, President Cleveland (in his second term) and Secretary of State Olney seemed content to use Andrade and Scruggs's propaganda and the press excitement as reinforcements for their aggressive position against the British, paralleling Harrison's approach to Chile. But background braggadocio soon became front-page demands for serious war preparations against a major world power. Cleveland and Olney were surprised at the intensity of the press and public outcry, as were Salisbury and Pauncefote. Not only did the press and public applaud the war preparations, but they heaped mockery and abuse on the British. After the crisis abated, Salisbury sent trusted political ally Stafford Northcote to the United States to examine the depth of anti-British feeling. Northcote interviewed Godkin, who saw deeply rooted resentment against Britain in U.S. schools and popular culture.[13] This attitude was borne via the information system onto the editorial pages of newspapers and into extended essays in popular magazines. President Cleveland told Northcote that the public demand for aggressive action had caught him by surprise and was a contributing factor in his use of patriotic hyperbole.[14] Even Olney, who had boasted of the "sovereignty" of the United States over the Western Hemisphere, later admitted his concerns about poorly informed public opinion and its consequences for national policy.[15]

The blundering public diplomacy attempted by George Smalley, an experienced foreign correspondent, was a revealing conclusion to this episode in which the impulsive and nonrational came to play large roles. Smalley's efforts may have been based on thoughtful deliberation at the start, but the pervasive jingoism he found himself confronted with made him act recklessly, and his diplomacy became an embarrassment to all concerned. The information system was permeated by patriotic excess peppered with journalistic snippets. The one coherent message that emerged was the sanctity of the Monroe Doctrine, which in its media-hyped version left little room for negotiation. Sensible public figures such as Cleveland, Olney, Salisbury, Pauncefote, Godkin, and even Smalley found that rational discourse had fallen victim to the rabidity in and rapidity of mass communication.

Eventually U.S. and British diplomats resolved the Venezuela dispute, but the cultural predisposition to confrontational diplomacy and the excitement of anticipated combat remained much in evidence in the press. Spain's inability to suppress the Cuban revolt for independence often appeared in newspapers alongside the Venezuelan story in 1895 and 1896. Spain's problems deepened, and the administration of President William McKinley (1897–1901) engaged in determined diplomacy to find a solution to the civil unrest and its threat to the lives and property of U.S. citizens in Cuba. Here again the two streams of information flow worked at cross-purposes. During the last half of 1897 the government worked for a peaceful solution, while the press increasingly touted war. Secretary of State John Sherman and Stewart Woodford, the minister to Madrid, used the sensationalist journalism typical of Hearst's *New York Journal* and Pulitzer's *New York World* to apply pressure to the Spanish government. Like a water hose that suddenly twists and changes directions, the newspaper press turned the full force of its media flow on the McKinley administration in early 1898. Indeed, McKinley himself became the object of press scorn as a result of his speech in response to the explosion of the *Maine*. Failure to take decisive action could bring dire political consequences—the defeat of the Republican Party's candidates for the Senate and House of Representatives in the fall election. The information system informed the public, and the eligible voters would go to the polls in November. The war that McKinley had worked to avoid became the means by which to allay press criticism.

Once the United States went to war against Spain, the purposes of press sensationalism and government policy coincided, but there were some notable exceptions. One was the publication of the anticipated time of departure of the invasion fleet from Tampa that would give Spain useful information. The U.S. government instituted censorship and took direct control of the submarine-cable offices. Given the heavy coverage of the war, it was especially galling when a high-ranking officer came under press attack. General William Shafter's hesitation in the land approach to Santiago inspired a day or two of nervous speculation that portrayed the army's commanding officer in Cuba as indecisive and, in some accounts, inept. Shafter's exchange of punches with journalist Sylvester Scovel on July 17, 1898, during the ceremonial surrender of the Spanish army in Santiago was a catharsis during which official frustration with unfavorable media coverage found rare public expression.

ELECTRIC WIRES AND WARLIKE WORDS: THE DOMINANCE OF POPULAR CULTURE

The various administrations in Washington and the propaganda campaigns run by Martínez, Andrade, Scruggs, and the Cuban junta managed the information flow only temporarily, and, in the long run, actually lost control of it. Was any institution or person able to exert enough influence to justify the claim that they successfully managed the output of the information system? In short, the answer to this question is in the negative. Rather it was the spontaneous expression of popular culture that repeatedly came to the forefront. The public got what it wanted—exciting accounts of diplomatic confrontations, military and naval preparations, and, in 1898, war on sea and land. At the same time, the technology had made advancements: these improvements included larger submarine cables, a shift from the mirror galvanometer to the siphon recorder (mainly in the North Atlantic), a more productive transmission process based on the duplex system, and sheathed cables on the ocean floor for greater protection. The nationwide telegraph network received international messages and fed them to the metropolitan editorial offices, and the Hoe presses used cheap newsprint to bring the public a new form of an old product: the newspaper that combined international news with patriotism and entertainment. This communications system, owned and operated by private-sector corporations, was profitable at every level.

Admittedly the prominent newspaper editors of the day exercised a great deal of influence over the choice of the topics to be featured on the front page and also the vocabulary, print format, and editorial slant of the presentation. Pulitzer and Hearst were the main contributors to the emergence of sensationalism or yellow journalism. When the focus of attention turns to particular episodes, however, it becomes apparent that even the editors with their considerable means could not determine the direction that newspapers across the United States would follow. Pulitzer's *New York World* was critical of the President Cleveland's provocative response to London in the Venezuela boundary dispute in 1895, a position that placed this powerful editor at odds with most of the U.S. press. Two years later Hearst lavished money and front-page coverage on the Evangelina Cisneros story only to discover that most of his colleagues in the metropolitan press were downplaying or ignoring it. For over two years smaller city newspapers as well as several metropolitan organs such as the *New York Tribune* rejected the sensationalism of Hearst and Pulitzer. Not until the destruction of the *Maine* in February 1898 did mainstream press adopt the approach of the yellow press.

The *New York Tribune*'s Whitelaw Reid came to grasp the diversity and fluidity in the nation's newspapers during his railroad journey from the West Coast back to New York. Reid was disappointed to see that the same publications that had denounced the methods of Hearst and Pulitzer in 1896 and 1897 had reversed themselves in early 1898. His comment is worth repeating from chapter 10: "It was an actual amazement to me, in crossing the continent, to notice how extensively the papers which have so denounced what is termed 'yellow journalism' on the editorial pages, were nevertheless repeating, with job type display, the *Journal*'s most sensational news."[16]

There was no single approach to news coverage or editorial comment in these years but rather a wide diversity in style and content. For example, Reid advocated the cause of the reform wing of the Republican Party and was therefore at odds with the practice of journalistic objectivity espoused by the *New York Times*. Adolph Ochs took over the editorship of the *Times* in 1896, when Hearst and Pulitzer were in the midst of their sensationalist competition. The *Times* published carefully assembled, factual news written in a low-key style, while Hearst and Pulitzer went in the opposite direction. Yet by the turn of the century all three of these newspapers and Reid's *Tribune* enjoyed wide circulation and financial success. Among the ten metropolitan newspapers emphasized in this study styles ranged from the sensationalistic exuberance of the *San Francisco Examiner* to the calmer style of that city's *Chronicle*. The *Boston Globe* broke with regional partisanship as a Democratic paper in a Republican region. This Republican preference was reflected in the pages of the *Philadelphia Inquirer*. The South's *Atlanta Constitution* and *New Orleans Picayune* held firm with the Democrats.

A shared perspective on international issues emerged in this diversity of press orientations. In part, this commonality resulted from the flow of news dispatches on foreign affairs that moved through the submarine cables and the national telegraph network. There was an anomaly in all this. The centralized wire network with connections to international cables fed a stream of information to this diverse group of newspapers, whose editors, transcending their ideological, political, regional, and editorial differences, espoused a jingoistic response to most of the current international issues. The new technology coupled with the infusion of revenue from advertisements gave each newspaper a sense of independence often reflected in editorials but also grounded in the reality that the choice of front-page stories would have a large impact on circulation and profits. Editors gauged the public mood in their cities and regions and selected information and opinion pieces to publish that they believed would hold their readers. Immersed in the national popular culture of the United States and confronted by a large and unfamiliar international environment, these readers turned to patriotism and aggression as sources of reassurance. The cultural mediation previously performed by regional and local newspapers turned into a disintermediation in which the national interest was paramount. What emerged was a varied landscape of newspapers scattered across the United States that gravitated toward a common approach to the news centered on national traditions such as the Monroe Doctrine and armed might in the form of new steel-hulled warships equipped with modern firepower.

The technological coalescence of wires and presses created an imagined community from Boston to Atlanta, from Philadelphia to San Francisco. The typical front-page format coupled with the impact of the content of stories such as the 1885 intervention in Panama, confrontation with Chile and Britain, revolution in Cuba, and the explosion of the *Maine* worked together to create an intensity of emotion that became a shared experience culminating in new bursts of patriotic sentiment. This chronology of events traces an ascending level of public excitement, each new episode seeming to build on the emotionality of the previous one. The presentation of the news became more dramatic in the 1890s at the same time that events became more intense. The typical front page changed from the eight-column, small-type format of the 1880s to bigger type and wider columns beneath banner headlines often accompanied by illustrations and editorial cartoons. Visual images and large, bold lettering captured the attention of an eager audience and incited visceral responses.

In this chronology there was an acceleration of events, as historian C. A. Bayly notes in the concluding chapter to his synthesis on the long nineteenth century, which he appropriately entitled, "The Great Acceleration, c. 1890–1914." In it, he stresses "the growing velocity of international connections."[17] The international cable network became the main conduit for this speedup and intensification in the flow of information. With the exceptions of the War of 1812 and the U.S.-Mexican War, the first six decades of the nineteenth century saw few international crises that had a direct impact on the United States. The *Virginius* affair of 1873, the dispute with Chile at the close of the War of the Pacific, and the 1885 intervention in Panama appeared at intervals of eight and two years and had appreciable but temporary impacts. In the decade of the 1890s, the acceleration became obvious. The *Itata* chase, the *Baltimore* affair, and the related crises with Chile became prominent news items in 1891–1892. Public interest in the Venezuela border dispute and consequent tensions with Britain persisted from 1894 to 1896. The Cuban revolution led by José Martí in 1895 was a growing concern and a subject of public discussion from 1895 to 1898, when the United States fought and defeated Spain. The coverage of these nearly continuous international crises in the 1890s represented a significant acceleration of international news on the front pages of newspapers across the United States when compared with previous decades. Equally important, the stories themselves reached the front pages more rapidly. The months and weeks required to convey information about the War of 1812 were shortened to weeks and days in the U.S.-Mexican War, which, in turn, were shortened to a day or two in the 1890s.

This acceleration was magnified by what Rebecca Edwards has termed "reach" or the movement of information as well as of people and goods across great distances.[18] The railroad became the most important carrier of passengers and freight, but improvements in engines and naval architecture also made large oceangoing vessels important parts of these technological breakthroughs. Electric telegraphy and submarine cables handled the movement of an increasingly larger volume of information. These innovations combined in the last decades of the nineteenth century to change the nature of commerce, employment, and business activity. Some newspaper readers in the United States were concerned about Cuban sugar mills and Chilean nitrate mines because trade and investments tied these sources of raw products to the U.S. economy. But related political and diplomatic issues in these foreign lands were of more immediate popular interest. The metropolitan press linked their readers to Havana, Santiago, Colón, Madrid, and London as well as Washington and New York.

The informational reach of the newspaper reader expanded greatly, but this reach also drew on familiar patterns of behavior and traditional ideas. Expansion into frontier areas was a main theme in U.S. history. Such "habits of empire" as discussed by Walter Nugent had a deeply ingrained appeal for the public.[19] Opportunities for expansion on the North American continent had largely disappeared by the 1890s, but the print media, relying on the electric-communications system, gave the public a vicarious sense of continuity with older forms of expansion. Print and visual images that combined the nation's self-proclaimed mission to civilize less-developed peoples with the might of steel-hulled warships and courageous riflemen fed the popular imagination in confrontations with old world powers and in the domination of less powerful lands of the Americas. Newspapers, magazines, and books refined and embellished the clashes with Britain and Spain and also the intrusions into Chile, Colombia, Panama, and Cuba and in the process developed an artificial (and faulty) logic that seemed to justify these aggressions as an extension of past expansionism and, in most cases, as a defense of the Monroe Doctrine.

This combination of acceleration, intensification, and reach expressed in more dramatic front-page presentations was exciting. It sold newspapers. It also created anxieties for some. The United States as depicted in the pages of its newspapers, popular magazines, and books had ventured out onto the international stage as an aspiring world power. This national audacity carried with it uncertain consequences. The USS *Charleston*'s chase of the Chilean *Itata* was a race much like a sporting event, but the subsequent confrontations between Washington and Santiago carried with them the possibility of armed conflict, loss of life, and national embarrassment. The confrontation with Great Britain raised the stakes considerably, as did the war with Spain. These conflicts produced an anxiety—what Tom Lutz calls "American nervousness"—an understandable reaction to a new and larger role in international politics and the uncertainty as to where that new role would land the country.[20] Admittedly, this nervousness is a general concept that should be explained by reference to the actions and writings of individuals (which Lutz does), but it also deserves attention as a larger social and political phenomenon. Several of the commentators cited in this study revealed a sense of anxiety and disappointment in the outcome of events. Representative James Blount's expressions of confusion at the rapid pace of information flow in the Chilean crisis of 1891–1892, Britishers Julian Pauncefote and Stafford Northcote's concerns about volatile public opinion in the Venezuelan boundary dispute, and E. L. Godkin's persistent critiques of the quality of reporting expressed uncertainty about not only the validity of the information made public to the masses of readers but also the expansive role of the United States in world affairs.

RESPONSES TO THE NEW SYSTEM

The information system did not cause these controversial, disruptive events, but it did provide the platform on which the various competitors engaged in their competitions. Therefore, in line with the holistic perspective used in this study, the information system became an essential part of these conflicts. The Pender and Scrymser companies and Western Union implemented technical improvements in telegraphy and cable deployment to accelerate the movement of messages and to enhance their reliability. Newspaper publishers and editors upgraded their printing presses and enlivened their front-page layouts. The cumulative impact of these changes was to increase the velocity, volume, and entertainment value of the information in circulation. What moved through the wires and onto the front pages drew from the traditional values of popular culture in the United States and found expression in aggressive U.S. foreign-policy options.

The controversies stirred up by the new information system between the 1870s and 1903 brought about responses in government (and also the business community) that led to an increasing reliance on new practices in the fields of propaganda and public relations. The first piecemeal attempts at shaping public opinion on foreign policy issues in this period took place before these fields were institutionalized by presidents and the State Department. The main actors in this study used both the more subtle, indirect methods later associated with public relations and the more forceful, ideological style typical of what came to be called as propaganda.[21] Presidents Harrison, Cleveland, and McKinley gave increasing attention to the newspaper press and began to devise means to deal with the information flow both in the public media and within the more confined space of diplomatic negotiations. Harrison and Cleveland prepared statements directed at the press and the general public to further their policy objectives. McKinley set up a war room in the White House to serve as a communications center during the Spanish-American War and engaged in a propaganda campaign to win public support for the U.S. counterinsurgency campaign in the Philippines. His secretary of state, John Hay, cultivated the favor of reporters with regular press conferences—personal meetings with small groups of influential reporters.[22]

While these early efforts mainly relied on policy statements to manage information flow, other more effective approaches to molding public opinion arose in the same period. The work of Chilean diplomat Marcial Martínez from 1879 to 1881 and Commander John Grimes Walker in 1885 foreshadowed the public-relations techniques and manipulative methods that emerged in the 1890s. The collaboration of William Scruggs and José de Andrade in *The Monroe Doctrine on Trial* surprised the heads of state in Washington and London, and the Cuban junta's propaganda frustrated the Spanish in the 1895–1898 period. The U.S. government, however, lagged behind in the effective use of propaganda and public relations until the twentieth century.

President Theodore Roosevelt took an activist role in upgrading the government's practices of dealing with the information system. The 1902–1903 diplomatic clash between the United States and Germany that grew out of Kaiser Wilhelm's aggressive approach to settling a debt question with Venezuela provided the occasion for Roosevelt to take advantage of the press uproar in the United States. Germany, with the reluctant cooperation of Britain, established a blockade of Venezuelan ports in December 1902. The situation became a crisis when Germany's warships sank two Venezuelan vessels and destroyed a Venezuelan fort at the mouth of Lake Maracaibo. U.S. newspaper editorials expressed outrage and concluded that Germany had initiated a campaign against the Monroe Doctrine with the goal of establishing a permanent colony in the Caribbean. While Berlin's intentions were limited to the modest aim of debt collection, commentators in the United States warned of the worst possible scenario. According to Nancy Mitchell, "The brief war scare was an American tempest, the American Navy and American press chasing each other like a cat after its tail, spinning into an absurd frenzy."[23] Roosevelt, by his own account in a 1906 letter to diplomat Henry White, exploited this frenzy by warning German Ambassador Theodor von Holleben that public opinion was so aroused against Germany "that any president might feel obliged to act."[24] While Roosevelt may not have given the ultimatum he later claimed he did, his use of press and public opinion to formulate a policy statement for Berlin fits the documentary evidence in a 1903 letter to secretary of State John Hay.[25] Given the furor in the U.S. press and Britain's reluctance, Germany decided to accept arbitration. Both Roosevelt and his German contacts responded to the press and public's devotion to the Monroe Doctrine.[26]

Roosevelt also took the initiative in expanding government contributions to the flow of information in the communications system. The steady release of progress reports on the construction of the Panama Canal was a smart public-relations move on that part of his administration. The U.S. government's institutionalization of its inputs into the information system improved its public relations and propaganda. The new information system had created a new media environment in which governments, corporations, and individuals attempted to build favorable images for themselves and the causes for which they stood. World War I witnessed an escalation of propaganda under President Woodrow Wilson's Committee on Public Information. The CPI and other information "services" ushered in a new level of information manipulation and control that intensified throughout the twentieth century. Information systems came to play a role via propaganda, public relations, and combinations of the two in the spread of communist movements across the globe, the course of Mexico's epic revolution, the campaign for independence in India, the rise of fascism in Italy, Germany, and Spain, the global conflicts in World War II, and various confrontations during the Cold War.[27] Historians have focused on propaganda in the twentieth century, but the evolution of the information system can best be understood by recognizing the increased acceleration, intensity, and reach in the flow of its messages in the period under study here.

The arrival of the Internet and cable television in the last decades of the twentieth century brought another revolution in communications that parallels the alignment of international cables, the U.S. domestic telegraph network, and mass-circulation newspapers that created new channels for expressions of patriotism and imperialism in the last decades of the nineteenth century. Since the 1980s the multiplication of cable news channels and the expansion of the internet plus talk radio in the United States has created openings for outpourings of conservative opinion that have contributed to the Tea Party movement. As media analyst Jane Chapman observes, the globalization of communications opened the way for the rise of Rupert Murdoch's News Corporation, led by its Fox News channel in the United States. The growth of this global-communications giant coincided with the rise of conservative talk radio and the grassroots Tea Party movement, both of which have received a push through Murdoch's Fox News. Elements of the left have also taken advantage of new opportunities. The Zapatista movement in Mexico in the 1990s achieved national and international recognition via the Internet. Personal media and the internet played an important role in the early stages of the antiauthoritarian revolts that swept through Tunisia, Egypt, Libya, and other Middle Eastern states in 2011. Innovations in media technology do not explain these phenomena, but a full understanding of their rise to prominence could address these technologies in a fashion similar to this study of the new information system of the late nineteenth century. These cases from the recent past approximate the pattern that emerged in the late 1800s. Changes in media technology and technique facilitate eruptions of radical, often extremist responses, to contemporary events.[28]

This experience of the late nineteenth marked a crucial chapter in the struggles of political and media institutions and the people who ran them to adjust to the changes in communications technology and their unpredictable effects on society. These struggles exemplify the venerable race between the utilization of new technical systems and the efforts of humans to keep pace with new ways of doing things. Diplomats and political leaders fell behind in the crises of the mid- and late 1890s. Journalists, editors, and propagandists were adept in exploiting the system to secure short-term advantages such as newspaper "scoops" and the placement of persuasive articles and essays in the press, but they were unable to accomplish their long-term goals because of the diversity of editorial approaches in the press and overriding power relations (such as in the case of the United States and Britain in the Venezuela boundary controversy). The dominance of U.S. popular culture in the information flow broke through in 1895–1898 as a mix of grassroots sentiment stirred and sustained by a jingoistic press and the vote-hungry patriotism of politicians. Neither the Cleveland nor the McKinley administration was able to deal effectively with the system. The interaction between the public and the stimulation effected by the press and politicians generated high levels of intensity across the United States through the new electric-communications network and the mass-circulation press that fed on this network during and after the Spanish-American War. The institutional structure for the powerful display of patriotic celebrations and nationalistic aggressiveness was in place for better or worse not only in the United States but also Europe, much of Latin America, and many nations and colonial territories across the globe.

Notes

CHAPTER ONE

1. William L. Harris, "Venezuela: Wars, Claims, and the Cry for a Stronger Monroe Doctrine," in *United States-Latin American Relations, 1850–1903: Establishing a Relationship*, ed. Thomas M. Leonard (Tuscaloosa: University of Alabama Press, 1999), 107–22; William E. Curtis, *Venezuela: A Land Where It's Always Summer* (New York: Harper and Brothers, 1896), 245–59.
2. William E. Curtis, *Handbook of the American Republics*, vol. 3 (Washington, D.C.: Bureau of the American Republics, 1893), 551, 557.
3. Harris, "Venezuela," 114–21; Paul Gibb, "Unmasterly Inactivity? Sir Julian Pauncefote, Lord Salisbury, and the Venezuelan Boundary Dispute," *Diplomacy and Statecraft* 16.1 (2005): 23–55; Marshall Bertram, *The Birth of Anglo-American Friendship: The Prime Facet of the Venezuelan Boundary Dispute: A Study of the Interaction of Diplomacy and Public Opinion* (Lanham, MD: University Press of America, 1992); Walter LaFeber, *The New Empire: An Interpretation of American Expansion, 1860–1898* (Ithaca, NY: Cornell University Press, 1963), 242–83; John A. S. Grenville and George Berkeley Young, *Politics, Strategy, and American Diplomacy, 1873–1917* (New Haven, CT: Yale University Press, 1966), 125–78; Charles C. Tansill, *The Foreign Policy of Thomas F. Bayard* (New York: Fordham University Press, 1940), 630–37. Bertram places emphasis on the role of the press in this controversy.
4. Gibb, "Unmasterly Activity?," 23–55; Bertram, *Birth*, 19–55. For more on the Monroe Doctrine, see Dexter Perkins, *A History of the Monroe Doctrine*

(Boston: Little, Brown, 1963); David W. Dent, *The Legacy of the Monroe Doctrine* (Westport, CT: Greenwood, 1999); Gretchen Murphy, *Hemispheric Imaginings: The Monroe Doctrine and Narratives of U.S. Empire* (Durham, NC: Duke University Press, 2005); and Alonso Aguilar, *Pan-Americanism from Monroe to the Present: A View from the Other Side* (New York: Monthly Review Press, 1968).

5. Bertram, *Birth*, 19–55; LaFeber, *New Empire*, 254–70.
6. *San Francisco Examiner*, December 20, 21, and 22, 1895, 1.
7. William L. Scruggs, *The Venezuelan Question: British Aggression in Venezuela; or, The Monroe Doctrine on Trial*, 3rd ed. (Atlanta, GA: Franklin Printing Company, 1896). On the preparation of the pamphlet, see William L. Scruggs to Ezequiel Rojas, October 30, 1895, William Lindsay Scruggs Papers, 1828–1925, pt. 2, Scrapbooks, reel 1, Manuscript Division, Library of Congress, Washington, D.C.; and José de Andrade to William L. Scruggs, June 28, 1895, July 24, 1895, and July 29, 1895, Scruggs Papers.
8. Bertram, *Birth*, 53.
9. Thomas Pakenham, *The Scramble for Africa* (New York: Archon Books, 1991), 487–90; William Langer, *The Diplomacy of Imperialism*, 2nd ed. (New York: Knopf, 1960), 213–58; Oron J. Hale, *Publicity and Diplomacy with Special Reference to England and Germany, 1890–1914* (Gloucester, MA: Peter Smith, 1960), 102–37.
10. Gibb, "Unmasterly Activity?," 23–55; Bertram, *Birth*, 75–139.
11. Daniel Headrick, *The Invisible Weapon: Telecommunications and International Politics, 1851–1945* (New York: Oxford University Press, 1991), 28–49; Peter Hugill, *Global Communications since 1844: Geopolitics and Technology* (Baltimore, MD: Johns Hopkins University Press, 1999), 5–51.
12. Henry Cabot Lodge, "England, Venezuela, and the Monroe Doctrine," *North American Review* 160 (June 1895): 651–58.
13. Julian Pauncefote to Lord Salisbury, December 18, 1895, in *British Documents in Foreign Affairs: Foreign Office Confidential Print, North America, 1837–1914* (Frederick, MD: University Publications of America, 1987), 289; H. W. Brands, *The Reckless Decade: America in the 1890s* (New York: St. Martin's, 1995), 301. See chapter 7 for a discussion of U.S. press coverage of this crisis.
14. John H. Schroeder, *Mr. Polk's War: American Opposition and Dissent, 1846–1848* (Madison: University of Wisconsin Press, 1973), 3–50; Thomas Hietala, *Manifest Design: Anxious Aggrandizement in Late Jacksonian America* (Ithaca, NY: Cornell University Press, 1985), 215–72; Walter Nugent, *Habits of Empire: A History Of American Expansion* (New York: Knopf, 2008), 130–236; William David Sloan, *The Media in America: A History* (Northport,

AL: Vision Press, 2011), 121–58; Kevin Barnhurst and John Nerone, *The Form of the News: A History* (New York: Guilford Press, 2001), 31–110.

15. Headrick, *Invisible Weapon*, 3–49; Dwayne Winseck and Robert M. Pike, *Communications and Empire: Media, Markets, and Globalization, 1860–1930* (Durham, NC: Duke University Press, 2007), 1–91.

16. William Hadfield, *Brazil and the River Plate* (London: Edward Stanford, 1877), 1–4, 7; Curtis, *Handbook of the American Republics*, 549; Robert Greenhill, "The Royal Mail Steam Packet Company and the Development of Steamship Links with Latin America," *Maritime History* 3.1 (April 1973): 67–91; Daniel Headrick, *The Tools of Empire: Technology and European Imperialism in the Nineteenth Century* (New York: Oxford University Press, 1981), 129–49, 165–79; Daniel Headrick, *The Tentacles of Progress: Technology Transfer in the Age of Imperialism, 1850–1940* (New York: Oxford University Press, 1988), 18–48; William Henry Flayhart, *The American Line (1871–1902)* (New York: Norton, 2000), 137–224.

17. James R. Beniger, *The Control Revolution: Technological Change and Economic Origins of the Information Society* (Cambridge, MA: Harvard University Press, 1986); Paul A. Alcorn, *Social Issues in Technology: A Format for Investigation* (Upper Saddle River, NJ: Prentice Hall, 2003); Carl J. Couch, *Information Technologies and Social Orders* (New York: Aldine de Gruyter, 1996); Asa Briggs and Peter Burke, *A Social History of the Media: From Gutenberg to the Internet* (Cambridge, UK: Polity Press, 2002); Melvin L. De Fleur and Sandra Ball-Rokeach, *Theories of Mass Communications* (Boston: Allyn and Bacon, 1898); Armand Mattelart, *The Invention of Communication* (Minneapolis: University of Minnesota Press, 1994); Armand Mattelart, *Networking the World, 1974–2000* (Minneapolis: University of Minnesota Press, 2000).

    On the development of the communications system in the United States, see Alfred D. Chandler Jr. and James Cortada, eds., *A Nation Transformed by Information: How Information Has Shaped the United States from Colonial Times to the Present* (Cambridge, MA: Harvard University Press, 2000); and Daniel Czitrom, *Media and the American Mind* (Chapel Hill: University of North Carolina Press, 1982). Paul Starr's *The Creation of the Media: Political Origins of Modern Communications* (New York: Basic Books, 2004) places the evolution of the United States system in a comparative context. JoAnne Yates discusses the use of the telegraph as a part of the information systems that developed within corporations in the United States from 1850 to 1920 in *Control through Communication: The Rise of System in American Management* (Baltimore, MD: Johns Hopkins University Press, 1989). Richard D. Brown's *Knowledge Is Power: The Diffusion of Information in Early America, 1700–1865* (New York: Oxford University Press, 1989) and David M. Henkin's *The Postal Age: The Emergence of Modern Communications in Nineteenth-Century America* (Chicago: University of Chicago Press, 2007) both deal with the U.S. Post Office as a communications system.

An overview of the varied approaches in U.S. journalism is found in J. Herbert Altschull, *From Milton to McLuhan: The Ideas Behind American Journalism* (White Plains, NY: Longman, 1990). For some historical perspectives on the internet of the late twentieth and early twenty-first centuries and its predecessors, see Earl Morrogh, *Information Architecture: An Emerging 21st Century Profession* (Upper Saddle River, NJ: Prentice Hall, 2003), esp. 3–32. On the history of systems in general and their emergence in the late nineteenth century, see Thomas P. Hughes, *American Genesis: A Century of Invention and Technological Enthusiasm* (New York: Viking, 1989); Ruth Schwartz Cowan, *A Social History of American Technology* (New York: Oxford, 1997), esp. 149–72; Alan J. Marcus and Howard P. Segal, *Technology in America: A Brief History* (Belmont, CA: Wadsworth Thomson, 1999), esp. 110–73; and Jacques Ellul, *The Technological Society* (New York: Vintage, 1964).

18. Vary T. Coats and Bernard Finn, *A Retrospective Technology Assessment: Submarine Telegraphy, the Transatlantic Cable of 1866* (San Francisco: San Francisco Press, 1979); Ken Beauchamp, *A History of Telegraphy: Its History and Technology* (New Brunswick, NJ: Institute of Engineering and Technology, 2001); Headrick, *Invisible Weapon*; Hugill, *Global Communications.*
19. Jill Hills, *The Struggle for the Control of Global Communications: The Formative Century* (Urbana: University of Illinois Press, 2002).
20. Winseck and Pike, *Communication and Empire.*
21. Tom Standage, *The Victorian Internet* (New York: Walker, 1998).
22. Jorma Ahvenainen, *The History of the Caribbean Telegraphs before the First World War* (Helsinki: Finnish Academy of Science and Letters, 1996); Jorma Ahvenainen, *The European Cable Companies in South America before the First World War* (Helsinki: Finnish Academy of Science and Letters, 2004).
23. David Paull Nickles, *Under the Wire: How the Telegraph Changed Diplomacy* (Cambridge, MA: Harvard University Press, 2002). See also Nickles's "Telegraph Diplomats: The United States' Relations with France in 1848 and 1870," *Technology and Culture* 40.1 (1999): 1–25.
24. Richard Bradford, *The* Virginius *Affair* (Boulder: Colorado Associated University Press, 1980); Bertram, *Birth*; Harris, "Venezuela"; John L. Offner, *An Unwanted War: The Diplomacy of the United States and Spain over Cuba, 1895–1898* (Chapel Hill: University of North Carolina Press, 1992).
25. The U.S. domestic economy is likewise the focus in Alexander James Field, "The Magnetic Telegraph, Price and Quantity Data, and the New

Management of Capital," *Journal of Economic History* 52.2 (1992): 401–13; and Richard DuBoff, "Business Demand and the Development of the Telegraph in the United States," *Business History Review* 54.4 (1986), 459–79. See also Steve Fraser, *Every Man a Speculator: A History of Wall Street in American Life* (New York: Harper Collins, 2005); John Steele Gordon, *The Scarlet Woman of Wall Street* (New York: Weidenfeld and Nicholson, 1988); Richard John, "Recasting the Information Infrastructure for the Industrial Age," in *A Nation Transformed by Information*, 55–105; and JoAnne Yates, "Business Use of Information and Technology during the Industrial Age," in *A Nation Transformed by Information*, 107–35.

26. H. S. Ferns, *Britain and Argentina in the Nineteenth Century* (Oxford, UK: Clarendon Press, 1960); H. S. Ferns, "The Baring Crisis Revisited," *Journal of Latin American Studies* 24.2 (1992): 241–73; Harold Blakemore, *British Nitrates and Chilean Politics, 1886–1896: Balmaceda and North* (London: Athlone Press, 1974); Alex Nalbach, "'Poisoned at the Source'? Telegraphic News Services and Big Business in the Nineteenth Century," *Business History Review* 77.4 (2003): 577–610.
27. Menahem Blondheim, *News over the Wires: The Telegraph and the Flow of Public Information in America, 1844–1897* (Cambridge, MA: Harvard University Press, 1994).
28. Richard Schwarzlose, *The Nation's Newsbrokers*, 2 vols. (Evanston, IL: Northwestern University Press, 1989–90); Ted Curtis Smythe, *The Gilded Age Press* (Westport, CT: Praeger, 2003).
29. Donald Read, *The Power of News: The History of Reuters* (New York: Oxford University Press, 1992); Simon Potter, *News and the British World: The Emergence of an Imperial Press System* (Oxford, UK: Clarendon Press, 2003); Winseck and Pike, *Communication and Empire*; Oliver Boyd-Barrett and Michael Palmer, *Le trafique de nouveles: Les agencies mondiales d'information* (Paris: Alain Moreau, 1981); Oliver Boyd-Barrett, *The International News Agencies* (London: Constable, 1980); Rhoda Desbordes, "Transnationales de l'information au Nouveau Monde: Quelques aspects de l'implementation de agences internationales de information en Amerique du Sud, 1874–1919" (PhD diss., Sorbonne Nouvelle, 2005).
30. Nalbach, "'Poisoned at the Source'?," 577–610.
31. Phillip Knightley, *The First Casualty: From the Crimea to Vietnam: The War Correspondent as Hero, Propagandist, and Myth Maker* (New York: Harcourt Brace Jovanovich, 1975); John Hohenberg, *Foreign Correspondence: The Great Reporters and Their Times* (New York: Columbia University Press, 1964); John Maxwell Hamilton, *Journalism's Roving Eye: A History of American Foreign Reporting* (Baton Rouge: Louisiana State University Press, 2009);

Brooke Kroeger, *Nellie Bly: Daredevil, Reporter, Feminist* (New York: Three Rivers Press, 1995).

32. Nickles, "Telegraph Diplomats"; Nickles, *Under the Wire*, 31–134.
33. Don M. Coerver, "Mexico: Conflicting Self-Interests," in *United States-Latin American Relations, 1850–1903*, 22–27; Thomas Schoonover, *Dollars over Dominion: The Triumph of Liberalism in Mexican–United States Relations, 1861–1867* (Baton Rouge: Louisiana State University Press, 1978); Robert R. Miller, *Arms Across the Border: United States Aid to Juárez during the French Intervention in Mexico* (Philadelphia: American Philosophical Society, 1973).
34. Amy Kaplan, *The Anarchy of Empire in the Making of U.S. Culture* (Cambridge, MA: Harvard University Press, 2002); Karen Hoganson, *Fighting for American Manhood: How Gender Politics Provoked the Spanish-American War and the Philippine-American Wars* (New Haven, CT: Yale University Press, 1998); Walter Hixson, *The Myth of American Diplomacy: National Identity and U.S. Foreign Policy* (New Haven, CT: Yale University Press, 2008); Robert Hannigan, *New World Power: American Foreign Policy, 1898–1917* (Philadelphia: University of Pennsylvania Press, 2002); Michael Hunt, *Ideology and U.S. Foreign Policy* (New Haven, CT: Yale University Press, 1987). For a helpful perspective on the connections between propaganda and public opinion, see Troy R. E. Paddock, "Introduction: Newspapers, Public Opinion, and Propaganda," in *A Call to Arms: Propaganda, Public Opinion, and Newspapers in the Great War*, ed. Troy R. E. Paddock (Westport, CT: Praeger, 2004), 1–13.
35. Most studies in Atlantic history focus on the first three centuries of American-European-African interactions, but it seems that the history of submarine cables and the information system they made possible in the late 1800s is a legitimate part of this field. Among the many interesting publications in Atlantic history are Bernard Bailyn, *Atlantic History: Concept and Contours* (Cambridge, MA: Harvard University Press, 2005); John R. Gillis, *Islands of the Mind: How the Human Imagination Created the Atlantic World* (New York: Palgrave Macmillan, 2004); Peter Coclanis, "*Drang Nach Osten*: Bernard Bailyn, the World-Island, and the Idea of Atlantic History," *Journal of World History* 13.1 (2002), 169–82; and Alison Games, "Atlantic History: Definitions, Challenges, and Opportunities," *American Historical Review* 111.3 (2006): 741–57.
36. David Bushnell and Neill Macauley, *The Emergence of Latin America in the Nineteenth Century* (New York: Oxford University Press, 1994); Stuart Voss, *Latin America in the Middle Period, 1750–1929* (Wilmington, DE: Scholarly Resources, 2002); Thomas O'Brien, *Making the Americas: The United States and Latin America from the Age of Revolutions to the Era of Globalization* (Albuquerque: University of New Mexico Press, 2007), chaps. 1–3; Thomas

O'Brien, *The Century of U.S. Capitalism in Latin America* (Albuquerque: University of New Mexico Press, 1999), chaps. 1–2; Marcello Carmangani, *El otro occidente: América Latina desde la invasión europa hasta la globalización* (Mexico City: Colegio de México, 2004), chaps. 1–4; Marcello Carmangani, ed., *Federalismos latinamericanos: México/Brasil/Argentina* (Mexico City: Colegio de México, 1993); Lester Langley, *The Americas in the Age of Revolution, 1750–1850* (Princeton, NJ: Princeton University Press, 1996); Lester Langley, *America and the Americas: The United States in the Western Hemisphere* (Athens: University of Georgia Press, 1998).

37. Matthew Brown, ed., *Informal Empire in Latin America: Culture, Commerce, and Capital* (Malden, MA: Blackwell, 2008); Alan Knight, "U.S. Imperialism/Hegemony and Latin American Resistance," in *Empire and Dissent: The United States and Latin America*, ed. Fred Rosen (Durham, NC: Duke University Press, 2008), 23–52; Emily Rosenberg, *Financial Missionaries to the World: The Politics and Culture of Dollar Diplomacy, 1900–1930* (Durham, NC: Duke University Press, 2003); Gilbert Joseph, "Close Encounters: Toward a Cultural History of U.S.–Latin American Relations," in *Close Encounters of Empire: Writing the Cultural History of U.S.–Latin American Relations*, ed. Gilbert M. Joseph, Catherine C. LeGrand, and Ricardo D. Salvatore (Durham, NC: Duke University Press, 1998), 3–46; O'Brien, *Century of U.S. Capitalism*; Hunt, *Ideology and U.S. Foreign Policy*.

38. Alan Knight, "Britain and Latin America," in *The Oxford History of the British Empire: The Nineteenth Century*, ed. Andrew Porter (New York: Oxford University Press, 1999), 122–45; Gad Heuman, "The British West Indies," in *Oxford History of the British Empire*, 470–94. The contributions to this volume discuss the overall structure and dynamics of the British Empire and help to place its role and image in the Americas in a global context. For examples, see Robert Kubicek, "British Expansion, Empire, and Technological Change," 247–69. See also Charles Jones, *El Reino Unido y América: Inversiones y influencia económica* (Madrid: Mapfre, 1992).

39. Rory Miller, *Britain and Latin America in the Nineteenth and Twentieth Centuries* (London: Longman, 1993), 47–178; Joseph Smith, *Illusions of Conflict: Anglo-American Diplomacy Toward Latin America, 1865–1896* (Pittsburgh, PA: University of Pittsburgh Press, 1979).

40. Carlos Marichel, *A Century of Debt Crises in Latin America: From Independence to the Great Depression, 1820–1930* (Princeton, NJ: Princeton University Press, 1989), 126–70; Ferns, *Britain and Argentina*, 435–84; Ferns, "Baring Crisis Revisited," 241–73. For other relevant studies, see chapter 7.

41. Alfred Jackson Hanna and Kathryn Hanna, *Napoleon III and Mexico* (Chapel Hill: University of North Carolina Press, 1971); Nancy Nichols Barber,

The French Experience in Mexico, 1821–1861: A History of Constant Misunderstanding (Chapel Hill: University of North Carolina Press, 1979); Robert H. Duncan, "Political Legitimization and Maximilian's Second Empire in Mexico, 1864–1867," *Mexican Studies* 12.1 (1996): 27–66.

42. Maron J. Simon, *The Panama Affair* (New York: Scribner, 1971); David McCullough, *The Path Between the Seas: The Creation of the Panama Canal, 1870–1914* (New York: Simon and Schuster, 1977), 45–203.
43. The question of who invented the term "Latin America" seems to be a matter of some dispute—see, for example, *Dictionary of Human Geography* (2011, s.v. "Latin America"): "John Phelan (1968) attributes the term to nineteenth-century French scholars, who positioned France as the leader of a Latin 'race' engaged in a struggle for domination against Anglo-Saxon and Slavic racial blocs. According to Phelan (1968, p. 296), the term 'Latin America' was baptized in 1861 in *La revue des races Latines,* a magazine 'dedicated to the cause of Pan-Latinism.' . . . However, the idea of pan-Latinism also circulated in Spain and Hispanic America in the mid-nineteenth century, as intellectuals from the Creole elite expressed anxiety about the imperial ambitions of Anglo-Saxon America (Ardao, 1992). Chilean scholar Miguel Rojas Mix (1986) attributes the term to Francisco Balbao (1823–65), an exiled Chilean writer living in Paris in the mid-nineteenth century. According to Mix, Balboa used the term in a speech in 1856."
44. Thomas Schoonover, *The French in Central America: Culture and Commerce, 1820–1930* (Wilmington, DE: Scholarly Resources, 2000); Thomas Schoonover, *Uncle Sam's War of 1898 and the Origins of Globalization* (Lexington: University Press of Kentucky, 2003), 25; Christian Herman, *La politique de la France en Amérique Latine, 1826–1850* (Bordeaux: Maison des pays iberique, 1996); Monica Quijada, "Sobre el origin y difusión del nombre América Latina (O una variación heterodoxa en torno el tema de la construción social de la verdad)," *Revista Indiana* 58.214 (1998): 595–616; Martha Barbance, *Vie commercial de la route de Cap Horn au xixe siecle: L'armament A. D. Bordes et fils* (Paris: SEVPEN, 1969); Jurgen Schneider, "Trade Relations Between France and Latin America, 1810–1850," in *América Latina en la época de Simón Bolívar,* ed. Reinhard Liehr (Berlin: Colloquium, 1989), 423–37; German Arceniegas, *Latin America: A Cultural History* (London: Barrie and Rockliff, 1969), 314–441.
45. Robert L. Scheina, *Latin America's Wars: The Age of the Caudillo, 1791–1899* (Washington, D.C.: Brassey's, 2003), 333–64; Louis A. Pérez, *Cuba between Empires, 1878–1902* (Pittsburgh, PA: University of Pittsburgh Press, 1983); Carmen Adams Fernández, *La América distorsionada: Una visión española de finales del siglo XIX* (Oviedo, ES: University de Oviedo, 1999); Lester

Langley, *Struggle for the American Mediterranean: United States–European Rivalry in the Gulf Caribbean, 1776–1904* (Athens: University of Georgia Press, 1982); Rodrigo Botero, *Ambivalent Embrace: America's Troubled Relationship with Spain from the Revolutionary War to the Cold War* (Westport, CT: Greenwood, 2001), chaps. 2–4; Alberto Elena and Javier Ordóñez, "Science, Technology, and the Spanish Colonial Experience in the Nineteenth Century," *Osiris* 15, s.2 (2000): 70–82.

46. Nancy Mitchell, *The Danger of Dreams: German and American Imperialism in Latin America* (Chapel Hill: University of North Carolina Press, 1999), 9–63; Thomas Schoonover, *Germany in Central America: Competitive Imperialism, 1829–1929* (Tuscaloosa: University of Alabama Press, 1998); Holger Herwig, *Germany's Vision of Empire in Venezuela, 1871–1914* (Princeton, NJ: Princeton University Press, 1986).
47. McCullough, *Path*, 124–241.
48. Ferns, *Britain and Argentina*; Blakemore, *British Nitrates and Chilean Politics.*
49. Herwig, *Germany's Vision*, 80–109.
50. For the application of the concepts of nationalism and "imagined communities" to Latin America, see *Beyond Imagined Communities: Reading and Writing the Nation in Nineteenth-Century Latin America*, ed. Sara Castro-Klarén and John Charles Chasteen (Baltimore, MD: Johns Hopkins University Press, 2003); *Nationalism in the New World*, ed. Don Doyle and Marco Antonio Pamplona (Athens: University of Georgia Press, 2006); *Constitucionalismo y orden liberal América Latina, 1850–1920*, ed. Marcello Carmagnani (Torino: Otto, 2000); *Construcciones impresas: Panfletos, diarios y revistas en la formación de los estados nacionales en América Latina, 1820–1920*, ed. Paula Alonso (Buenos Aires: Fondo de cultura económica, 2003); *Nación y nacionalismo en América Latina*, ed. Jorge Enrique González (Buenos Aires: Consejo latinoamericano de ciencias sociales, 2007); and *Ensayos sobre la nueva historia política de América Latina, siglo XIX*, ed. Guillermo Palacios (Mexico City: Colegio de México, 2007). For the seminal work in this area, see Benedict Anderson, *Imagined Communities: Reflections on the Origin and Spread of Nationalism*, 2nd ed. (London: Verso, 1991).
51. Michael Adas, *Dominance by Design: Technological Imperatives and America's Civilizing Mission* (Cambridge, MA: Harvard University Press, 2006); Headrick, *Tentacles of Progress*. For a different perspective on this issue, see David Landes, *The Wealth and Poverty of Nations* (New York: Norton, 1998); and David Landes, *The Unbound Prometheus: Technological Change and Industrial Development in Western Europe from 1750 to the Present* (Cambridge, MA: Cambridge University Press, 1972).

52. Among the many informative studies of the United States in the late nineteenth century, see Jack Beatty, *The Age of Betrayal: The Triumph of Money in America, 1865–1900* (New York: Vintage, 2007); Rebecca Edwards, *New Spirits: Americans in the Gilded Age, 1865–1925* (New York: Oxford University Press, 2006); Mark Wahlgren Summers, *The Gilded Age; or, The Hazards of New Functions* (Upper Saddle River, NJ: Prentice Hall, 1997); and Robert Wiebe, *The Search for Order, 1877–1920* (New York: Hill and Wang, 1967).
53. John, "Recasting the Information Infrastructure"; Yates, "Business Use"; Brown, *Knowledge Is Power*; Henkin, *The Postal Age.*
54. John, "Recasting the Information Infrastructure," 73–75.
55. Hohenberg, *Foreign Correspondence*, 38–44; Hamilton, *Journalism's Roving Eye*, 48–56; George Wilkins Kendall, *Dispatches from the Mexican War* (Norman: University of Oklahoma Press, 1999); Robert Johannsen, *To the Halls of Montezuma: The Mexican War in the American Imagination* (New York: Oxford University Press, 1985). Johannsen discusses the images of the Mexican War in U.S. newspapers and magazines.
56. Blondheim, *News over the Wires*, 11–140.
57. Brown, *Knowledge Is Power*, 258.
58. John J. Pauly, "The Great Chicago Fire as a National Event," *American Quarterly* 36.5 (1984): 677.
59. Pauly, "The Great Chicago," 682.
60. Hohenberg, *Foreign Correspondence*, 74–80; Hamilton, *Journalism's Roving Eye*, 57–75; Knightley, *First Casualty*, 3–50; George W. Smalley, *Anglo-American Memories* (New York: Putnam's, 1911), 220–52; *New York Times*, August 24, 1870, 5; Geoffrey Wawro, *The Franco-Prussian War: The German Conquest of France* (Cambridge: Cambridge University Press, 2003). Michéle Martin's *Images at War: Illustrated Periodicals and Constructed Nations* (Toronto: University of Toronto Press, 2006) examines the coverage of the war in the illustrated presses of England, France, Germany, and Canada.
61. Beatty, *Age of Betrayal*, 154–57; Knightley, *First Casualty*, 49–50.
62. Gerald Baldasty, *The Commercialization of News in the Nineteenth Century* (Madison: University of Wisconsin Press, 1992); Paulette Kilmer, "The Press in Industrial America, 1865–1883," in *The Media in America: A History*, ed. William David Sloan (Newport, AL: Vision Press, 2005), 199–222.
63. W. Joseph Campbell, "The Age of New Journalism, 1883–1890," in *Media in America*, 223–48; Smyth, *Gilded Age Press*, 123–24.
64. Smyth, *Gilded Age Press*, 136–27; Campbell, "Age of New Journalism," 236; Bingham Duncan, *Whitelaw Reid: Journalist, Politician, Diplomat* (Athens: University of Georgia Press, 1975), 97–98.
65. Smythe, *Gilded Age Press*, 93–96, 77–81, 133–38, 173–93; Kilmer, "The Press

in Industrial America"; Campbell, "Age of New Journalism"; Duncan, *Whitelaw Reid*, 51–195; David Nasaw, *The Chief: The Life of William Randolph Hearst* (Boston: Houghton Mifflin, 2000), 67–94; Harold E. Davis, "Henry W. Grady, Master of the Atlanta Ring—1880–1886," *Georgia Historical Quarterly* 69.1 (1988): 1–16; E. J. Edwards, "Charles H. Taylor of the Boston 'Globe,'" *McClure's Magazine*, July 1896, 144–47. An extensive survey of U.S. press opposition to the Spanish-American War is found in Piero Gleijeses, "1898: The Opposition to the Spanish-American War," *Journal of Latin American Studies* 35.4 (2003): 681–719. The *Literary Digest* occasionally categorized the political inclinations of the newspapers it quoted. For example, see *Literary Digest*, April 6, 1895, 1–2, May 4, 1895, 1–2, and May 11, 1895, 1–3.

66. U.S. Department of the Interior, Census Office, *Abstract of the Eleventh Census, 1890* (1894; repr., New York: Arno Press, 1974), 3, 25–28. The population figures for the adjacent states of New Jersey and Connecticut are added to those of New York, and the figures for Vermont, New Hampshire, and Rhode Island are added to the total for Massachusetts.
67. Murphy, *Hemispheric Imaginings*, 97–144; Perkins, *Monroe Doctrine*, 149–91.
68. Michael Kammen, *Mystic Chords of Memory: The Transformation of Tradition in American Culture* (New York: Vintage, 1991), 179.
69. Cecilia Elizabeth O'Leary, *To Die For: The Paradox of American Patriotism* (Princeton, NJ: Princeton University Press, 1999), 155. See also chapters 9 and 10 in O'Leary's book.
70. Kilmer, "The Press in Industrial America," 200. Some immigrants were literate in their native languages but not in English, as suggested by the large number of foreign-language newspapers published in New York, Chicago, St. Louis, and other midwestern and northeastern cities. See Jean Folkerts and Dwight L. Teeter, *Voices of a Nation: A History of Mass Media in the United States* (Boston: Allyn and Bacon, 2002), 283–88; and Denis Brian, *Pulitzer: A Life* (New York: Wiley, 2001), 5–39.
71. Hamilton, *Journalism's Roving Eye*, 89–109; David R. Spencer, *The Yellow Journalism: The Press and America's Emergence as a World Power* (Evanston, IL: Northwestern University Press, 2007); W. Joseph Campbell, *Yellow Journalism: Puncturing the Myths, Defining the Legacies* (Westport, CT: Praeger, 2003); John D. Stevens, *Sensationalism and the New York Press* (New York: Columbia University Press, 1991), 55–100; Barnhurst and Nerone, *The Form of the News*.
72. Qtd. in Clarence Brown, *The Correspondents' War* (New York: Scribner, 1967), 389.

CHAPTER TWO

1. For a comprehensive history of the British and other European cable companies in Latin America, see two books by Ahvenainen: *European Cable Companies* and *History of the Caribbean Telegraphs*. In addition to the ten companies discussed in this chapter, there were several other companies that operated in Latin America, but they were small, and many did not survive for more than a few years. For more detail, see the two works by Ahvenainen.
2. Winseck and Pike, *Communication and Empire*, 61–91. Ahvenainen examines the contrasting financial results for several of the British cable companies and their frequently acrimonious relationships. For examples, see Ahvenainen, *History of the Caribbean Telegraphs*, 46–93; and *European Cable Companies*, 154–87, 201–7.
3. William Hadfield, *Brazil, the River Plate, and the Falkland Islands* (London: Longman, Brown, Green, and Longmans, 1854), 1–4; William Hadfield, *Brazil and the River Plate* (London: Edward Stanford, 1877), 1–4, 9. For historical studies of transatlantic shipping in the nineteenth century, see Walter Hubbard and Richard F. Winter, *North Atlantic Mail Sailings, 1840–75* (Canton, OH: U.S. Philatelic Society, 1988); and Flayhart, *American Line*.
4. *Brazil and River Plate Mail*, January 7, 1864, 61, and August 7, 1867, 12; Greenhill, "Royal Mail Steam Packet Company." See also A. J. Arnold and Robert Greenhill, "Mail Contracts, Subsidies, and Profits: The Royal Mail Steam Packet Company, 1842–1905," in *Management, Finance and Industrial Relations in the Maritime Industry*, ed. Simon P. Ville and David M. Williams (St. John's, Newfoundland: International Maritime Economic History Association, 1994), 47–60; and A. J. Arnold and Robert Greenhill, "Business, Government, and Profit: The Management and Regulation of Overseas Mail Contracts, 1836–90," *Accounting, Business and Financial History* 5.3 (1995): 334–59.
5. *Brazil and River Plate Mail*, January 7, 1864, 60.
6. *South American Journal* (hereafter *SAJ*), November 7, 1871, 15. The *South American Journal* was formerly the *Brazil and River Plate Mail*.
7. Rene De La Pedraja, *Oil and Coffee: Latin American Merchant Shipping from the Imperial Era to the 1950s* (Westport, CT: Greenwood Press, 1998), 17–22; Claudio Veliz, *Historia de la marina mercante de Chile* (Santiago: Universidad de Chile, 1961).
8. William Eleroy Curtis, *Trade and Transportation Between the United States and Spanish America* (Washington, D.C.: U.S. Government Printing Office, 1889); Leonard Swann, *John Roach: Maritime Entrepreneur* (Annapolis, MD: U.S. Naval Institute Press, 1965), 95–124; David M. Pletcher, *The Diplomacy of Trade and Investment: American Economic Expansion in the Hemisphere, 1865–1900* (Columbia: University of Missouri Press, 1998), 180–215.

9. Headrick, *Invisible Weapon*, 11–27; Robert J. Cain, "Telegraphs in the British Empire, 1850–1900" (PhD diss., Duke University, 1971); Winseck and Pike, *Communication and Empire*, 16–42.
10. F. H. C. Tarver, "The Associated Companies in South America," unpublished manuscript, April 30, 1934, Cable and Wireless Archive, Porthcurno Telegraph Museum, Cornwall, UK.
11. Ahvenainen, *European Cable Companies*, 31–39; Horacio Reggini, *Sarmiento y las telecomunicaciones: La obsesión del hilo* (Buenos Aires: Ediciones Galápago, 1996), 83–103; *SAJ*, February 6, 1867, 15–16; Tarver, "The Associated Companies," 1–3.
12. Tarver, "The Associated Companies," 1–3.
13. Ahvenainen, *History of the Caribbean Telegraphs*, 9–22.
14. *New York Times*, July 29, 1868, 1; Ahvenainen, *History of the Caribbean Telegraphs*, 18–19, James Scrymser, *Personal Reminiscences of James Scrymser in Times of Peace and War* (Easton, PA: Eschenbach Printing Company, 1915), 70–75; Scrymser obituary, *New York Times*, January 15, 1917, 4.
15. "Annual Report to the Stockholders of the Western Union Company," October 7, 1873, 13–14, Western Union Telegraph Company Records, Lemelson Center, National Museum of American History, Washington, D.C.; William Smith to Polo de Bernabe, December 5, 1872 (report on cable repair), Western Union Telegraph Company Records; *New York Times*, July 29, 1868, 1; Ahvenainen, *History of the Caribbean Telegraphs*, 9–22.
16. Louis A. Pérez Jr., *Cuba and the United States: Ties of Singular Intimacy* (Athens: University of Georgia Press, 1990), 29–54; Pérez, *Cuba between Empires*; Oscar Zanetti and Alejandro Garcia, *Sugar and Railroads: A Cuban History*, trans. Franklin Knight and Mary Todd (Chapel Hill: University of North Carolina Press, 1998), 78–193.
17. *Bankers' Magazine*, July 1870, 9–12.
18. Jorma Ahvenainen, *The Far Eastern Telegraphs: The History of Telegraphic Communications between the Far East, Europe, and America before the First World War* (Helsinki: Suomalainen Tiedeakatemia, 1981); Hugh Barty-King, *Girdle Round the Earth: The Story of Cable and Wireless* (London: Heinemann, 1979), 3–109; Headrick, *Invisible Weapon*, 3–72; Cain, "Telegraphs in the British Empire."
19. The circumstances surrounding Scrymser's loss of International Ocean remain elusive. See his *Personal Reminiscences*, 75; and Ahvenainen, *History of the Caribbean Telegraphs*, 23–25.
20. Ahvenainen, *History of the Caribbean Telegraphs*, 23–25.
21. Ibid., 26–30.
22. West India and Panama, directors' annual reports, November 23, 1870,

2, November 22, 1871, 1–2, and November 17, 1873, 2, Cable and Wireless Archive.

23. Ahvenainen, *History of the Caribbean Telegraphs*, 42.
24. "Memoria del director general del telégrafos," in *Memoria del ministerio del interior presentada del congreso nacional en 1874* (Santiago: Imprenta nacional, 1874), 409–19; Santiago Marin Vicuña, *Los hermanos Clark* (Santiago: Balcells, 1929), 58–68; John J. Johnson, *Pioneer Telegraphy in Chile, 1852–1876* (Stanford, CA: Stanford University Press, 1948), 99–104.
25. Johnson, *Pioneer Telegraphy*, 104–5.
26. Ibid.
27. Ibid.
28. Ibid.; *Times* (London), September 2, 1872, 5; Tarver, "The Associated Companies," 3–4.
29. *Memoria presentada por el director general de telégrafos nacionales al ministerio del interior correspondiente a 1872*, vol. 1 (Buenos Aires: Imprenta de la Union, 1873), 16–25; *Memoria presentada por el director general de telégrafos nacionales al ministerio del interior correspondiente al ano 1873*, vol. 2 (Buenos Aires: Kidd, 1874), 52–59; Reggini, *Sarmiento*, 105–95; Walter B. L. Bose, "Historia de las comunicaciones," in *Historia Argentina contemporanea, 1862–1930*, vol. 3 of *Historia económica* by Horacio Juan Cuccorese et al. (Buenos Aires: Libreria El Ateno, 1966), 582–603; John E. Hodge, "The Role of the Telegraph in the Consolidation of the Argentine Republic," *The Americas* 41.1 (1984): 59–80.
30. *Memoria presentada por el director*, 1:56–59; Bose, "Historia," 588–94; Hodge, "Role of the Telegraph."

venainen, *European Cable Companies*, 59–96.

32. *SAJ*, November 7, 1871, 14, and November 22, 1871, 7.
33. William Abbot, letter to the editor, *Times* (London), July 19, 1872, 10.
34. Winseck and Pike, *Communication and Empire*, 17–19, 80–84; Barty-King, *Girdle*, 3–22.
35. *SAJ*, November 7, 1871, 14–15, and November 22, 1871, 6–7, 13–14.
36. *SAJ*, December 22, 1871, 13, January 22, 1872, 13, and April 22, 1872, 13–14; *Times* (London), April 16, 1872, 10; Ahvenainen, *European Cable Companies*, 13–28, 92–96.
37. *SAJ*, January 23, 1873, 13, and February 7, 1873, 2. See also *Times* (London), February 6, 1873, 6, and October 15, 1873, 15.
38. *SAJ*, March 23, 1873, 13; *Times* (London), March 22, 1873, 13; Ahvenainen, *European Cable Companies*, 47–91.
39. *SAJ*, February 7, 1873, 3, and February 24, 1873, 13–14.
40. *SAJ*, September 23, 1873, 13–14; Ahvenainen, *European Cable Companies*, 73–76.

41. *SAJ*, September 23, 1873, 13–14, February 24, 1873, 20, March 8, 1873, 12, and October 8, 1873, 6–8; *Times* (London), January 2, 1873, 6, January 16, 1873, 6, and February 22, 1873, 16.
42. *SAJ*, March 7, 1874, 7, 14.
43. *SAJ*, July 8, 1874, 18; Reggini, *Sarmiento*, 137–95.
44. Brazilian Submarine, directors' reports, 1874–79, Cable and Wireless Archive; *Times* (London), April 7, 1875, 7, October 28, 1875, 7, March 31, 1876, 10, October 6, 1876, 8.
45. *Times* (London), December 16, 1875, 6, December 31, 1875, 6, February 8, 1877, 7, April 5, 1877, 7, January 28, 1879, 7, January 6, 1880, 7, and January 8, 1880, 5.
46. On Pender's influence in international telegraphy and in British politics, see Headrick, *Invisible Weapon*, 33–37; Barty-King, *Girdle*, 3–63; and "Death of Sir John Pender," *Times* (London), July 8, 1896, 10.
47. Ahvenainen, *History of the Caribbean Telegraphs*, 91–109.
48. Winseck and Pike in *Communication and Empire* emphasize the financial connections and managerial cooperation between the British and U.S. cable firms, but Scrymser and some of his contemporaries saw these relationships as mainly competitive. See Herbert Laws Webb, "A New Chilean Cable," *Electrical World*, June 22, 1889, 356, Herbert Laws Webb, "A Boom in Submarine Cables," *Electrical World*, June 29, 1889, 374, and Herbert Laws Webb, "Puck's Girdle Round the Earth," *Electrical World*, March 1, 1890, 166–68. See also Headrick, *Invisible Weapon*, 38–39; John A. Britton and Jorma Ahvenainen, "Showdown in South America: James Scrymser, John Pender, and United States–British Cable Competition," *Business History Review* 78.1 (2004): 1–27; and Ahvenainen, *European Cable Companies*, 215–30, 389–409.
49. Charles Bright, *Submarine Telegraphs: Their History, Construction, and Workings* (London: Cosby, Lockwood, and Sons, 1898), 211–522; Coates and Finn, *Retrospective Technology Assessment*, 1–65; Beauchamp, *History of Telegraphy*, 134–79.
50. Bright, *Submarine Telegraphs*, 592–603; "The Great Transatlantic Cable," *The American Experience*, season 17, episode 7, directed by Peter Jones, aired April 11, 2005 (New York: PBS, 2005), DVD.
51. William J. Maver Jr., "Ocean Telegraphy," pt. 3, *Electrical World*, June 14, 1884, 191; "Magnet Lab: National High Magnetic Field Laboratory," www.magnet.fsu.edu. This process was demonstrated to me John Packer, honorary curator at the Porthcurno Telegraph Museum in Cornwall, UK, in 2001.
52. Lazare Weiller, "The Annihilation of Distance," *Living Age*, October 15, 1895, 175. Packer also demonstrated this part of the process to me.
53. Webb, "Puck's Girdle Round the Earth," 166.
54. Webb, "Puck's Girdle Round the Earth," 166; Tarver, "The Associated Companies," 18, 98–102.

55. Barty-King, *Girdle*, 59–60; Beauchamp, *History of Telegraphy*, 168.

CHAPTER THREE

1. *New York Times*, April 12, 1885, 6.
2. *SAJ*, February 6, 1867, 15–16.
3. On the diplomatic and political crises, see Smith, *Illusions of Conflict*; LaFeber, *New Empire*; Thomas Schoonover, *The United States in Central America, 1860–1911: Episodes of Social Imperialism and Imperial Rivalry in the World System, 1860–1911* (Durham, NC: Duke University Press, 1991); Robert N. Burr, *By Reason or Force: Chile and the Balancing of Power in South America, 1830–1905* (Berkeley: University of California Press, 1967); Thomas F. McGann, *Argentina, the United States, and the Inter-American System, 1880–1914* (Cambridge, MA: Harvard University Press, 1957).

   On economic issues, see Pletcher, *Diplomacy of Trade*; Miller, *Britain and Latin* America; Marichel, *Century of Debt Crises*; Victor Bulmer-Thomas, *The Economic History of Latin America since Independence* (Cambridge: Cambridge University Press, 1994); Steven Topik, *Trade and Gunboats: The United States and Brazil in the Age of Empire* (Stanford, CA: Stanford University Press, 1996); Steven Topik and Allen Wells, eds., *The Second Conquest of Latin America: Coffee, Henequen, and Oil during the Export Boom, 1850–1930* (Austin: University of Texas Press, 1998); John Coatsworth and Alan M. Taylor, eds., *Latin America and the World Economy since 1800* (Cambridge, MA: Rockefeller Center for Latin American Studies, Harvard University, 1998); and Carmagnani, *El otro occidente*.
4. Bradford, Virginius *Affair*, 25–54; Jay Sexton, "The United States, the Cuban Rebellion, and the Multilateral Initiative of 1875," *Diplomatic History* 30.3 (2006): 335–50; Andrés Cisneros, Carlos Escudé, et al., *Historia general de las relaciones exteriores de la República Argentina* (2000), www.argentina-rree.com, chaps. 33–34; Harold Peterson, *Argentina and the United States, 1810–1960* (New York: State University of New York Press, 1964), 240–46.
5. Bradford, Virginius *Affair*, 25–54.
6. West India and Panama, directors' report, September 17, 1873, Cable and Wireless Archive, Porthcurno Telegraph Museum, Cornwall, UK; Ahvenainen, *History of the Caribbean Telegraphs*, 9–45.
7. West India and Panama, directors' report, November 17, 1873, Cable and Wireless Archive; *Times* (London), November 18, 1873, 7.
8. Headrick, *Invisible Weapon*, 33–39.
9. Bradford, Virginius *Affair*, 44–54.

10. Ibid.; Allan Nevins, *Hamilton Fish: The Inner History of the Grant Administration*, 2 vols. (New York: Unger, 1957), 2:667.
11. Bradford, Virginius *Affair*, 57–108; Nevins, *Hamilton Fish*, 2:678–87.
12. Hamilton Fish, diary entry, November 28, 1873, qtd. in Nevins, *Hamilton Fish*, 2:686. Fish's problems were complicated by the State Department's reliance on an awkward and often confusing security code for the transmission of telegrams. See Ralph E. Weber, *United States Diplomatic Codes and Ciphers* (Chicago: Precedent Publishing, 1979), 234–37.
13. *New York Times*, November 8, 1873, 3; Bradford, Virginius *Affair*, 57–62.
14. Journalists supplied Fish with the first telegraphic information on the executions of Ryan and the three Cubans and later with an erroneous report concerning more mass executions. See Bradford, Virginius *Affair*, 57–58 and 61–62.
15. *New York Times*, November 9, 1873, 5, November 10, 1873, 8, and November 13, 1873, 3. The *New York Herald* accompanied its coverage with several editorials calling for public support of the Cuban rebels and also the intervention of the United States government against Spain. In her eulogistic *Life of Capt. Joseph Fry: The Cuban Martyr* (Hartford, CT: J. B. Burr, 1875), Jeanie Mort Walker quoted extensively from the *Herald*'s editorials. See especially 301–13. The *New York Tribune* also provided extensive coverage of the *Virginius* affair, generally on its front page, from November 10 to 25, 1873, but did not adopt the sensationalist approach of the *New York Herald*.
16. *New York Times*, November 16, 1873, 1, November 21, 1873, 1, and November 26, 1873, 5; *New York Tribune*, November 12, 1873, 1.
17. *New York Tribune*, November 20, 1873, 1; Bradford, Virginius *Affair*, 62–73; Nevins, *Hamilton Fish*, 2:672–73.
18. *Boston Globe*, November 13, 1, November 14, 1, and November 15, 1873, 1. See more generally the *Globe*'s front page between November 12 and 15, November 17 and 22, and November 24 and 29.
19. Bradford, Virginius *Affair*, 83–85.
20. *New York Times*, January 1, 1874, 1, and January 3, 1874, 4. For examples of the coverage in the metropolitan dailies, see *Philadelphia Inquirer*, November 11–25, 1873; *Atlanta Constitution*, November 7–16, 1873; *New Orleans Picayune*, November 7–16, 1873; *Chicago Tribune*, November 9–16; *New York Tribune*, November 10–25; *San Francisco Examiner*, November 7 and November 10–12, 1873; and *San Francisco Chronicle*, November 9 and November 11–13, 1873. For the administration's view of the press coverage, see Nevins, *Hamilton Fish*, 2:680–91.
21. María Cristina San Román and Guillermo Horacio Gasió, "La presidencia de

Avellaneda," in *La Argentina del ochenta al centenario,* ed. Gustavo Ferrari and Ezequiel Gallo (Buenos Aires: Editorial Sudamerica, 1989), 35–44; Natalio R. Botana, "1880: La federalización de Buenos Aires," in *La Argentina,* 107–122.

22. *Hansard's Parliamentary Debates,* vol. 253 (London: Cornelius Buck, 1880), 183; Ferns, *Britain and Argentina,* 388. On the lack of knowledge in Great Britain about affairs in Argentina, see *Times* (London), July 1, 1880, 8.
23. Edwin Egerton to Earl Granville, June 4, 11, and 24, 1880, FO 6/369, Public Record Office, London; Ferns, *Britain and Argentina,* 389–95; Ezequiel Gallo, "Argentina: Society and Politics, 1880–1916," in *The Cambridge History of Latin America,* vol. 5, *c. 1870–1916,* ed. Leslie Bethel (Cambridge: Cambridge University Press, 1986), 359–63; San Román and Gasió, "La presidencia de Avellaneda," 35–44; Botana, "La federalización," 107–22.
24. Bose, "Historia de las comunicaciones," 582–96; *Memoria presentada por el director del telégrafos nacionales al ministerio del interior correspondiente al ano 1873,* vol. 2 (Buenos Aires: Kidd, 1874), 52–59.
25. Edwin Egerton to Earl Granville, August 1, 1880, FO 6/360; Ferns, *Britain and Argentina,* 392–93.
26. For example, see Nicolas Bouwer to Barings, December 7, 18, 23, and 31, 1879, January 8, 1880, and February 19, 1880, HC 4.1.65, Baring Archive, London.
27. Nicolas Bouwer to Barings, February 2, 1880, and April 12, 1880, HC 4.1.65, Baring Archive. See also Bouwer letter to Barings, February 27, 1880, HC 4.1.65, Baring Archive.
28. Nicolas Bouwer to Barings, June 5, 1880, HC 4.1.65, Baring Archive.
29. Nicolas Bouwer to Barings, June 18 and 21, 1880. Confirmed by letter of June 23, 1880, HC 4.1.65, Baring Archive.
30. Nicolas Bouwer to Barings, July 2, 1880, August 16 and 25, 1880, HC 4.1.65, Baring Archive.
31. *Times* (London), June 15, 1880, 10, June 17, 1880, 5, and June 19, 1880, 7.
32. Ibid., June 15, 1880, 10, June 17, 1880, 5, and June 19, 1880, 7. According to Jorma Ahvenainen, River Plate Telegraph seldom had technical difficulties but did experience interruptions due to political unrest. See Ahvenainen, *European Cable Companies,* 31–34, 76, 115–16.
33. *Times* (London), June 22, 1880, 5, June 26, 1880, 7, June 28, 1880, 8, June 29, 1880, 5, and June 30, 1880, 5.
34. Ibid., July 6, 1880, 11.
35. Ibid., July 5, 1880, 11, and July 6, 1880, 11.
36. Ibid., July 5, 1889, 11. A similar late report that obviously arrived by ship to Lisbon from Buenos Aires announced the June 30 reopening of the port of Buenos Aires on July 30. See *Times* (London), July 30, 1880, 5.

37. George Rauch, *Conflict in the Southern Cone: The Argentine Military and the Boundary Dispute with Chile, 1870–1902* (Westport, CT, Praeger, 1999), 35–56; Geoffrey S. Smith, "The Role of José M. Balmaceda in Preserving Argentine Neutrality in the War of the Pacific," *Hispanic American Historical Review* 49.2 (1969): 254–67. For the background of this episode, see these useful syntheses: Joseph Tulchin, "Argentina: Clash of Global Visions I," in *United States-Latin American Relations, 1850–1903: Establishing a Relationship*, ed. Thomas M. Leonard (Tuscaloosa: University of Alabama Press, 1999), 147–68; and William F. Sater, "Chile: Clash of Global Visions II," in *United States-Latin American Relations*, 169–96.
38. Rauch, *Conflict*, 35–165; Cisneros, Escudé, et al., *Historia general*, chaps. 33–34; Smith, "The Role of José M. Balmaceda." See also Bruce W. Farcau, *The Ten Cents War: Chile, Peru, and Bolivia in the War of the Pacific, 1879–1884* (Westport, CT: Praeger, 2000); Richard O. Perry, "Argentina and Chile: the Struggle for Patagonia, 1843–1881," *The Americas* 36.3 (1980): 347–63; Burr, *By Reason or Force*, 153–55; and Mario Barros, *Historia diplomatica de Chile, 1541–1930* (Barcelona: Ediciones Ariel, 1970), 306–450. Submarine-cable communications played an important role in the War of the Pacific and its aftermath. For an assessment of this topic, see chapter 4.
39. Reggini, *Sarmiento*, 155–74.
40. The telegraph lines in both countries were operational. See "Memoria del director jeneral de telegrafías," in *Memoria del ministerio del interior presentada del congreso nacional en 1881* (Santiago: Imprenta nacional, 1881), 173–81; and "Memoria del director jeneral de telegrafías," in *Memoria del ministerio del interior presentada del congreso nacional en 1883* (Santiago: Imprenta nacional, 1883), xviii–xxi. The participants on Argentine side of the negotiations made use of the new facilities in the large and up-to-date Correos and Telégrafos building located in the heart of Buenos Aires. See Bose, "Historia de las comunicaciones," 588–98.
41. Cisneros, Escudé, et al, *Historia general*, chap. 34; Peterson, *Argentina and the United States*, 240–41; Burr, *By Reason or Force*, 153–55.
42. Cisneros, Escudé, et al, *Historia general*, chap 34; Peterson, *Argentina and the United States*, 242.
43. Cisneros, Escudé, et al, *Historia general*, chap. 34; Peterson, *Argentina and the United States*, 240–46.
44. The Venezuela boundary controversy of 1895–1896 elicited an outpouring of anti-British jingoism in the United States that drew the two countries toward hostilities. See Bertram, *Birth of Anglo-American Friendship*; Harris, "Venezuela"; and Grenville and Young, *Politics, Strategy, and American Diplomacy*, 125–87. On the tensions between the United States and

Spain in the Cuban crises of the 1890s and the Spanish-American War, see Offner, *Unwanted War*; Joyce Milton, *The Yellow Kids: Foreign Correspondents in the Heyday of Yellow Journalism* (New York: Harper and Row, 1989); Campbell, *Yellow Journalism*; and Pérez, *Cuba and the United States*, 55–112.

45. Joyce Goldberg, *The* Baltimore *Affair* (Lincoln: University of Nebraska Press, 1986); William F. Sater, *Chile and the United States: Empires in Conflict* (Athens: University of Georgia Press, 1990); John A. Britton, "'The Confusion Provoked by Instantaneous Discussion': The New International Communications Network and the Chilean Crisis of 1891–92 in the United States," *Technology and Culture: The International Quarterly of the Society for the History of Technology* 48.2 (2007): 729–57.

46. R. L. Thomson to A. F. Walter, July 3, 1890, *Times* of London Archive, London. For examples of the Argentine reporting from London, see *La Prensa*, September 11, 1891, 4, September 13, 1891, 4, September 14, 1891, 3, October 5, 1891, 3, and November 7, 1891, 4. For examples of British coverage of Argentina, see the *Times* (London), July 5, 1890, 7, July 10, 1890, 5, July 11, 1890, 5, and July 19, 1890, 7.

47. For a preliminary assessment of the role of communications in these crises, see John A. Britton, "International Communications and the Anglo-Argentine Crises of the 1890s," paper presented at the annual meeting of SECOLAS, March 3, 2001, Veracruz, Mexico. Among the many works on the diplomatic and economic aspects of these events are Ferns, *Britain and Argentina*, 397–491; Ferns, "Baring Crisis Revisited"; Andrew Thompson, "Informal Empire? An Exploration in the History of Anglo-Argentine Relations, 1810–1914," *Journal of Latin American Studies* 24.2 (1992): 419–36; A. G. Hopkins, "Informal Empire in Argentina: An Alternative View," *Journal of Latin American Studies* 26.2 (1994): 469–84; Cisneros, Escudé, et al., *Historia general*, chap. 40; Gerardo della Paolera and Alan Taylor, *Straining at the Anchor: The Argentine Currency Board and the Search for Macroeconomic Stability, 1880–1935* (Chicago: University of Chicago Press, 2001), 37–95; Marichel, *Century of Debt Crises*, chaps. 5 and 6; Carlos F. Díaz Alejandro, "La economía argentina durante el período 1880–1913," in *La Argentina*, 369–76; A. G. Ford, "Comercio exterior y inversiones extranjeras," in *La Argentina*, 497–512; and Philip Ziegler, *The Sixth Great Power: A History of One of the Greatest of All Banking Families, the House of Barings, 1762–1929* (New York: Knopf, 1988), 229–90.

48. Britton and Ahvenainen, "Showdown in South America."

CHAPTER FOUR

1. *New York Times*, January 7, 1866, 8; Sater, *Chile and the United States*, 26–30. For an overview of Vicuña Mackenna's role in Chilean history, see José Luis Rénique, "Benjamín Vicuña Mackenna: exilio, historia y nación," in *La república peregrina: Hombres de armas y letras en América del Sur, 1800–1884*, ed. Carmen McEvoy and Ana María Stuven (Lima: Instituto francés de estudios andinos, 2007), 487–529.
2. Sater, *Chile and the United States*, 26.
3. *New York Times*, January 7, 1866, 8.
4. Ibid., January 7, 1866; Sater, *Chile and the United States*, 26–30.
5. *New York Times*, February 10, 1866, 2, and February 15, 1866, 2; Sater, *Chile and the United States*, 27–28.
6. Sater, *Chile and the United States*, 28; William C. Davis, *The Last Conquistadores* (Athens: University of Georgia Press, 1950), 291–310; Rodrigo Fuenzalida Bade, *La armada de Chile*, vol. 2 (Valparaíso: Talleres empresa periodista, 1978), 628–30; Scheina, *Latin America's Wars*, 333–40.
7. Jay Monaghan, *Chile, Peru, and the California Gold Rush* (Berkeley: University of California Press, 1973).
8. Knightley, *The First Casualty*, 3–63; Dora Neill Raymond, *British Policy and Opinion during the Franco-Prussian War* (New York: AMS Press, 1967).
9. Read, *Power of News*, 40–110; Desbordes, "Transantionales de l'information"; Ken Ward, *Mass Communications and the Modern World* (Chicago: Dorsey Press, 1989), 46–49.
10. Schwarzlose, *Nation's Newsbrokers*; Victor Rosewater, *History of Coöperative News-Gathering in the United States* (New York: Appleton, 1930), 1–189; Hohenberg, *Foreign Correspondence*, 35–150. On reportage of the U.S. Civil War, see J. Cutler Andrews, *The North Reports the Civil War* (Pittsburgh, PA: University of Pittsburgh Press, 1955). U.S. press coverage of the Cuban revolt against Spain, known as the Ten Years War, was sporadic at best, editorial opinion taking primacy rather than coverage in the field. See Gerald E. Poyo, *"With All, and for the Good of All": The Emergence of Popular Nationalism in the Cuban Communities of the United States, 1848–1898* (Durham, NC: Duke University Press, 1989), 20–51; Bradford, Virginius *Affair*; and Nevins, *Hamilton Fish*, 1:176–200, 2: 667–94, 861–87.
11. Miller, *Britain and Latin America*, 69–178; Larry Clayton, *Peru and the United States* (Athens: University of Georgia Press, 1999), 51–53; Heraclio Bonilla, *Gran Bretana y el Peru* (Lima: Instituto de estudios peruanos, 1997), 4–12; Marichel, *Century of Debt Crises*, 81–90.
12. Clayton, *Peru and the United States*, 53–60; Larry Clayton, *Grace: W. R. Grace*

and Company, 1850–1930 (Ottowa, IL: Jameson Books, 1985), 32–83; Pletcher, *Diplomacy of Trade and Investment*, 195–200.

13. There are many studies of this war in both Spanish and English. The following list represents only a beginning: William F. Sater, *Chile and the War of the Pacific* (Lincoln: University of Nebraska Press, 1986); Gonzalo Bulnes, *La Guerra del Pacífico*, 3 vols. (1911–1919; repr., Santiago: Editorial del Pacifico, 1955); Roberto Querejazu-Calvo, *Guano, salitre, y sangre* (La Paz: Editorial los amigos del libro, 1979); Jorge Basadre, *Reflexiones en torno a la guerra de 1879* (Lima: Campodonico, 1979); and Farcau, *Ten Cents War.*
14. West Coast of America, "Report of the Proceedings at the Extraordinary General Meeting," September 1, 1890, 3, Cable and Wireless Archive, Porthcurno Telegraph Museum, Cornwall, UK; *Times* (London), June 20, 1879, 10, April 24, 1880, 8, July 30, 1880, 11, and August 9, 1880, 7.
15. *Times* (London), October 4, 1879, 5, October 6, 1879, 5, October 10, 1879, 7, October 13, 1879, 10, October 14, 1879, 9, October 20, 1879, 11, and October 21, 1879, 5.
16. Qtd. in ibid., May 30, 1879, 10. A paraphrase of the quote was printed in *New York Times*, May 30, 1879, 1.
17. *New York Times*, May 30, 1879, 1.
18. Ibid., May 30, 1879, 1. The dateline for this story was May 30, which is most probably incorrect.
19. Ibid., June 17, 1879, 7. On June 25 the *Times* carried another fairly accurate, brief account from Valparaíso via Lisbon datelined May 31 (on page 7). The *New York Times* gave a detailed account of the battle on June 16, 1879 (page 1), based on information obtained from sources on the U.S.-registered steamship *Adams* and from the *South Pacific Times.*
20. *Times* (London), October 10, 1879, 7. Greater detail about the capture of the *Huascar* appeared on November 3, 1879, 6, November 13, 1879, 10, and November 18, 1879, 5. The *Times* and Reuters shared news gathering on the War of the Pacific with Havas, the French agency. Havas usually noted the date and point of origin of the coverage and also had a delay of about thirty days. See the analysis in Desbordes, "Transantionales de l'information au Nouveau Monde," chap. 9.
21. *New York Times*, November 4, 1879, 5.
22. *Times* (London), November 11, 1879, 3, and November 12, 1879, 12. More detailed accounts appeared in the *Times* on December 5, 1879, 5, and December 11, 1879, 6. For examples of rumors and "wild stories," see *New York Times*, August 30, 1879, 5, November 11, 1879, 1, and December 4, 1879, 1; and *Times* (London), June 6, 1879, 5, August 23, 1879, 5, August 28, 1879, 5, and December 10, 1879, 5.

23. *New York Times,* November 11, 1879, 1.
24. *Times* (London), December 5, 1879, 5. The dateline reads "Buenos Aires and Valparaíso, December 3, 1879."
25. Ibid., December 29, 1879, 4.
26. Ibid., December 23, 1879, 3.
27. *New York Times,* December 14, 1879, 1, and December 24, 1879, 5; *New York Tribune,* December 28, 1881, 1; *Boston Globe,* December 27, 1881, 1; *Washington Post,* December 25, 1881, 1; *Chicago Tribune,* December 26, 1881, 5; *San Francisco Examiner,* December 27, 1881, 3; *San Francisco Chronicle,* December 27, 1881, 3.
28. For these circulars and other propaganda in the United States, see Herbert Millington, *American Diplomacy and the War of the Pacific* (New York: Columbia University Press, 1948), 26–31.
29. *New York Times,* July 29, 1879, 5. An interesting pro-Peruvian account of the Lavalle mission to Chile can be found in Victor M. Maurtua, *The Question of the Pacific,* trans. F. A. Pezet (Philadelphia: Lasher, 1901), 43–51.
30. *New York Times,* August 25, 1879, 3. The identity of the publisher of the *Peruvian Mail* was not indicated but he was obviously sympathetic to the Peruvian government.
31. Ibid., August 30, 1879, 5. The U.S. press tended to favor Peru and Bolivia in the early stages of the war. See Millington, *American Diplomacy,* 30–31. For examples of comments on rumors and other dubious information from a variety of sources, see *Times* (London), June 6, 1879, 5, August 23, 1879, 5, and December 10, 1879, 5; and *New York Times,* August 30, 1879, 5, November 11, 1879, 1, and December 4, 1879, 1.
32. Clements R. Markham, *The War between Peru and Chile, 1879–1882* (London: Sampson Low, Marston, Searle, and Rivington, 1882), 272–73; Clements Markham to *Times* (London), October 15, 1883, 12. On Markham's lengthy career as a writer on Peruvian topics, see Juan J. R. Villarías Robles, "El peruanismo de Sir Clements Markham (1830–1916)," in *Los americanistas del siglo XIX: La construcción de una comunidad cienífica internacional,* ed. Leoncio López-Ocón, Jean-Pierre Chaumeil, and Ana Verde Casanova (Madrid: Iberoamericana, 2005), 111–43.
33. For example, see *Times* (London), June 3, 1879, 5, June 25, 1879, 7, November 3, 1879, 6, November 14, 1879, 5, December 3, 1879, 5, and January 19, 1880, 5.
34. Marciál Martínez to Ministerio de relaciones exteriores, November 6, 1881, qtd. in Fredrick Pike, *Chile and the United States, 1880–1962* (Notre Dame, IN: University of Notre Dame Press, 1963), 53. For an extensive analysis of

Blaine's role in the War of the Pacific, see David Healy, *James G. Blaine and Latin America* (Columbia: University of Missouri Press, 2001), 54–119.

35. Marciál Martínez to Ministerio de relaciones exteriores, October 8, 1881, October 26, 1881, November 6, 1881, and December 18, 1881, qtd. in Pike, *Chile and the United States*, 53–55 and 323–25. Martínez was also aware of the furor raised in the Chilean press concerning the openly pro-Peruvian statements by U.S. minister to Peru Stephen Hurlbut virtually from the time of his arrival in Lima in August 1881. Stephen Hurlbut's brother William was editor of the pro-Peru, pro-Blaine *New York World*. See Healy, *James G. Blaine*, 85–86; and Millington, *American Diplomacy*, 88–93.
36. Marcial Martínez, *General Considerations in Relation to the Difficulty Between Chile, on the One Hand, and Peru and Bolivia on the Other* (Philadelphia: Times Printing House, 1881), 5. Martínez explained the organization of his propaganda campaign in a letter to Eulogio Altamirano, November 7, 1883, in his *Opusculos sobre asuntos internacionales*, vol. 3 of *Obras completas* (Santiago: La ilustracíon, 1919), 333–44. See also his *Memorandum Delivered by the Minister of Chile to the Secretary of State, Stating the Actual Condition of Affairs on the Pacific Coast* (Washington, D.C.: n.p., 1881). This memorandum was published with three other Chilean government documents and a thirty-five-page explanatory text in *General Considerations*. Martínez wrote an account of his experience in Washington from 1880 to 1881 in the prologue to a Spanish edition of this publication that appeared in 1914. See Marcial Martínez, *Cuestión Chileno-Peruana* (Santiago: La ilustración, 1914), 4–18.
37. Millington, *American Diplomacy*, 26–31.
38. Martínez, *General Considerations*, 27; Martínez, *Cuestión Chileno-Peruano*, 6–9; Francisco Javier Ovalle Castillo, *La personalidad de don Marcial Martínez* (Santiago: Imprenta universitaria, 1918), 86–91; *New York Times*, November 18, 1881, 4, and November 26, 1881, 1.
39. *Boston Globe*, November 25, 1881, 1; *New Orleans Picayune*, November 26, 1881, 1; *Atlanta Constitution*, November 25, 1881, 1; *San Francisco Chronicle*, November 27, 1881, 8; *Chicago Tribune*, November 29, 1881, 1.
40. *Atlanta Constitution*, November 27, 1881, 4 (editorial); *New Orleans Picayune*, November 26, 1881, 1 (article); *San Francisco Examiner*, November 30, 1881, 2 (editorial); *New York Times*, November 18, 1881, 4, and November 26, 1881, 4 (editorials); *New York Tribune*, November 28, 1881, 4 (editorial); *Chicago Tribune*, November 29, 1881, 1 (article). The *Chicago Tribune*'s editorial defended Blaine but faulted the work of Hurlbut in Peru (*Chicago Tribune*, December 2, 1881, 4). The generally Republican *Washington Post* had taken a somewhat reluctant pro-Blaine editorial position on November 25 (page 2) but also testified to the importance of Martínez by mentioning him in this editorial.

41. Martínez, *General Considerations*, 4–5, 13–19, 18, 22–35, and 64–70; Martínez to Altamirano, November 7, 1883, in Martínez, *Opusculos sobre asuntos internacionales*, 339. See also Bulnes, *Guerra del Pacífico*, 3:78–80; Pike, *Chile and the United States*, 47–62; Sater, *Chile and the United States*, 35–50; and Healy, *James G. Blaine*, 83–87. For newspaper coverage, see *New York Times*, November 26, 1881, 1, 6; *New Orleans Times-Picayune*, November 26, 1881, 1; *San Francisco Examiner*, November 27, 1881, 8, and November 30, 2 (editorial); *Chicago Tribune*, November 29, 1881, 1, and December 2, 1881, 4 (editorial).
42. Clayton, *Peru and the United States*, 71–73. Maurtua, *Question of the Pacific*, 61–163, provides extensive quotations from diplomatic documents and a discussion of the efforts by the United States to influence the outcome of the peace negotiations.
43. Knightley, *First Casualty*, 45–50; Raymond, *British Policy and Opinion*.
44. Sater, *Chile and the United States*, 43–50; Bulnes, *Guerra del Pacífico*, 3:80–121; Healy, *James G. Blaine*, 54–119.
45. Martínez, *Opusculos sobre asuntos internacionales*, 335–91; Ovalle Castillo, *La personalidad*, 91–103; Russell Bastert, "Diplomatic Reversal: Frelinghuysen's Opposition to Blaine's Pan-American Policy," *Mississippi Valley Historical Review* 42.4 (1956): 653–70; Sater, *Chile and the United States*, 42–46; José Miguel Yrarrázaval Larrain, *El Presidente Balmaceda*, 2 vols. (Santiago: Editorial nascimiento, 1940), 1:223–32; Bulnes, *Guerra del Pacífico*, 111–21; Edward Crapol, *James G. Blaine: Architect of Empire* (Wilmington, DE: Scholarly Resources, 2000), 70–74; Millington, *American Diplomacy*, 121–43; Healy, *James G. Blaine*, 106–8.

CHAPTER FIVE

1. *New York Times*, September 27, 1883.
2. Ivan Musicant, *The Banana Wars: A History of United States Military Intervention in Latin America from the Spanish-American War to the Invasion of Panama* (New York: Macmillan, 1990), 81–96; Kenneth Hagan, *American Gunboat Diplomacy and the Old Navy, 1877–1889* (Westport, CT: Greenwood, 1973), 160–87; Daniel H. Wicks, "Dress Rehearsal: United States Intervention on the Isthmus of Panama, 1885," *Pacific Historical Review* 49.4 (1980): 581–605. For a valuable perspective on the U.S. Navy in the 1880s see Wicks's "New Navy and New Empire: The Life and Times of John Grimes Walker" (PhD diss., University of California, Berkeley, 1979), esp. 38–207 and 376–428. For the number of U.S. Marines and sailors on shore, see the *New York Tribune*, April 22, 1885, 5; and Robert Coontz, *From the Mississippi to the Sea* (Philadelphia: Dorrance, 1930), 92. The *New York Times* account (September 4, 1885, 3) said that Prestán was suspended by a derrick over a railroad car.

3. Stephen J. Randall, *Colombia and the United States: Hegemony and Interdependence* (Athens: University of Georgia Press, 1992), 69.
4. *New York Times,* January 15, 1880, and May 11, 1881.
5. Ferdinand de Lesseps to James Scrymser, March 31, 1880, qtd. in Scrymser, *Personal Reminiscences,* 90. De Lesseps also praised Scrymser's intentions to connect South America with Australia and New Zealand and thereby reach Asia through John Pender's Eastern Telegraph Company and its affiliates, but this part of the plan never came into being.
6. Ahvenainen, *History of the Caribbean Telegraphs;* Ahvenainen, *European Cable Companies.* For the global context, see Headrick, *Invisible Weapon,* 3–92. On the French efforts in Panama, see McCullough, *Path Between the Seas,* 124–81; Gerstel Mack, *The Land Divided: A History of the Panama Canal and Other Isthmian Projects* (1944; repr., New York: Octagon, 1974), 281–414; Simon, *Panama Affair,* 1–123; and Schoonover, *The French in Central America,* 55–82.
7. *New York Times,* October 1, 1882.
8. Ibid., December 21, 1881, and October 21, 1882.
9. Ahvenainen, *History of the Caribbean Telegraphs,* 50–55, 80–86; Britton and Ahvenainen, "Showdown in South America"; West India and Panama, board minutes, May 30 and September 26, 1883, Cable and Wireless Archive, Porthcurno Telegraph Museum, Cornwall, UK.
10. *New York Times,* June 4, 1884. According to Ahvenainen, by comparison West India and Panama saw a sharp decline in profits from £46,871 in 1883 to £28,745 in 1886 but returned to the £40,000 range in the late 1880s (*History of the Caribbean Telegraphs,* 84).
11. David McCreery, *Development and the State in Reforma Guatemala* (Athens: Ohio University, Center for International Studies, 1983); Ralph Lee Woodward, *Central America: A Nation Divided* (New York: Oxford University Press, 1999), chaps. 5–6; Jim Handy, *Gift of the Devil: A History of Guatemala* (Boston: South End Press, 1984), chap. 3; Thomas Karnes, *The Failure of Union: Central America, 1824–1960* (Chapel Hill: University of North Carolina Press, 1961), chap. 7.
12. Karnes, *Failure of Union,* 156–57.
13. Ibid., 158.
14. Ibid., 158–60. The *New York Times* reported from Mexico City via Scrymser's cable that the telegraph had prevented war in Central America. The public assumed that Díaz's announcement would stop the Barrios offensive into El Salvador, and so the mood in Mexico City was celebratory: "The opinion is prevalent here that if not for the existence of the cable companies an old-fashioned war would have occurred among the Central America republics"

(*New York Times*, March 13, 1885). This opinion proved to be premature. See *New York Times*, March 28, 1885; and Karnes, *Failure of Union*, 158–60.

15. Karnes, *Failure of Union*, 160; Jürgen Buchenau, *In the Shadow of the Giant: The Making of Mexico's Central American Policy, 1876–1930* (Tuscaloosa: University of Alabama Press, 1996), 35.
16. *New York Times*, March 8 and 13, 1885, and April 1, 1885; *San Francisco Chronicle*, April 5, 1885, 1; Karnes, *Failure of Union*, 160.
17. Buchenau, *Shadow of the Giant*, 36; *New York Times*, March 18, 1885.
18. Karnes, *Failure of Union*, 161.
19. *New York Times*, March 17 and 18, 1885; Hagan, *American Gunboat Diplomacy*, 171.
20. *New York Times*, March 18, 1885.
21. *New York Times*, April 2, 1885. See accounts in the *Washington Post, Chicago Tribune, Atlanta Constitution, New Orleans Picayune*, and *San Francisco Chronicle*, April 2–6, 1885; and also Karnes, *Failure of Union*, 162; Buchenau, *Shadow of the Giant*, 30–39; and Paul Burgess, *Justo Rufino Barrios: A Biography* (Quetzaltenango, Guatemala: El noticiero evangelico, 1946), 226–33.
22. Thomas M. Leonard, "Central America: The Search for Economic Development," in *United States-Latin American Relations, 1850–1903: Establishing a Relationship*, ed. Thomas M. Leonard (Tuscaloosa: University of Alabama Press, 1999), 93–95. For additional information on European business influences in Latin America, see Walther L. Bernecker, "Las relaciones entre Europa y Latinoamerica durante el sigle XIX: Ofensivas comerciales e intereses economicos," *Hispania* 53.183 (1993): 177–212.
23. Schoonover, *United States in Central America*, 68–69; Schoonover, *Germany in Central America*, 75–76. For additional information on German influence in the Western hemisphere in this period, see Ian Forbes, "German Informal Imperialism in South America before 1914," *Economic History Review* 31.3 (1978): 384–98.
24. Schoonover, *United States in Central America*, 62–69; *New York Times*, March 23, 1878, and April 24, 1878; *Times* (London), March 30, 1878.
25. Schoonover, *United States in Central America*, 55.
26. Qtd. in ibid., 55.
27. Schoonover, *French in Central America*, 1–82; Hanna and Hanna, *Napoleon III and Mexico*; Brian Hammett, *Juárez* (London: Longman, 1994).
28. Ahvenainen, *History of the Caribbean Telegraphs*, 94–109.
29. McCullough, *Path*, 124–81; Mack, *Land Divided*; Simon, *Panama Affair*. The use of the telegraph to signal the beginning of construction is described in Mack, *Land Divided*, 319.
30. Pletcher, *Diplomacy of Trade*, 132–33.

31. Ibid., 131–32.

32. John Kasson, "The Monroe Doctrine in 1881," *North American Review* 133 (December 1881): 530–31; Ari Hoogenboom, *Rutherford B. Hayes, Warrior and President* (Lawrence: University Press of Kansas, 1995), 414–21; Healy, *James G. Blaine*, 40–53; McCullough, *Path*; 45–152; David M. Pletcher, *The Awkward Years: American Foreign Relations Under Garfield and Arthur* (Columbia: University of Missouri Press, 1961), 22–39, 102–15, 270–83; Smith, *Illusions of Conflict*, 81–116.

33. Helen Delpar, "Colombia: Troubled Friendship," in *United States-Latin American Relations, 1850–1903*, 58–60; Randall, *Colombia and the United States*, 26–42; Mercedes Chen Daley, "The Watermelon Riot: Cultural Encounters in Panama City, April 15, 1856," *Hispanic American Historical Review* 70.1 (1990): 85–108; Aims McGuiness, *Path of Empire: Panama and the California Gold Rush* (Ithaca, NY: Cornell University Press, 2009); John H. Kemble, *The Panama Route* (New York: Da Capa, 1972).

34. McCullough, *Path*, 170–74; Mack, *Land Divided*, 342; Delpar, "Colombia," 70–71.

35. Randall, *Colombia and the United States*, 69.

36. Musicant, *Banana Wars*, 82–96; Hagan, *American Gunboat Diplomacy*, 160–87; McCullough, *Path*, 153–81.

37. Delpar, "Colombia," 70–74; Musicant, *Banana Wars*, 86–86; McCullough, *Path*, 175–77.

38. Lewis Clark to William E. Chandler, June 18, 1885, in *Military Forces of the United States in Colombia*, Senate documents, vol. 4, 58th Congress, 2nd sess., serial 4589 (Washington, D.C.): U.S. Government Printing Office, 1904), 53; Robert K. Wright Jr. to Lewis Clark, January 18, 1885, in *Military Forces*, 54; Thomas Adamson to William Hunter, *Military Forces*, 55–57.

39. Hagan, *American Gunboat Diplomacy*, 171; Alfred Thayer Mahan, *The Letters and Papers of Alfred Thayer Mahan*, 3 vols., ed. Robert Seager II and Doris Maguire (Annapolis, MD: Naval Institute Press, 1975), 1:594–601. See especially Mahan to William C. Whitney, April 27, 1885.

40. Hagan, *American Gunboat Diplomacy*, 172–73.

41. Paraphrases of the telegram in *New York Tribune*, April 2, 1885. See also the *New York Times*, April 2, 1885.

42. *New York Tribune*, April 2, 1885; Mark D. Hirsch, *William C. Whitney: Modern Warwick* (New York: Dodd, Mead, 1948), 270–72.

43. *New York Tribune*, April 3, 1885, 1, and April 7, 1885, 1; *Report of Commander McCalla upon the Naval Expedition to the Isthmus of Panama* (Washington, D.C.: U.S. Government Printing Office, 1885), 3–11; Bowman Hendry

McCalla, "Memoirs of a Naval Career," typescript, 1910, chap. 13, Manuscript Division, Library of Congress, Washington, D.C. See also Paolo E. Coletta, *Bowman Hendry McCalla: A Fighting Sailor* (Washington, D.C.: University Press of America, 1979), 42–44; and Hagan, *American Gunboat Diplomacy*, 172–82. On the limitations involving the use of the telegraph to transmit detailed commands from Washington to Panama, see Wicks, "Dress Rehearsal," 583.

44. *Report of Commander McCalla*, 11–16; *New York Times*, April 25, 1885, 1; *New York Tribune*, April 25, 1885, 1; Hagan, *American Gunboat Diplomacy*, 182–87.
45. *New York Tribune*, April 26, 1885, 1; McCalla, "Memoirs of a Naval Career," chap. 13; Wicks, "Dress Rehearsal," 593; Coletta, *Bowman Hendry McCalla*, 44–46; Hagan, *American Gunboat Diplomacy*, 177.
46. *New York Tribune*, April 27, 1885, 1. McCalla defended his decision to occupy the CSA cable station in central Panama City in his "Memoirs of a Naval Career," chap. 13.
47. William C. Whitney to James E. Jouett, April 3, 1885, in *Military Forces*, 135; Jouett to Whitney, April 17, 1885, in *Military Forces*, 106–11; Bowman McCalla, "Notice to the Public," April 24, 1885, in *Military Forces*, 120; *Report of Commander McCalla*, 12–18. Jouett's telegram to the Navy Department received wide circulation in the U.S. press. "It was absolutely necessary to occupy Panama to protect transit and American property. The safety of Panama and American interests are secured by this move. Your instructions were published in the Panama papers. No misunderstanding will result. Government officials will arrive soon, when everything will be turned over to them and we will withdraw." See *New York Times*, April 26, 1885, 1; *New York Tribune*, April 26, 1885, 1; and *Chicago Tribune*, April 26, 1885, 9. See also Hagan, *American Gunboat Diplomacy*, 184–87.
48. John G. Walker to Bowman H. McCalla, April 6, 1885, John G. Walker Papers, Manuscript Division, Library of Congress, Washington, D.C. This letter is also quoted in part in Hagan, *American Gunboat Diplomacy*, 181. See also Wicks, "Dress Rehearsal," 590–92. McCalla expressed his interest in the U.S. press and public reactions to the intervention in his "Memoirs of a Naval Career," chapter 13, 15–16.
49. Walker to McCalla, April 6, 1885; Wicks, "Dress Rehearsal," 593. For examples of King's reporting, see *New York Tribune*, April 11, 22, 23, 25, 26, 27, 28, 29, 30, and May 1 and 2, 1885. The potential for the release of information in the press to cause problems was exemplified by the unexplained publication of a terse telegram from Commander Kane of the *Galena*, dated April 2, to the

State Department in Washington stating his concern that Colombian authorities would promptly release any of the rebels captured by U.S. forces in the event that he or any other officer turned their prisoners over to local officials. The publication of this telegram led to hostile exchanges between Colombian and U.S. diplomats for several days. See Hagan, *American Gunboat Diplomacy*, 177–78.

50. *Chicago Tribune*, April 25, 1885, 3; *New York Times*, April 25, 1885, 1; *Washington Post*, April 25, 1885, 1; *San Francisco Examiner*, April 25, 1885, 1.
51. *New Orleans Picayune*, April 26, 1885, editorial page, and April 28, 1885, editorial page; *San Francisco Chronicle*, April 25, 1885, 3, April 26, 1885, 3, and April 27, 1885, 3. See also *San Francisco Chronicle*, April 2, 1885, 2.
52. *New York Tribune*, April 29, 1885, 1. For King's supportive coverage of U.S. forces, see the *New York Tribune*, April 11, 1885, 1, and April 22–May 2, 1885. Most of his articles appear on page 1. Historian Mercedes Chen Daley argues that the 1857 uprising was an anti-foreign (largely anti-U.S.) movement with an extensive social base of support. See her "Watermelon Riot," 85–108.
53. *New York Times*, September 4, 1885, 3; Ricardo Becerra and Jorge Holguin to Thomas F. Bayard, July 31, 1885, in *Papers Relating to the Foreign Policy of the United States, 1885* (Washington, D.C.: Government Printing Office, 1886), 279; Ricardo Becerra to T. F. Bayard, August 28, 1885, in *Papers*, 280.
54. Coontz, *From the Mississippi*, 92. For photographic corroboration of the Coontz version, see McCullough, *Path*, 96. For differing accounts, see *New York Times*, September 4, 1885, 3; and *New York Tribune*, September 4, 1885.
55. Gustave Anguizola, *Philippe Bunau-Varilla: The Man behind the Panama Canal* (Chicago: Nelson Hall, 1980), 65; Ovidio Díaz Espino, *How Wall Street Created a Nation: J. P. Morgan, Teddy Roosevelt, and the Panama Canal* (New York: Four Walls Eight Windows, 2001); McCullough, *Path*, 329–402.
56. Delpar, "Colombia," 64, 79–80. The *New York Times* (April 11, 1885) reported that telegraphic communications between Bogotá and the United States had been interrupted for three months before being restored on April 10, 1885.
57. Robert Seager II, "Alfred Thayer Mahan: Christian Expansionist, Navalist, Historian," in *Makers of the American Naval Tradition*, ed. James C. Bradford (Annapolis, MD: Naval Institute Press, 1990), 33. See also Seager's (ed.) *Alfred Thayer Mahan*, 146–49; William E. Livezey, *Mahan on Sea Power* (Norman: University of Oklahoma Press, 1947), 93–174; and Alfred Thayer Mahan, *The Influence of Seapower upon History* (New York: Sagamore Press, 1957), esp. 29–30 and 72–73.
58. Seager, *Alfred Thayer Mahan*, 211–307; Warren Zimmermann, *First Great Triumph* (New York: Farrar, Straus and Giroux, 2002), 85–122. Mahan

mentions communications in *Influence* on page 30, which may be read as a reference to transportation. Walker was supportive of Mahan and his historical and strategic research and writing. See Wicks, "New Navy and New Empire," 138–52.

59. Walker to McCalla, April 6, 1885; McCalla, "Memoirs of a Naval Career," chap. 13, 15–16.

60. Wicks, "Dress Rehearsal," 591, 596. Wicks also extensively discusses Walker's interest in public relations and the press coverage of the intervention in his "New Navy and New Empire," 59–60, 387–427. Mahan generally did not include telegraphic communications in his analyses of naval strategy. For example, in his 1890 contingency plan for war between the United States and Great Britain, Mahan did not consider the role of telegraphy. In this plan, according to Seager, Mahan "failed to take into account the impact that telegraphic communications were likely to have on fleet operation. (He still thought of fleet communications in terms of telescope-fitted observation balloons and carrier pigeons.)" See Seager, *Alfred Thayer Mahan*, 230. By contrast, Walker saw telegraphic contact with U.S. forces on sea and land in the Panama intervention, as well as telegraphed newspaper reports from the scene, as crucial.

61. Navy Department, General Order No. 324, October 3, 1884, microcopy 984, roll 1, National Archives, Washington, D.C.

62. See letters from John G. Walker to B. F. Stevens, December 7, 1891, and February 1, Walker Papers. See also telegrams from him dated April 15 and 27 and May 15, 16, and 18, 1892, Walker Papers.

63. Navy Department, General Orders No. 356, July 9 1886, No. 377, October 31, 1889, No. 417, August 19, 1889, No. 428, July 24, 1894, No. 451, July 8, 1895, and No. 464, July 23, 1896, microcopy 984, roll 1, National Archives.

64. Nickles, *Under the Wire*; Arthur Lubow, *The Reporter Who Would Be King: A Biography of Richard Harding Davis* (New York: Scribner, 1992); Milton, *Yellow Kids*; Brown, *Correspondents' War*; Campbell, *Yellow Journalism*; Knightley, *First Casualty*, 55–78; Thomas Pakenham, *The Boer War* (New York: Random House, 1979), 171–72, 289–91, 378–85.

65. *New York Tribune*, April 29, 1885, 1.

CHAPTER SIX

1. Headrick, *Tools of Empire*, 157–64; Headrick, *Invisible Weapon*, 28–72.

2. Western and Brazilian, directors' report, June 18, 1880, Cable and Wireless Archive, Porthcurno Telegraph Museum, Cornwall, UK.

3. Western and Brazilian, directors' report, February 28, 1887, Cable and Wireless Archive.

4. Western and Brazilian, directors' report, March 11, 1881, December 30, 1881, January 19, 1885, and January 13, 1886, Cable and Wireless Archive. For a discussion of sheathed cables, see Bright, *Submarine Telegraphs,* 403–91.
5. *New York Times,* December 21, 1881, 8, March 8, 1885, and January 12, 1889, 1.
6. Bright, *Submarine Telegraphs,* 604–34.
7. Charles R. Flint to William E. Curtis, April 20, 1892, James G. Blaine Papers, Manuscript Division, Library of Congress, Washington, D.C.; Williams E. Curtis to Charles R. Flint, April 12, 1892, James G. Blaine Papers.
8. West Coast of America, minute books, March 23, March 27, April 13, April 17, and April 26, 1882, Cable and Wireless Archive.
9. West Coast of America, minute books, November 22, 1882, December 20, 1882, April 2, 1883, and August 2, 1883, Cable and Wireless Archive.
10. *Times* (London), January 23, 1885, 11, and March 9, 1887, 11.
11. Ibid., June 12, 1890; Britton and Ahvenainen, "Showdown in South America."
12. Sater, *Chile and the United States,* 35–68; Pike, *Chile and the United States,* 55–59.
13. Sater, *Chile and the United States,* 51–53.
14. Burr, *By Reason or Force,* 175–96.
15. Walter R. Herrick, *The American Naval Revolution* (Baton Rouge: Louisiana State University Press, 1966); B. Franklin Cooling, *Benjamin Franklin Tracy: Father of the Modern American Fighting Navy* (Hamden, CT: Archon, 1973); Bradford, ed., *Makers of the American Naval Tradition.*
16. Simon Collier and William F. Sater, *A History of Chile, 1808–1994* (Cambridge: Cambridge University Press, 1996), 149–54
17. Collier and Sater, *History of Chile,* 149–54.
18. West Coast of America, minute books, March 29, April 17 and 26, October 4, November 22, and December 20, 1882, and July 19, 1883, Cable and Wireless Archive; "Agreement for Working (with CSA)," July 27, 1883, Cable and Wireless Archive. For an overview, see Britton and Ahvenainen, "Showdown in South America."
19. *New York Times,* May 11, 1881, 5, October 1, 1881, 2, December 21, 1881, 8; James Scrymser to Frederick T. Frelinghuysen, December 9, 1884, in *Papers Relating to the Foreign Relations of the United States, 1885* (Washington, D.C.: U.S. Government Printing Office, 1886), 1–2.
20. For the West Coast of America's version of Scrymser's policies see that company's minute books, January 29, February 13 and 26, April 23 and 30, and June 4 and 18, 1890, Cable and Wireless Archive; as well as West Coast

of America, "Report of the Extraordinary General Meeting, September 1, 1890," 4–5, Cable and Wireless Archive.

21. Britton and Ahvenainen, "Showdown in South America," 17–23.
22. Scrymser to Frelinghuysen, December 9, 1884, 1–2; *New York Times*, January 15, 1880, 1, and May 11, 1881, 5.
23. *New York Times*, December 21, 1881, 8.
24. *Papers*, 54–60; Britton and Ahvenainen, "Showdown in South America," 13–14.
25. Britton and Ahvenainen, "Showdown in South America," 17–27; Ahvenainen, *European Cable Companies*, 96–114, 215–18.
26. Goldberg, Baltimore *Affair*, 47–49. For Scrymser's defense of his actions, see *New York Times*, November 3, 1891, 2; and Scrymser, *Personal Reminiscences*, 84.
27. Charles W. Calhoun, *Benjamin Harrison* (New York: Times Books, 2005), 83–129; R. Hal Williams, *Years of Decision: American Politics in the 1890s* (Prospect Heights, IL: Waveland Press, 1993), 47–68.
28. Goldberg, Baltimore *Affair*, 76–123; Healy, *James G. Blaine*, 221–34.
29. Healy, *James G. Blaine*, 205–15; Goldberg, Baltimore *Affair*, 20–32; Sater, *Chile and the United States*, 59–60.
30. Goldberg, Baltimore *Affair*, 78.
31. *New York Times*, May 5, 1891, 5.
32. See *Chicago Tribune, San Francisco Examiner, San Francisco Chronicle, New Orleans Picayune*, and *New York Times*, May 6, 7, and 8, 1891. Most of the coverage appears on the front page.
33. Osgood Hardy, "The *Itata* Incident," *Hispanic American Historical Review* 5.2 (1922), 204–12.
34. *New York Times*, May 8, 1891, 1.
35. Ibid., 1891, 1.
36. Ibid., May 8, 1891, 1.
37. Hardy, "*Itata*," 212; *San Francisco Examiner*, May 12, 1891, 1.
38. *San Francisco Chronicle*, May 17, 1891, 1; *New York Times*, May 16, 1891, 1.
39. *New York Times*, May 18, 1891, 1.
40. Ibid., May 16, 1891, 1. See the *New York Tribune, Boston Globe, Philadelphia Inquirer, New Orleans Picayune, Atlanta Constitution, Chicago Tribune, Washington Post, San Francisco Examiner*, and *San Francisco Chronicle* from May 10, 1891, to June 5, 1891, for coverage of the chase. Most of the stories on the *Itata* affair appear on the front page or the first page that carries news items. The *Boston Globe* headline is from May 17, 1891, 1.
41. *San Francisco Chronicle*, May 17, 1891, 1. This was also quoted in *New York Times*, May 18, 1891, 1.
42. *San Francisco Examiner*, June 5, 1891, 1; *New York Times*, June 5, 1891, 2; *New*

Orleans Picayune, June 5, 1891, 1; *Washington Post,* June 5, 1891, 1. For similar coverage see the *Atlanta Constitution, Chicago Tribune, New York Tribune,* and *San Francisco Chronicle,* June 5 and June 6, 1891.

43. Goldberg, Baltimore *Affair,* 1–19.
44. *New York Times,* October 18, 1891, 1.
45. Pike, *Chile and the United States,* 73–74.
46. *Times* (London), November 2, 1891, 13.
47. Ibid., November 2, 1891, 13.
48. *New York Times,* November 3, 1891, 2.
49. Goldberg, Baltimore *Affair,* 48.
50. *New York Times,* November 20, 1891, 1; Goldberg, Baltimore *Affair,* 47–51; Bravo Valdivieso, Bulnes Serrano, and Vial Correa, *Balmaceda y la guerra civil* (Santiago: Editorial fundación, 1991), 353–54; Pike, *Chile and the United States,* 68.
51. Pike, *Chile and the United States,* 71.
52. Goldberg, Baltimore *Affair,* 94.
53. Ibid., 82.
54. Ibid., 81–83.
55. James Soley to Benjamin Tracy, December 23 and 28, 1891 (telegrams), and December 24, 1891 (letter), box 11, Benjamin Tracy Papers, Manuscript Division, Library of Congress, Washington, D.C.; Sater, *Chile and the United States,* 64–65.
56. Goldberg, Baltimore *Affair,* 84–86.
57. For accounts datelined from Chile, see *Chicago Tribune,* December 14, 1891, 1; and *San Francisco Examiner,* December 13, 1891, 1, and December 14, 1891, 1. For accounts from Chile via London and Panama, see *New York Tribune,* December 15, 1891, 1; and *New York Times,* December 13, 1891, 1, December 14, 1891, 1, December 23, 1891, 1, and December 25, 1891, 1.
58. *New York Tribune,* January 17, 1892, 1.
59. *San Francisco Examiner,* January 19, 1892, 1.
60. C. P. Huntington to Benjamin Tracy, January 17, 1891 (letter), and January 18 and 19, 1891 (telegrams), box 11, Tracy Papers; Benjamin Tracy to C. P. Huntington, January 18 and 20, 1891 (telegrams), box 26, Tracy Papers; Cooling, *Benjamin Franklin Tracy,* 119–22.
61. *Philadelphia Inquirer,* January 20, 1892, 1; *Chicago Tribune,* January 26, 1892, 3.
62. *Chicago Tribune,* January 25, 1892, 4.
63. *Washington Post,* January 25, 1892, 1.
64. *San Francisco Examiner,* January 19, 1892, 2.
65. *New Orleans Picayune,* January 24, 1892, 1.
66. *State* (Columbia, SC), January 24, 1892, 1.
67. *San Francisco Examiner,* January 19, 1892, 2, and January 21, 1892, 1.

68. *New York Tribune*, January 25, 1892, 1.
69. *Washington Post*, January 14, 1892, 1; *Boston Globe*, January 20, 1892, 1.
70. *Boston Globe*, January 18, 1891, 1.
71. Austin Corbin to Benjamin Tracy, January 18, 1892, box 11, Tracy Papers.
72. Ross to the Chilean minister of foreign relations, January 22, 1892, qtd. in Pike, *Chile and the United States*, 78. For Ross's experiences in producing Chilean propaganda in Europe, see Augusto Matte and Agustín Ross, *Memoria presentada a la excma junta de gobierno* (Paris: Imprimerie et librarie administratives et critiques Paul Dupont, 1892).
73. Ross to the Chilean minister of foreign relations, January 22, 1892, qtd. in Pike, *Chile and the United States*, 78.
74. Britton, "Confusion Provoked by Instantaneous Discussion"; *Philadelphia Inquirer*, January 24, 1892, 1.
75. Historian David Paull Nickles notes that U.S. diplomats in the 1880s and 1890s were instructed to use the telegraph only in emergency situations (*Under the Wire*, 178). Harrison emerged as an effective practitioner of telegraph diplomacy in comparison to the diplomats and politicians discussed in Nickles's study. On the *Virginius* affair see Bradford, Virginius *Affair*, 25–85; and Nevins, *Hamilton Fish*, 2:607–87. For the Samoan crisis, see Paul Kennedy, *The Samoan Tangle: A Study in Anglo-German-American Relations, 1878–1900* (New York: Harper and Row, 1974), 51–97. For the Bering Sea controversy and a helpful summary of Harrison's diplomacy in general, see Calhoun, *Benjamin Harrison*, 119–33. Calhoun also summarizes the U.S.-Italian dispute regarding the deaths of several Italians in a New Orleans riot in 1891. The diplomatic negotiations in this case continued for a year. Another relevant case involved the communications between Colombia and the United States regarding the 1885 U.S. intervention in Panama. These exchanges spread out over the month of April, and a number of them were conducted via telegraphy. See Delpar, "Colombia," 58–80.

    Secretary of State Blaine instructed the Chilean minister in Washington that Harrison would not accept "any further diplomatic delays." See *New York Daily Tribune*, January 24, 1892, 1. The Chileans apparently responded on January 23, but the precise time that Harrison saw this response remains uncertain. See Goldberg, Baltimore *Affair*, 128–29; and Healy, *James G. Blaine*, 228–29.
76. Goldberg, Baltimore *Affair*, 128–29. All of the following metropolitan papers carried the January 25 report from Santiago on their front pages on January 26: *New York Times, New York Tribune, Boston Globe, Philadelphia Inquirer, Washington Post, Atlanta Constitution, New Orleans Picayune, Chicago Tribune, San Francisco Examiner*, and *San Francisco Chronicle*.

77. Goldberg, Baltimore *Affair*; Healy, *James G. Blaine*; Emilio Meneses Ciuffardi, *El factor naval en las relaciones entre Chile y los Estados Unidos (1885–1951)* (Santiago: Ediciones pedagógicas chileanas, 1989); José Miguel Barros Franco, *El caso del* Baltimore: *Apuntes para la historia diplomatica de Chile* (Santiago: Universidad de Chile, 1950); Pike, *United States and Chile.*
78. *Congressional Record,* 52nd Congress, 1st sess., vol. 23 (Washington, D.C.: U.S. Government Printing Office, 1893), 551.
79. *New York Times,* November 1, 1891, 1.
80. Ibid., November 8, 1891, 5.
81. *Times* (London), November 11, 1891, 1.
82. E. L. Godkin, "The Secret of Warriorism," *Nation,* March 24, 1892, 222–23. See also Godkin's "The Ultimatum," *Nation,* January 28, 1892, 64–66. For more on Godkin, see two books by William M. Armstrong: *E. L. Godkin and American Foreign Policy, 1865–1900* (New York: Bookman, 1957); and *E. L. Godkin: A Biography* (Albany: State University of New York Press, 1978). For a useful interpretive article, see Edward Caudill, "E. L. Godkin and His (Special and Influential) View of 19th Century Journalism," *Journalism Quarterly* 69.4 (1992): 1039–49.
83. Godkin, "Secret of Warriorism," 222–23.
84. *Washington Post,* January 14, 1892, 1.

CHAPTER SEVEN

1. *New York Times,* September 17, 1891, 5, and September 20, 1891, 2.
2. Ibid., December 25, 1891, 1, and December 26, 1891, 1.
3. Elliot Young, *Catarino Garza's Revolution on the Texas-Mexico Border* (Durham, NC: Duke University Press, 2004), 194.
4. Pletcher, *Diplomacy of Trade and Investment,* 77–113; David M. Pletcher, *Rails, Mines, and Progress: Seven American Promoters in Mexico, 1867-1911* (Ithaca, NY: Cornell University Press, 1958); Fernando Rosenzweig, "El comercio exterior," in *Historia moderna de Mexico,* ed. Daniel Cosio Villegas (Mexico City: Editorial Hermes, 1974), 635–729; Luis Nicolau D'Olwer, "Las inversiones extranjeras," in *Historia moderna,* 973–1185; Stephen Haber, *Industry and Underdevelopment: The Industrialization of Mexico, 1890–1940* (Stanford, CA: Stanford University Press, 1989).
5. *New York Times,* March 25, 1881, 8, and May 11, 1881, 5.
6. Young, *Catarino Garza's Revolution,* chaps. 1–2; José Luis Navarro Burciaga, "Catarino Garza, periodista opositor a Porfirio Díaz en Tamaulipas," in *Porfirio Díaz frente al descontento popular regional (1891–1893),* ed. Frederick Katz and Jane-Dale Lloyd (Mexico: Universidad Iberoamericana, 1986), 59–78.

7. *New York Times*, December 25, 1891, 1, and December 26, 1891; *Washington Post*, December 26, 1891, 1.
8. *New York Times*, December 28, 1891, 1; Rebecca Robbins Raines, *Getting the Message Through: A Branch History of the U.S. Army Signal Corps* (Washington, D.C.: Center for Military History, 1996), 82.
9. *San Francisco Chronicle*, January 3, 1892, 11. See also *San Francisco Chronicle*, January 4, 1892, 1 and 4; and *New York Times*, January 4, 1892, 1.
10. Young, *Catarino Garza's Revolution*, 207–8.
11. *New Orleans Picayune*, January 3, 1892, 2.
12. Ibid., January 5, 1892, 1.
13. Ibid., January 11, 1892, 1.
14. *New York Tribune*, January 9, 1892, 3.
15. *New York Times*, January 1, 3, 4, 7, 10, 11, and 20, 1892. Most of the coverage appears on page 1 or 2. The *Chicago Tribune* carried a map of the border uprising; see January 12, 1892, 2. See also the *Boston Globe, Philadelphia Inquirer, Atlanta Constitution, New York Tribune, Washington Post, Chicago Tribune*, and *San Francisco Examiner* in the same period, as well as Charlotte Roehl, "Porfirio Díaz in the Press of the United States" (PhD diss., University of Chicago, 1953), 14–17, 61–64. Garza received reports on the coverage of his movement in the U.S. press. See a letter to Garza reproduced in Navarro Burciaga, "Catarino Garza," 79–80.
16. H. W. Allen, "President Díaz and the Mexico of Today," *Review of Reviews* 61 (January 1893): 678.
17. *New York Times*, February 9, 1892, 2.
18. Young, *Catarino Garza's Revolution*, 268–69.
19. Paul Garner, *Porfirio Díaz* (London: Longman, 2001), chap. 6; Mauricio Tenorio-Trillo, *Mexico at the World's Fairs: Crafting a Modern Nation* (Berkeley: University of California Press, 1996); Adolfo Duclós Salinas, *The Riches of Mexico and Its Institutions* (St. Louis, MO: Nixon-Jones Printing Company, 1893). Duclós Salinas prepared his volume for the Columbian Exposition.
20. *New York Times*, December 24, 1891, 1
21. Young, *Catarino Garza's Revolution*, 198–99.
22. Ibid., 201–3.
23. *New York Times*, December 31, 1891, 1. See also *New York Times*, February 25, 1892, 1. Midwestern newspapers such as the *St. Louis Republic* and the *Cleveland Leader* took pro-Díaz positions. See Roehl, "Porfirio Díaz in the Press," 14–17.
24. *New York Tribune*, January 9, 1892, 3.
25. Matías Romero, *Articulos sobre México publicado en los Estados Unidos de América por Matías Romero en 1891–1892* (México: Oficina impresora de

estampillas, 1892). On Romero's lengthy career in Washington, see Harry Bernstein, *Matías Romero, 1837–1898* (Mexico: Fondo de cultura económica, 1973). Romero explained the public relations and propaganda methods he used in the United States in his preface to *Mexico and the United States: A Study of Subjects Affecting their Political, Commercial, and Social Relations Made with a View toward their Promotion.* (New York: G. P. Putnam's Sons, 1898), iii–ix.

26. Matías Romero, "The Garza Raid and Its Lessons," *North American Review* 155 (September 1892): 333–34.
27. Ibid., 335–36.
28. Ibid., 336.
29. *Literary Digest*, January 23, 1892, 331.
30. Allen, "President Díaz," 678.
31. John G. Bourke, "An American Congo," *Scribner's* (May 1894): 590–610; Young, *Catarino Garza's Revolution*, 240–67.
32. Young, *Catarino Garza's Revolution*, 238; Richard Harding Davis, *The West from a Car-Window* (New York: Harper and Brothers, 1892). See also Richard Harding Davis, "The Garza Revolution," *Harper's Weekly*, January 16, 1892, 51–53, and "$300,000 Reward," *Harper's Weekly*, January 23, 1892, 91–93.
33. Richard Harding Davis, *Three Gringos in Venezuela and Central America* (New York: Harper and Brothers, 1896), 143. On the background of Davis's Central America trip, see Lubow, *Reporter*, chap. 5.
34. Davis, *Three Gringos*, 144, 147.
35. *Review of Reviews* 3 (June 1891): 493.
36. Lubow, *Reporter*, 123.
37. Karen Roggenkamp, *Narrating the News: New Journalism and Literary Genre in Late Nineteenth-Century American Newspapers and Fiction* (Kent, OH: Kent State University Press, 2005), 53. See also Milton, *Yellow Kids*; Campbell, *Yellow Journalism*; and Brown, *Correspondents' War.*
38. Romero, "The Garza Raid," 336.
39. According to Lubow, Davis drew on his visit to Cuba as a youth for many of the images of places and people in *Soldiers of Fortune* and on his travels in Mexico, Central America, and Venezuela as an adult for portraying Latin American politics (*Reporter*, 123–25, 374–75).
40. Clayton, *Peru and the United States*, 53–57; Joseph Schott, *Rails Across Panama* (New York: Bobbs-Merrill, 1967); Marshall Eakin, *British Enterprise in Brazil: The St. John d'el Rey Mining Company and the Morro Velho Gold Mine, 1830–1960* (Durham, NC: Duke University Press, 1989), 37–46; O'Brien, *Century of U.S. Capitalism*, 14–16, 36–38.

41. Marcus and Segal, *Technology in America*, 132–35; Schwartz Cowan, *Social History of American Technology*, 138–47; Burton Bledstein, *The Culture of Professionalism* (New York: Norton, 1976), 93–195. By the 1890s, universities were graduating engineers trained in these areas and the public had developed considerable respect for these professions.
42. Arthur L. Pearse, "Gold and Silver in South America," *Engineering Magazine* 7 (May 1894): 182–92; Emilio M. Amores, "Trade and Industry of South America," *Engineering Magazine* 10 (February 1896): 811–20.
43. Richard Harding Davis, *Soldiers of Fortune* (New York: Charles Scribner's Sons, 1897), 127, 128.
44. Ibid., 15, 20, 25–26, 127–28, 168.
45. Ibid., 59.
46. Ibid., 205.
47. Ibid., 101–2.
48. Ibid., 243–48.
49. Ibid., 309–29.
50. Ibid., 336.
51. Especially helpful on Davis's writing is Murphy, *Hemispheric Imaginings*, 119–44. See also Lars Schoultz, *Beneath the United States: A History of U.S. Policy toward Latin America* (Cambridge, MA: Harvard University Press, 1998), 78–175; Frederick Pike, *The United States and Latin America: Myths and Stereotypes of Civilization and Nature* (Austin: University of Texas Press, 1992), 113–92; Hoganson, *American Manhood*; and Paul T. McCartney, *Power and Progress: American Identity, the War of 1898, and the Rise of American Imperialism* (Baton Rouge: Louisiana State University Press, 2006).
52. See the following books by William Eleroy Curtis: *The Capitals of Spanish America* (New York: Harper and Brothers, 1888); *Trade and Transportation*; *The United States and the Foreign Powers* (Meadville, PA: Flood and Vincent, 1892); *The Relics of Columbus: A Illustrated Description of the Historical Collection in the Monastery of La Rabida* (Washington, D.C.: Lowdermilk, 1893); *Venezuela*; and *Between the Andes and the Ocean* (Chicago: H. S. Stone and Company, 1900). In these books, Curtis advocated the late nineteenth-century commercial expansion of the United States in Latin America. For historical assessments of this period, see Pletcher, *Diplomacy of Trade*; Walter LaFeber, *New Empire*; and Schoultz, *Beneath the United States*, esp. 78–175.
53. Joseph Smith, "The Latin American Trade Commission of 1884–5," *Inter-American Economic Affairs* 24.4 (1971): 3–24; McGann, *Argentina*, 121–22.
54. *Commission on Relations with Latin America, 49th Congress, First Session,*

House Executive Document 50 (Washington, D.C.: Government Printing Office, 1886).

55. Curtis, *Capitals*, 1–113, 171–95, 257–97.
56. Ibid., 114–70.
57. Ibid., 387–402.
58. Ibid., 454.
59. Ibid., 472.
60. Ibid., 551–89.
61. Ibid., 552.
62. Ibid., 590.
63. *New York Times*, August 23, 1889, 2, October 3, 1889, 5, October 4, 1889, 5, November 17, 1889, 5.
64. McGann, *Argentina*, 130–64.
65. *New York Times*, March 11, 1890, 9.
66. Matías Romero, "The Pan-American Conference," pt. 1, *North American Review* 151 (September 1890): 354–69; Matías Romero, "The Pan-American Conference," pt. 2, *North American Review* 151 (October 1890): 407–21.
67. Curtis, *Trade and Transportation*; *New York Times*, October 16, 1889, 5.
68. Topik, *Trade and Gunboats*, 45.
69. William Eleroy Curtis, "United States to Dominate Hemisphere," www.learner.org/workshops/primarysources/corporations/docs/curtis.

CHAPTER EIGHT

1. Richard Olney, "International Isolation of the United States," *Atlantic Monthly* 81 (May 1898): 588.
2. Jay Sexton, *The Monroe Doctrine: Empire and Nation in Nineteenth-Century America* (New York: Hill and Wang, 2011), 199–211.
3. Judith Ewell, *Venezuela and the United States: From Monroe's Hemisphere to Petroleum's Empire* (Athens: University of Georgia Press, 1996), 63–90; Benjamin Frankel, *Venezuela y los Estados Unidos* (Caracas: Fundación Boulton, 1977), 292–361; Mario Picon-Salas, Augusto Mijares, et al., *Venezuela independiente* (Caracas: Fundación Eugenio Mendoza, 1962), 126–43; Armando Rojas, *Las misiones diplomaticas de Guzmán Blanco* (Caracas: Monte Avila, 1971); Armando Rojas, *Historia de las relaciones diplomaticas entre Venezuela y los Estados Unidos*, vol. 1, *1810–1899* (Caracas: Ediciones de la presidencia de la república, 1979), 255–328. See also José Andrade to Walter Q. Gresham, March 31, 1894, accompanied by Andrade's "Memorandum on the Boundary Question between Venezuela and Great Britain," in *Papers Relating to the Foreign Relations of the United States, 1894* (Washington, D.C.:

U.S. Government Printing Office, 1895), 810–12 and 845–46; R. A. Humphreys, *Tradition and Revolt in Latin America and Other Essays* (London: Weidenfeld and Nicholson, 1969), 195.

4. William Scruggs to Ezequiel Rojas, October 30, 1895, William Lindsay Scruggs Papers, 1828–1925, pt. 2, Scrapbooks, reel 1, Manuscript Division, Library of Congress, Washington, D.C.; José Andrade to William Scruggs, June 28, 1895, July 24, 1895, and July 29, 1895, Scruggs Papers, pt. 1, reel 1; José Andrade to William Scruggs, June 25 and July 2, 1895, Scruggs Papers, pt. 1, reel 1; William Scruggs to José Andrade, October 20, 1895, Scruggs Papers, pt. 2, Scrapbooks, reel 1; Thomas F. Bayard to Grover Cleveland, April 5 and May 11, 1895, reel 90, Grover Cleveland Papers, 1743–1925, Manuscript Division, Library of Congress, Washington, D.C.
5. Harris, "Venezuela," 107–22; Bertram, *Birth*; Grenville and Young, *Politics, Strategy, and American Diplomacy*, 125–57; Humphreys, "Anglo-American Rivalries," 186–215; Theodore D. Jervey, "William Lindsay Scruggs—A Forgotten Diplomat," *South Atlantic Quarterly* 27.3 (1928): 292–309. Scruggs's pamphlet appeared in four editions from 1894 to 1896. For the last edition, see Scruggs, *Venezuelan Question*.
6. *Times* (London), November 30, 1872, 9. See also *New York Times*, December 14, 1872, 2.
7. Humphreys, *Tradition and Revolt*, 207–8.
8. Cecil Spring Rice to his sisters, March 12, 1895, in *The Letters and Friendships of Sir Cecil Spring Rice: A Record*, 2 vols., ed. Stephen Gwynn (Boston: Houghton Mifflin, 1929), 1:173–75.
9. Cecil Spring Rice to Francis Villiers, April 12, 1895, in *Letters*, 1:175–76.
10. *Washington Post*, October 19, 1895, 1, 3; *New York Times*, October 16, 1895, 5, and October 19, 1895, 1; *San Francisco Examiner*, October 19, 1895, 1.
11. *New York Times*, October 19, 1895, 1; see also *New York Times*, October 16, 1895, 5. On the activities of Venezuela and Great Britain in the disputed area, see Enrique Bernardo Núñez, *Tres momentos en la controversia de límites de Guyana: El incidente de Yuruán* (Caracas: Imprenta nacional, 1962).
12. See the following newspapers from October 19 through October 22, 1895: *Washington Post, New York Times, New York Tribune, Boston Globe, Philadelphia Inquirer, Chicago Tribune, New Orleans Times-Picayune*, and the *San Francisco Examiner.*
13. *New York Times*, December 18, 1895, 1–2.
14. Ibid., December 24, 1895, 9; *Literary Digest*, December 28, 1895, 1–3; *Public Opinion*, December 25, 1895, 837; Bertram, *Birth*, 39–58.
15. See the *Washington Post, New York Tribune, Boston Globe, Chicago Tribune, San*

Francisco Examiner, and *New Orleans Picayune* for the period from December 18 to December 23, 1895. See also *Boston Globe*, December 19, 1895, 1.

16. *San Francisco Examiner*, December 22, 1895, 1. See also *San Francisco Examiner*, December 20 and 21, 1895, 1.
17. Lord Selborne to Joseph Chamberlain, December 18, 1895, qtd. in Humphreys, *Tradition and Revolt*, 208.
18. The best account of Pauncefote's problems in Washington and Salisbury's misguided diplomacy is Gibb, "Unmasterly Inactivity?." See also R. B. Mowat, *The Life of Lord Pauncefote* (Boston: Houghton Mifflin, 1929), 122–202, and Pauncefote to Salisbury, December 18, 1895, FO 80/364, Public Records Office, London.
19. Julian Pauncefote to Eric Barrington, December 17, 1895, A/139, Salisbury Papers, Hatfield House, Hertfordshire, UK.
20. Julian Pauncefote to Lord Salisbury, October 25, 1895, A/139, Salisbury Papers.
21. Julian Pauncefote to Lord Salisbury, November 8, 1895, A/139, Salisbury Papers. Pauncefote mentioned another story that aroused anti-British feelings in the U.S. press in the same period: the Marlborough-Vanderbilt wedding.
22. Pauncefote to Salisbury, December 18, 1895, FO 80/364.
23. Julian Pauncefote to Lord Salisbury, December 20, 1895, A/139, Salisbury Papers. See also Gibb, "Unmasterly Inactivity?," 36–38.
24. "Confidential Memorandum," December 20, 1895, A/139, Salisbury Papers.
25. Julian Pauncefote to Lord Salisbury, December 24, 1895, FO 80/364.
26. Stafford Northcote to Lord Salisbury, March 28, 1895, and April 8, 1896, A/139, Salisbury Papers.
27. Stafford Northcote to Lord Salisbury, April 24, 1896, A/139, Salisbury Papers.
28. Henry M. Stanley, "The Issue Between Great Britain and America," *Nineteenth Century* 39 (January 1896): 1–2.
29. Stanley, "Issue," 3–4. About the same time that Stanley's essay appeared in print, Salisbury received a letter from the Raysons of New York City that contained reinforcement for the notion of a rampant Anglophobia in the United States. See A. and C. E. Rayson to Lord Salisbury, December 24, 1895, FO 80/364.
30. G. D. H. Gossip, "England in Nicaragua and Venezuela from the American Point of View," *Fortnightly Review* 64 (December 1895): 838; G. D. H. Gossip, "Venezuela before Europe and America from an American Point of View," *Fortnightly Review* 65 (January 1896): 403–4.
31. Gossip, "Venezuela before Europe," 406.
32. Ibid., 406–7; Gossip, "England in Nicaragua," 838.

33. Gossip, "England in Nicaragua," 839; and see also 835–42.
34. W. T. Stead, "Jingoism in America," *Contemporary Review* 67 (September 1895): 341. On Stead's work in the British press, see Jane Chapman, *Comparative Media History* (Malden, MA: Polity Press, 2005), 75–78.
35. Edward Dicey, "Common Sense and Venezuela," *Nineteenth Century* 39 (January 1896): 10.
36. Dicey, "Common Sense," 11.
37. Henry Booth to Lord Salisbury, February 22, 1896, FO 80/368. For the arrival of this publication at the U.S. State Department, see *New York Times*, March 10, 1896, 5.
38. Harry R. Whates, *The Third Salisbury Administration, 1895–1900* (London: Westminster Press, 1900), 65.
39. William James, *Richard Olney and His Public Service* (Boston: Houghton Mifflin, 1923), 122.
40. Gerald G. Eggert, *Richard Olney: Evolution of a Statesman* (University Park: Penn State University Press, 1974), 247; Grover Cleveland to Thomas F. Bayard, December 29, 1895, reel 92, Cleveland Papers.
41. Grover Cleveland to Bayard, December 29, 1895, reel 92, Cleveland Papers; Richard E. Welch Jr., *The Presidencies of Grover Cleveland* (Lawrence: University Press of Kansas, 1988), 185–92; H. W. Brands, *TR: The Last Romantic* (New York: Basic Books, 1997), 287–90; Bertram, *Birth*, 59–74.
42. Stafford Northcote to Lord Salisbury, March 22, 1896, A/139, Salisbury Papers.
43. Charles C. Tansill, *The Congressional Career of Thomas Francis Bayard, 1869–85* (New York: Fordham University Press, 1940), 746.
44. Charles Eliot Norton, "Some Aspects of Civilization in America," *Forum* 20 (February 1896): 641–51. Pauncefote included a copy of this article in a letter he sent to Salisbury on February 14, 1896 (FO 80/368).
45. E. L. Godkin to Charles Eliot Norton, December 29, 1895, in *The Gilded Age Letters of E. L. Godkin*, ed. William M. Armstrong (Albany: State University of New York Press, 1974), 475. Godkin's criticism of the press extended far beyond this controversy. See Caudill, "E. L. Godkin."
46. E. L. Godkin to James Bryce, January 9, 1896, in *Gilded Age Letters*, 476–78.
47. Richard Olney to James Storrow, June 12, 1896, box 55, reel 20, Richard Olney Papers, 1830–1928, Manuscript Division, Library of Congress, Washington, D.C.; Eggert, *Richard Olney*, 237–42.
48. Olney, "International Isolation of the United States," 577–78.
49. J. A. S. Grenville, *Lord Salisbury and Foreign Policy: The Close of the Nineteenth Century* (London: Athlone Press, 1964), 3–23; Mowat, *Life*, 112–85.

50. *New York Times,* January 6, 1896, 5; and Eggert, *Richard Olney,* 225–26.
51. Henry Norman to Richard Olney, January 6, 1896, January 9, 1896, and January 10, 1896, box 42, reel 15, Olney Papers.
52. *Daily Chronicle* (London), qtd. in *New York Times,* January 1, 1896, 5, January 4, 1896, 1, January 5, 1896, 5, January 6, 1896, 5, January 12, 1896, 15, and January 13, 1896, 5. See also Eggert, *Richard Olney,* 225–26.
53. *New York Times,* January 6, 1896, 5. In a letter to Olney, Norman commented with pride on the reprinting of his work in the *New York Times* and other U.S. newspapers (January 6, 1896, box 42, reel 15, Olney Papers).
54. Julian Pauncefote to Lord Salisbury, January 3, 1896, A/139, Salisbury Papers.
55. *New York Times,* January 6, 1896, 5, and January 7, 1896, 1. Norman stated his goal of immediate reconciliation in a letter to Olney: "A few days of such work will have [the] conflict straightened out" (January 9, 1896, box 42, reel 15, Olney Papers). See also Henry Norman, "Anglo-American Arbitration," *Review of Reviews* 15 (January 1897): 75–76.
56. George W. Smalley, *Anglo-American Reminiscences* (New York: Putnam, 1911); Bradford Perkins, *The Great Rapprochement* (New York: Atheneum, 1968), 22–24. On Smalley's early career in journalism, see Hamilton, *Journalism's Roving Eye,* 57–75.
57. *Times* (London), December 5, 1895, 5, and December 6, 1895, 5.
58. Ibid., December 18, 1895, 6, December 19, 1895, 5, and December 20, 1895, 5; see also December 21, 1895, 5.
59. George Smalley to Richard Olney, January 22, 1896, box 43, reel 16, Olney Papers.
60. George Smalley to Richard Olney, January 25, 1896, box 43, reel 16, Olney Papers; Joseph Matthews, "Informal Diplomacy in the Venezuelan Crisis of 1896," *Mississippi Valley Historical Review* 50.2 (1963): 200–201.
61. Matthews, "Informal Diplomacy" 204–6; George Smalley to Richard Olney, January 27, 1896, box 43, reel 156, and January 28, 1896, box 44, reel 16, Olney Papers.
62. *Times* (London), February 20, 1896, 5.
63. Ibid., February 21, 1896, 5.
64. George Smalley to Richard Olney, January 30, 1896, box 44, reel 16, Olney Papers.
65. George Smalley to Richard Olney, January 31, 1896, and February 1, 1896, box 44, reel 16, Olney Papers.
66. Matthews, "Informal Diplomacy," 206.
67. Joseph Chamberlain to Lord Salisbury, February 21, 1896, A/139, Salisbury Papers.

68. Grenville, *Lord Salisbury*, 69.
69. *Nation*, February 22, 1896, 169.
70. Grenville, *Lord Salisbury*, 65–73; Welch, *Presidencies of Grover Cleveland*, 177–92; Bertram, *Birth*, 95–141; Ernest R. May, *Imperial Democracy: The Emergence of America as a Great Power* (New York: Harper and Row, 1961), 43–66.
71. Pakenham, *Scramble for Africa*, 487–503; Hale, *Publicity and*, esp. 102–37; Langer, *Diplomacy of Imperialism*, 213–58. News of the Jameson raid and the Kruger telegram received extensive coverage in newspapers in the United States. See the *New York Times*, *Chicago Tribune*, *San Francisco Chronicle*, and the seven other metropolitan newspapers emphasized in this study from January 3 to January 8, 1896. Between January 13 and January 15, the *New York Times* and the *San Francisco Chronicle* noted an implicit connection between Britain's willingness to negotiate on the Venezuelan boundary issue and the growing hostility of Germany regarding South Africa.
72. Welch, *Presidencies of Grover Cleveland*, 192.
73. Edwards, *New Spirits*, esp. 37–59; Wiebe, *Search for Order*, 11–78, 224–56; Summers, *Gilded Age*; Williams, *Years of Decision*; Brands, *Reckless Decade*.

CHAPTER NINE

1. Hamilton, *Journalism's Roving Eye*, esp. 89–109; Campbell, *Yellow Journalism*; Stevens, *Sensationalism*, 55–100; Milton, *Yellow Kids*; W. A. Swanberg, *Pulitzer* (New York: Scribner, 1967); W. A. Swanberg, *Citizen Hearst* (New York: Scribner, 1961); Nasaw, *Chief*, 95–167. On the long-term U.S. interest in Cuba, see Louis Pérez, *Cuba in the American Imagination* (Chapel Hill: University of North Carolina Press, 2008), 24–95.
2. Pérez, *Cuba between Empires*, 3–18.
3. Louis Pérez, *On Becoming Cuban: Identity, Nationality, and Culture* (Chapel Hill: University of North Carolina Press, 1999), 16–95; Pérez, *Cuba and the United States*, 55–81.
4. Western Union Company, annual reports, October 7, 1873, 11–14, October 14, 1874, 8–9, October 13, 1875, 9–13, October 11, 1876, 10–12, and October 10, 1877, 12–13, Western Union Telegraph Company Records, Lemelson Center, National Museum of American History, Washington, D.C.; Scrymser, *Personal Reminiscences*, 68–75; Ahvenainen, *History of the Caribbean Telegraphs*, 12–26.
5. Cuba Submarine, directors' report, February 12, 1895, 2, and August 7, 1895, 2, Cable and Wireless Archive, Porthcurno Telegraph Museum, Cornwall, UK. The profitability calculations for Cuba Submarine are derived from the

directors' reports for the years 1895–98, various issues of the *Times* (London), and Ahvenainen, *History of the Caribbean Telegraphs*, 123–45.

6. West India and Panama, directors' report, May 19, 1897, 2–3, Cable and Wireless Archive. Profitability calculations for West India and Panama are based on the company's directors' reports for the years 1895–98, various issues of the *Times* (London), and Ahvenainen, *History of the Caribbean Telegraphs*, 123–45.
7. West India and Panama, board minute books, December 20, 29, and 30, 1897, and January 12, 1898, Cable and Wireless Archive; *New York Times*, December 20, 1897, 2, and December 27, 1897, 6. On the French company, see *Electrical World*, October 10, 1896, 413–14, November 21, 1896, 620, and December 12, 1897, 715; and *New York Times*, October 2, 1896, 8. Jorma Ahvenainen has a thorough discussion of the complexities in the early history of these companies in his *History of the Caribbean Telegraphs*, 94–122. On the work of reporters see John R. Spears, "Afloat for the News in War Times," *Scribner's Magazine* 24 (October 1898): 501–4; Ray Stannard Baker, "How the News of the War Is Reported," *McClure's Magazine* 11 (September 1898): 491–95, and Milton, *Yellow Kids*, 337–38.
8. *New York Times*, May 27, 1895, 5, June 10, 1896, 5, July 2, 1896, 5; Pérez, *Cuba between Empires*, 39–53.
9. Milton, *Yellow Kids*, 57–66.
10. Brown, *Correspondents' War*, 44; George Bronson Rea, *Facts and Fakes about Cuba* (New York: George Munro's Sons, 1897), 21–24.
11. Brown, *Correspondents' War*, 50–53.
12. Milton, *Yellow Kids*, 79–82.
13. Brown, *Correspondents' War*, 22–25.
14. *New York Times*, January 8, 1896, 5, and January 9, 1896, 6.
15. Brown, *Correspondents' War*, 24–25.
16. Milton, *Yellow Kids*, 90–91; Grover Flint, *Marching with Gomez* (New York: Lamson and Wolfe 1898).
17. Brown, *Correspondents' War*, 45–50; Milton, *Yellow Kids*, 70–77; Rea, *Facts and Fakes*, xvii. For an extended discussion of the press coverage of the Cuban revolution see Brown, *Correspondents' War*, 22–128.
18. Pérez, *On Becoming Cuban*, 17–95, esp. 43–44; Pérez, *Cuba in the American Imagination*, 24–94; Nydia Sarabia, *Noticias confidenciales sobre Cuba, 1870–1895* (Havana: Editorial política, 1985); Gonzalo de Quesada, *Paginas escogidas* (Havana: Ediciones políticas, 1968), 5–7; Poyo, *"With All"*; John M. Kirk, *José Martí: Mentor of the Cuban Nation* (Gainesville: University Presses of Florida, 1983).
19. Not all Cubans in the United States agreed with the Estrada Palma-Quesada

approach of appealing for direct U.S. involvement. Many Cubans in Key West and Tampa were concerned about the possibility of U.S. annexation. See Poyo, *"With All,"* esp. 112–37; Pérez, *Becoming Cuban*, 17–95. For insights into the Junta's views of the U.S. government, see vols. 1 and 2 of *Correspondencia diplomática de la delegacíon cubana en Nueva York* (Havana: Archivo nacional de Cuba, 1943).

20. *New York Times*, April 28, 1896, 5.
21. Marcus Wilkerson, *Public Opinion and the Spanish-American War: A Study in War Propaganda* (1932; repr., New York: Russell and Russell, 1967), 54–61; George Auxier, "The Propaganda Activities of the Cuban Junta in Precipitating the Spanish-American War, 1895–1898," *Hispanic American Historical Review* 19.3 (1939): 288–305; George Auxier, "Middle Western Newspapers and the Spanish-American War, 1895–1898," *Mississippi Valley Historical Review* 26.4 (1939): 523–34. See also Martin M. Rosenberg and Thomas P. Ruff, *Indiana and the Coming of the Spanish-American War* (Muncie, IN: Ball State University, 1976), 5–9; and Richard F. Hamilton, *President McKinley, War, and Empire*, vol. 1, *President McKinley and the Coming of War, 1898* (New Brunswick, NJ: Transaction, 2006), 112–13, 216–17.
22. Auxier, "Propaganda Activities," 299–303; Wilkerson, *Public Opinion*, 29–61; Joseph Wisan, *The Cuban Crisis as Reflected in the New York Press (1895–1898)* (New York: Columbia University Press, 1934), 187–236. On the black legend in the nineteenth century, see Richard L. Kagan, "Prescott's Paradigm: American Historical Scholarship and the Decline of Spain," *American Historical Review* 101.2 (1996): 423–46.
23. See Swanberg, *Citizen Hearst*, 137–57; and Nasaw, *Chief*, 128–32, on Hearst and Evangelina Cossío Cisneros.
24. Gonzalo de Quesada and Henry D. Northrup, *The War in Cuba, Being a Full Account of Her Struggle for Freedom* (New York: Liberty Publishers, 1896).
25. Gabriel Cardona and Juan Carlos Losada, *Weyler, nuestro hombre en la Habana* (Barcelona: Editorial planeta, 1997), 154–98; Hilario Martín Jiménez, *Valeriano Weyler: Capitán-general, duque de Rubi, marqués de Tenerife y grande de España, de su vida y personalidad, 1838–1930* (Santa Cruz de Tenerife: Ediciones del umbral, 1998), 159–188; Pérez, *Cuba between Empires*, 39–56.
26. Cardona and Losada, *Weyler*, 199–216; Pérez, *Cuba between Empires*, 53–56, 151–58.
27. Charles Michelson, *The Ghost Talks* (New York: Putnam's, 1944), 85–88; Milton, *Yellow Kids*, 83–87.
28. Translation of these decrees as they appeared in the *New York Times*, February 17, 1896, 5. See also Martín Jiménez, *Valeriano Weyler*, 174–76; and

Cardona and Losada, *Weyler*, 199–201. For the newspaper coverage and comment, see *Chicago Tribune*, February 17, 1896, 3, and February 18, 1896, 6; *San Francisco Examiner*, February 17, 1896, 1; *San Francisco Chronicle*, February 18, 1896, 1; *Atlanta Constitution*, February 17, 1896, 1; *New Orleans Picayune*, February 17, 1896, 2; *New York Tribune*, February 17, 1896, 1; *Washington Post*, February 17, 1896, 1–2; and *Boston Globe*, February 20, 1896, 6. The *Philadelphia Inquirer* apparently gave no coverage to Weyler in this period.

29. *New York Times*, February 18, 1896, 5.
30. Ibid., May 16, 1896, 5. The Michelson story appeared in the *San Francisco Examiner*, February 17, 1898, 1.
31. *New York Times*, July 2, 1896, 5.
32. Milton, *Yellow Kids*, 139–52. See also Brown, *Correspondents' War*, 85–87.
33. *New York Times*, September 11, 1896, 5; Brown, *Correspondents' War*, 65; Rea, *Facts and Fakes*, 178–79.
34. Brown, *Correspondents' War*, 88–89. A third correspondent identified as Woodward apparently died in a combat situation in May 1895. See *New York Times*, May 27, 1895, 5.
35. Brands, *Reckless Decade*, 254–86; H. Wayne Morgan, *William McKinley and His America* (Syracuse, NY: Syracuse University Press, 1963), 103–207.
36. Milton, *Yellow Kids*, 102.
37. *Chicago Tribune*, November 13, 1896, 1, November 14, 1896, 1, and November 17, 1896, 2; *San Francisco Examiner*, November 13, 1896, 1, 2, and November 14, 1896, 1, 6; *Washington Post*, November 13, 1896, 1; *New Orleans Picayune*, November 14, 1896, 1; *Boston Globe*, November 13, 1896, 1, November 14, 1896, 7, and November 15–16, 1896, 1; *Philadelphia Inquirer*, November 12–15, 1896, 1; Wisan, *Cuban Crisis*, 167–69.
38. *New York Times*, November 14, 1896, 1.
39. Wisan, *Cuban Crisis*, 167–69.
40. *New York Times*, November 17, 1896, 1; see also November 14, 1896, 1.
41. Ibid., November 13, 1896, 2, November 14, 1896, 1, and November 17, 1896, 1; *New York Tribune*, November 14, 1896, 1, and November 15, 1896, 1; *Washington Post*, November 13, 1896, 1; *Chicago Tribune*, November 13, 1896, 1, and November 14, 1896, 1 and 12; *New Orleans Daily Picayune*, November 14, 1896, 1; *San Francisco Examiner*, November 13, 1896, 1, and November 14, 1896, 1; *San Francisco Chronicle*, November 13, 1896, 1, and November 14, 1896, 1; *Atlanta Constitution*, November 14, 1896, 1.
42. Pérez, *Cuba between Empires*, 39–88.
43. *New York Journal*, qtd. in Wisan, *Cuban Crisis*, 206. See also Wisan, *Cuban Crisis*, 187–236; Wilkerson, *Public Opinion*, 29–55; Milton, *Yellow Boys*, 89–152; and Brown, *Correspondents' War*, 22–62.

44. Qtd. in Wilkerson, *Public Opinion*, 32.
45. Ibid., 46–53.
46. Ibid., 42.
47. Ibid., 41–42, 54–56.
48. *New York Times*, February 18, 1896, 5; and *Chicago Tribune*, February 17, 1896, 3.
49. *New York Times*, February 18, 1896, 5; *Chicago Tribune*, February 17, 1896, 3; *Washington Post*, February 17, 1896, 1, 2; *New York Tribune*, February 17, 1896, 1; *Atlanta Constitution*, February 17, 1896, 1; *San Francisco Chronicle*, February 18, 1896, 1; *San Francisco Examiner*, February 17, 1896, 1; *New Orleans Daily Picayune*, February 17, 1896, 2.
50. *New York Tribune*, December 12, 1896, 1.
51. *New York Times*, May 14, 1897, 1.
52. *New York Tribune*, June 8, 1897, 1; *New York Times, New Orleans Picayune, Chicago Tribune, Atlanta Constitution, Washington Post, San Francisco Chronicle*, and *San Francisco Examiner*, June 8–9, 1897. The coverage appears on page 1 or 2.
53. *New York Times*, supplement, January 17, 1896, 3; *New York Times*, January 22, 1897, 2.
54. Cardona and Losada, *Weyler*, 154–231; Martín Jiménez, *Valeriano Weyler*, 171–88. For critical evaluations of Weyler, see Francisco Pérez Guzmán, *Herida profunda* (Havana: Unión de escritores y artistas de Cuba, 1998); and Emilio Roig de Leuchsenring, *Weyler en Cuba: Un precusor de la barbarie fascista* (Havana: Páginas, 1947).
55. Qtd. in Lubow, *Reporter*, 135–39. Davis preferred this description of Weyler in a letter to his mother, Rebecca Harding Davis, on January 10, 1897.
56. Lubow, *Reporter*, 141–42.
57. Ibid., 142–43; and Brown, *Correspondents' War*, 80–83. See also *New York Journal*, February 12, 1897, 1.
58. Lubow, *Reporter*, 143–44; Brown, *Correspondents' War*, 80–82.
59. *San Francisco Examiner*, February 12, 1897, 1, February 13, 1, and February 14, 1897, 1. See also *Washington Post*, February 14, 1897, 1; *New Orleans Picayune*, February 12, 1897, 7; *Chicago Tribune*, February 12, 1897, 1; *New York Times*, February 12–20, 1897 (the *Olivette* did not appear in the *New York Times* index for January 1–June 30, 1897); and *San Francisco Chronicle*, February 15, 1897, 1. The *New York Tribune* apparently did not cover this story. See issues from February 12, 1897–February 19, 1897.
60. James Creelman, *On the Great Highway* (Boston: Lothrop, 1901), 179–80. The *New York Journal* also exploited the story of the death of Ricardo Ruiz, a Cuban-born, naturalized U.S. citizen who died in a Spanish jail in Cuba

after his arrest on charges of rebel activity. The Ruiz case received wire-service coverage and became a nationwide story in February and March 1897. See Wilkerson, *Public Opinion*, 83–87.

61. Creelman, *On the Great Highway*, 180.
62. Swanberg, *Citizen Hearst*, 137–57; Nasaw, *Chief*, 128–32; Creelman, *On the Great Highway*, 179–82.
63. Wilkerson, *Public Opinion*, 87–91; Wisan, *Cuban Crisis*, 329–32; Swanberg, *Citizen Hearst*, 149–53; Brown, *Correspondents' War*, 101–2; Pérez, *Cuba in the American Imagination*, 77–80. For a sampling of the publicity work of the *New York Journal*, see Evangelina Cisneros, *The Story of Evangelina Cisneros Told by Herself* (New York: Continental Publishing Company, 1898).
64. *New York Tribune*, October 8–16, 1897; *San Francisco Chronicle*, October 8, 1897, 1, October 10, 1897, 13, and October 14, 1897, 2; *New Orleans Picayune*, October 8, 1897, 1, October 9, 1897, 2, October 13, 1897, 2, and October 15, 1897, 1; *Washington Post*, October 8, 1897, 1, October 9, 1897, 1, October 10, 1897, 1, October 12, 1897, 1, October 14, 1897, 7, and October 15, 1897, 1.
65. *New York Times*, October 17, 1897, 5. See also the editorial on page 6.
66. Campbell, *Yellow Journalism*, 34–38.
67. Campbell, *Yellow Journalism*, 97–147; Offner, *Unwanted War*, 38–39; Lewis L. Gould, *The Spanish-American War and President McKinley* (Lawrence: University Press of Kansas, 1980), 23–24.

CHAPTER TEN

1. Peggy Samuels and Harold Samuels, *Remembering the* Maine (Washington, D.C.: Smithsonian Institution Press, 1995), 110. See also Samuels and Samuels, *Remembering*, 109–21; Offner, *Unwanted War*, 122–26; and Margaret Leech, *In the Days of McKinley* (New York: Harper, 1959), 166–72.
2. Samuels and Samuels, *Remembering*, 111–12.
3. Headrick, *Invisible Weapon*, 28–39; Herbert Laws Webb, "With a Cable Expedition," *Scribner's Magazine* 7 (October 1890): 399–415; Henry Muir, "The Telegraph Systems of the World," *McClure's Magazine* 2 (July 1895): 99–112; Winseck and Pike, *Communication and Empire*, 16–91, 142–57.
4. Bright, *Submarine Telegraphs*, 635–75; Coates and Finn, *Retrospective Technology Assessment*, 15–173.
5. West Indian and Panama, directors' report, May 10, 1881, Cable and Wireless Archive, Porthcurno Telegraph Museum, Cornwall, UK. See also West India and Panama, directors' reports, May 21, 1874, October 21, 1874,

October 11, 1876, October 13, 1877, October 12, 1878, April 30, 1880, and October 15, 1889, Cable and Wireless Archive.

6. West India and Panama, directors' report, October 14, 1882, Cable and Wireless Archive.
7. West India and Panama, directors' report, May 12, 1886, Cable and Wireless Archive.
8. Francis Lambert, "On Grapnels for Raising Submarine Cables in Deep Water," *Society of Telegraph Engineers and Electricians*, November 13, 1878, 417–33; Bright, *Submarine Telegraphs*, 96–99; West Indian and Panama, directors' report, May 12, 1886, Cable and Wireless Archive.
9. West Indian and Panama, directors' reports, April 26, 1889, November 4, 1891, November 1, 1894, May 8, 1895, May 5, 1897, and May 4, 1898, Cable and Wireless Archive.
10. Campbell, "Age of New Journalism," 236; Smythe, *Gilded Age Press*, 125–28.
11. Barnhurst and Nerone, *Form of the News*, 185–202; Smythe, *Gilded Age Press*, 123–48.
12. These figures were derived from a survey of the ten metropolitan newspapers emphasized in this study from the following periods: early January 1892, late December 1895, February 10–11, 1898, and February 16–18, 1898.
13. Spencer, *Yellow Journalism*, 205–24; Smythe, *Gilded Age Press*, 173–201; John J. Johnson, *Latin America in Caricature* (Austin: University of Texas Press, 1980).
14. Creelman, *On the Great Highway*, 187.
15. *New York Times*, March 20, 1896, 5. For a portrait of Dupuy de Lôme, see Henry Cabot Lodge, *The War with Spain* (New York: Harper, 1899), 27.
16. Wisan, *Cuban Crisis*, 118–19.
17. *New York Times*, November 14, 1896, 1, March 2, 1897, 1, and March 11, 1897, 2; Brown, *Correspondents' War*, 87.
18. Enrique Dupuy de Lôme to Varina Jefferson Davis, April 24, 1897, qtd. in *New York Times*, August 27, 1897, 2. Dupuy de Lôme sent a ciphered telegram to Madrid in January 1898, again expressing his concern about U.S. public opinion. See G. J. A. O'Toole, *The Spanish War: An American Epic* (New York: Norton, 1984), 114.
19. *New York Times*, April 21, 1896, 5.
20. *New York World*, December 14, 1896, qtd. in Rea, *Facts and Fakes*, 217–18.
21. Cardona and Losada, *Weyler*, 199–200, 228; Martín Jiménez, *Valeriano Weyler*, 167–68, 174–76.
22. Creelman, *On the Great Highway*, 158–59.
23. Ibid., 161–62.

24. Offner, *Unwanted War*, 39; Stephen Ponder, *Managing the Press: Origins of the Media Presidency* (New York: St. Martin's Press, 1999), 1–15; Robert C. Hilderbrand, *Power and the People: Executive Management of Public Opinion in Foreign Affairs, 1897–1921* (Chapel Hill: University of North Carolina Press, 1981), 8–29; John Maxwell Hamilton, Renita Coleman, Bettye Grable, and Jaci Cole, "An Enabling Environment: A Reconsideration of the Press and the Spanish-American War," *Journalism Studies* 7.1 (2006): 78–93; Hamilton, *Journalism's Roving Eye*, 89–109. For the other perspective, see Hamilton, *President McKinley*, vol. 1, *President McKinley and the Coming of the War*; and Campbell, *Yellow Journalism*, 50–147. For evidence that McKinley and his staff kept abreast of newspaper coverage, see George Cortelyou diary, March 1, 18, 28, and 29, and April 2, 1898, George L. Cortelyou Papers, 1871–1948, Manuscript Division, Library of Congress, Washington, D.C.; and also newspapers clippings in scrapbooks, 1897–1901, "Current Comment" and foreign affairs, reels 94–98, in William McKinley Papers, ca. 1847–1945, Manuscript Division, Library of Congress, Washington, D.C. John Maxwell Hamilton's research team studied the *New York Journal, New York World, New York Times, Washington Post, Atlanta Constitution, Chicago Tribune, St. Louis Post Dispatch, Denver Post, Los Angeles Times*, and the *Seattle Post-Intelligence* ("An Enabling Environment," 78–93). Richard Hamilton in *President McKinley and the Coming of the War* argues that mass demonstrations, which were often reflective of the jingoism of the yellow press, were not important in this period and did not play a role Washington's decision to declare war against Spain. On McKinley's concern about newspaper coverage, see Gerald F. Linderman, *The Mirror of War: American Society and the Spanish-American War* (Ann Arbor: University of Michigan Press, 1974), 9–36. See also Gould, *Spanish-American War*, 35–42; and George Herring, "Imperial Tutor: William McKinley, the War of 1898, and the New Empire, 1898–1902," in *Selling War in a Media Age: The Presidency and Public Opinion in the American Century*, ed. Kenneth Osgood and Andrew K. Frank (Gainesville: University Press of Florida, 2010), 18–47. See also Pérez, *Cuba in the American Imagination*, 1–94, for the long-term U.S. preoccupation with Cuba.
25. John Sherman to Enrique Dupuy de Lôme, June 26, 1897, *Papers Relating to the Foreign Relations of the United States, 1897* (Washington, D.C.: U.S. Government Printing Office, 1898), 507–8.
26. John Sherman to Stewart L. Woodford, July 16, 1897, *Papers*, 558–61; Stewart L. Woodford to William McKinley, August 10, 1897, August 23, 1897, and October 17, 1897, RG 59, roll 123, State Department Records, National Archives and Records Administration, Washington, D.C. See also Offner, *Unwanted War*, 56–58.

27. Stewart L. Woodford to John Sherman, October 2, 1897, RG 59, roll 123, State Department Records. See also Duke of Tetuán to Stewart L. Woodford, September 24, 1897, Stewart L. Woodford to Duke of Tetuán, September 27, 1897, Stewart L. Woodford to State William R. Day, October 6 and October 23, 1897, RG 59, roll 123, State Department Records; and Offner, *Unwanted War*, 64. On Spain's unsuccessful efforts to gain European support to fend off the United States in Cuba see Orestes Ferrara, *The Last Spanish War: Revelations in Diplomacy* (New York: Paisley Press, 1937).
28. Stewart L. Woodford to William McKinley, January 17, 1898, qtd. in Offner, *Unwanted War*, 90. For Woodford's comments on Spanish government officials' views of the Cuban junta and the U.S. press, see Stewart L. Woodford to John Sherman, February 14, 1898, RG 59, roll 123, State Department Records. See also Offner, *Unwanted War*, 90–92.
29. Offner, *Unwanted War*, 92.
30. Ibid., 89. Britain played a large role in stopping Germany and France from answering Spain's call for support in its dealings with the United States. See R. G. Neale, *Britain and American Imperialism* (Brisbane: University of Queensland Press, 1965), 1–34.
31. Hannis Taylor, "A Review of the Cuban Question in Its Economic, Political, and Diplomatic Aspects," *North American Review* 165 (November 1897): 610–35; Offner, *Unwanted War*, 77–78.
32. Offner, *Unwanted War*, 28, 82, 95.
33. *New York Times*, January 14, 1898, 1; *Chicago Tribune*, January 13, 1898, 1; *New Orleans Picayune*, January 13, 1898, 7, and January 14, 1898, 1; *New York Tribune*, January 15, 1898, 2. The *New York Times* editorial appeared on January 15, 1898, 6. The coverage in the ten metropolitan newspapers ran from January 13 to January 17, 1898, usually on the front page. For the remarks by Adams and Dupuy de Lôme, see Offner, *Unwanted War*, 95–97.
34. Offner, *Unwanted War*, 99–100, 113.
35. Woodford, the U.S. minister in Spain, frequently reported on the Spanish press and public opinion. See Stewart L. Woodford to William McKinley, December 11 and December 18, 1897, and January 17, January 28, February 4, and March 31, 1898, RG 59, roll 123, State Department Records. For historical studies of the Spanish press, see Angel Smith, "The People and the Nation: Nationalist Mobilization and the Crisis of 1898 in Spain," in *The Crisis of 1898: Colonial Redistribution and Nationalist Mobilization*, ed. Angel Smith and Emma Davila-Cox (London: Macmillan, 1999), 152–79; David Ortiz, *Paper Liberals: Press and Politics in Restoration Spain* (Westport, CT: Greenwood, 2000); Sebastian Balfour, *The End of the Spanish*

Empire (Oxford, UK: Clarendon Press, 1997); and Melchor Fernández Almagro, *Historia política de la España contempóranea*, vol. 2 (Madrid: Alianza, 1968–1969), 233–398.

36. Smith, "The People and the Nation."
37. Félix Santos, *1898: La prensa y la guerra de Cuba* (Bilbao: Asociación Julián Zugazagoita, 1998), 96–122; Fernández, *La América distorsionada*, 97–115.
38. Enrique Dupuy de Lôme to José Canalejas, undated, *Papers*, 1007. The letter was probably written in mid-December 1897.
39. Dupuy de Lôme to Canalejas, undated, 1008.
40. Ibid.
41. Offner, *Unwanted War*, 122.
42. *Washington Post*, February 9, 1898, 1, and February 10, 1898, 1; *San Francisco Examiner*, February 9, 1898, 1–2, and February 10, 1898, 1–2. The *Boston Globe, Philadelphia Inquirer, New Orleans Picayune, New York Times, New York Tribune, Atlanta Constitution, San Francisco Chronicle*, and *Chicago Tribune* for February 10 and 11, 1898, all begin coverage on page 1. See also Stewart L. Woodford to John Sherman, February 11, 1898, and Stewart L. Woodford to Pio Gullón, February 14, 1898, *Papers*, 1009–13; and Ivan Musicant, *Empire by Default: The Spanish-American War and the Dawn of the American Century* (New York: Henry Holt, 1998), 135–36.
43. Offner, *Unwanted War*, 28–29; Gerald Eggert, "Our Man in Havana: Fitzhugh Lee," *Hispanic American Historical Review* 47.4 (1967): 463–85.
44. *New Orleans Daily Picayune*, January 13, 1898, 7, and January 14, 1898, 1.
45. *San Francisco Examiner*, January 13, 1898, 1.
46. Ibid., January 14, 1898, 1, 2, and 6 (editorial), and January 15, 1898, 1, 2. See also the *Chicago Tribune*, January 13, 1898, 1, January 14, 1898, 2, 6, and January 15, 1898, 1, 2; *New Orleans Daily Picayune*, January 13, 1898, 7, January 14, 1898, 2, and January 15, 1898, 1; and *Washington Post*, January 13, 1898, 1, and January 14, 1898, 1. See also Offner, *Unwanted* War, 94–97.
47. *San Francisco Examiner*, January 13, 1898, 1; Offner, *Unwanted War*, 98–99; Samuels and Samuels, *Remembering*, 57–64; O'Toole, *Spanish War*, 117–19.
48. Brown, *Correspondents' War*, 114–16; Milton, *Yellow Kids*, 217–21.
49. Brown, *Correspondents' War*, 114–21; Milton, *Yellow Kids*, 119–20; Samuels and Samuels, *Remembering*, 109–15.
50. Samuels and Samuels, *Remembering*, 115–16; Brown, *Correspondents' War*, 125.
51. Brown, *Correspondents' War*, 129–32.
52. Qtd. in Ibid., 121.
53. *Chicago Tribune*, February 17, 1898, 1. For similar coverage, see the *New*

Orleans Daily Picayune, February 17 and 18, 1898, 1; *Washington Post*, February 17 and 18, 1898, 1; and *San Francisco Examiner*, February 17 and 18, 1898, 1.

54. Wisan, *Cuban Crisis*, 390; Santos, *1898*, 65–92.
55. Wisan, *Cuban Crisis*, 391.
56. Ibid.
57. For analyses of the *World*'s coverage, see Wisan, *Cuban Crisis*, 392–96, and Milton, *Yellow Kids*, 225–28.
58. *New York Times*, February 20, 1898, 18, and February 23, 1898, 6; *New York Tribune*, March 1, 1898 1, and March 6, 1898, 5. See also *Literary Digest*, March 5, 1898, 271–73, March 19, 1898, 331–33, and April 2, 1898, 413–14; Campbell, *Yellow Journalism*, 116–19; and Wisan, *Cuban Crisis*, 392–99. In addition, see the *Washington Post, Chicago Tribune, Boston Globe, Philadelphia Inquirer, Atlanta Constitution, New Orleans Picayune, and San Francisco Chronicle* from February 17 to February 28, 1898.
59. Offner, *Unwanted War*, 125–26.
60. Hilderbrand, *Power and the People*, 18–19; John A. Porter to John D. Long, February 16, 1898, McKinley Papers.
61. Gould, *Spanish-American War*, 35–42.
62. Offner concludes that McKinley was reluctant to confront U.S. public opinion ("United States Politics and the 1898 War Over Cuba" in *Crisis of 1898*, 18–44). Hilderbrand views McKinley as a competent activist in dealing with the press and public opinion (*Power and the People*, 8–29). Hamilton also details McKinley's close attention to the press (*Journalism's Roving Eye*, 89–105).
63. Offner points out that "the *Maine* disaster did not generate intense diplomatic activity" (*Unwanted War*, 125).
64. Wilkerson, *Public Opinion*, 101; Santos, *1898*, 37–92.
65. Senator Henry Cabot Lodge, an outspoken advocate of war, commented on the tension and excitement in the general population in the weeks between the *Maine* explosion and the decision to declare war. See his *The War with Spain*, 29–35.
66. Samuels and Samuels, *Remembering*, 163–64; Wisan, *Cuban Crisis*, 404.
67. Brown, *Correspondents' War*, 140; Wisan, *Cuban Crisis*, 402–3.
68. Samuels and Samuels, *Remembering*, 175.
69. Stewart L. Woodford to McKinley, March 2, 1898, RG 59, roll 123, State Department Records, RG 59, roll 123; *San Francisco Examiner*, February 27, 1898, 1; Wilkerson, *Public Opinion*, 102. Woodford included the articles from the *San Francisco Examiner* in his letter to McKinley.
70. *Chicago Tribune*, February 27, 1898, 1, 2. The *Washington Evening Star*

reported the same conspiracy on February 28, 1898. See Samuels and Samuels, *Remembering*, 175.

71. *Chicago Tribune*, February 27, 1898, 1, 2.
72. *New York Times*, March 6, 1898, 1.
73. *Washington Post*, March 6, 1898, 1; *San Francisco Examiner*, March 6, 1898, 1.
74. Brown, *Correspondents' War*, 136–39.
75. *New York Tribune*, March 3, 1898, 1.
76. Whitelaw Reid to William McKinley, March 8, 1898, reel 72, Reid Family Papers, Manuscript Division, Library of Congress, Washington, D.C.; Whitelaw Reid to Donald Nicholson (telegram), March 7, 1898, reel 72, Reid Family Papers.
77. *San Francisco Chronicle*, March 6, 1898, 1.
78. *Atlanta Constitution*, March 7, 1898, 1; *Chicago Tribune*, January 13, 1892, 6.
79. Rosenberg and Ruff, *Indiana*, 25–30.
80. George H. Gibson, "Attitudes in North Carolina Regarding the Independence of Cuba, 1868–1898," *North Carolina Historical Review* 43.1 (1966): 62; Marvin Olasky, "Hawks or Doves? Texas Press and the Spanish American War," *Journalism Quarterly* 64.1 (1987): 207. See also Campbell, *Yellow Journalism*, 116–19; Hilderbrand, *Power and the People*, 22–27; Auxier, "Middle Western Newspapers"; Hamilton, *President McKinley and the Coming of the War*, 171–211; and Mark M. Welter, "The 1895–1898 Crisis in Minnesota Newspapers: Testing the 'Yellow Journalism' Theory," *Journalism Quarterly* 47.4 (1970): 719–24. Gibson, Olasky, Rosenberg and Ruff, Auxier, Hamilton, and Welter emphasize local editors' reluctance to adopt the more aggressive tone of the New York newspapers until early to mid-March.
81. Whitelaw Reid to Donald Nicholson, March 12, 1898, reel 72, Reid Family Papers.
82. Offner, *Unwanted War*, 129–30; Wisan, *Cuban Crisis*, 407–10.
83. Linderman, *Mirror of War*, 37–59; Offner, *Unwanted War*, 131–35; *New York Times*, March 19, 1898, 6.
84. Brown, *Correspondents' War*, 143–44.
85. Hoganson, *American Manhood*, 88–106; Offner, *Unwanted War*, 136. See also Wisan, *Cuban Crisis*, 422–28.
86. *New York World*, April 7, 1898, qtd. in Wilkerson, *Public Opinion*, 118. See also Wilkerson, *Public Opinion*, 98–120; and for excerpts from several prowar newspapers, see *Literary Digest*, March 5, 1898, 271–73, and March 19, 1898, 331–33. For an antiwar newspaper that shifted to the prowar camp, see the *New York Herald*, March 30, 1898, 5, April 2, 1898, 8, and April, 6, 1898, 8. The *Chicago Times-Herald*, the *Commercial Advertiser*, and the *Chicago*

Chronicle followed a similar pattern. See Gleijeses, "1898," for a thorough examination of antiwar advocacy in the U.S. press.

87. Offner, "United States Politics and the 1898 War Over Cuba," 29–35.
88. Offner, *Unwanted War,* 154–58.
89. Woodford to McKinley, March 2, 1898; *New York Times,* April 13, 1898, 1; *New Orleans Picayune,* April 13, 1898, 1, and April 12, 1898, 8; *Philadelphia Inquirer,* April 12, 1898, 7, and April 13, 1898, 1; *Boston Globe,* April 12, 1898, 1. Richard Hamilton argues in his study of the decision for war that public demonstrations—the masses in the streets—were not frequent or strong enough to push McKinley and Congress in that direction. The street demonstrations in Savannah, Richmond, and Washington seem to offer evidence that by April such demonstrations had begun to play a role in national politics, although it should be mentioned that many of the essentials for the decision for war were already in place. See Hamilton, *President McKinley and the Coming of the War,* 149–238. Setting aside this moot point, the big-city press tied together by electric wires did generate the impression of a nationwide demand for war against Spain. Politicians in Washington found that the press had a large impact and that its views seemed to approximate public opinion. Newspapers created the forum and set the agenda for public discussion. In this electronic and print environment, politicians and editors reached their conclusions. The print record—usually the front pages of newspapers—testifies to the information and thought processes that went into the decision-making in this critical period.
90. *New Orleans Picayune,* April 13, 1898, 1, and April 12, 1898, 8. See also *Washington Post,* April 12 and 13, 1898, 1; *New York Times,* April 13, 1898, 1; *New Orleans Picayune,* April 14, 1898, 1; *Atlanta Constitution,* April 12, 13, and 14, 1898, 1; *Chicago Tribune,* April 13, 1898, 1; and Eggert, "Fitzhugh Lee," 463–85. The *San Francisco Examiner* of April 12, 1898, placed Lee's coverage on page 3.
91. As early as February 26, 1898, Lee was claiming there was a "Spanish plot." See *San Francisco Examiner,* February 27, 1898, 1. Spain asked for his recall in March, but McKinley refused (*New Orleans Picayune,* March 7, 1898, 1). See also Offner, *Unwanted War,* 183–84; and Eggert, "Fitzhugh Lee," 463–85.
92. Offner, *Unwanted War,* 182.
93. Hilderbrand, *Power and the People,* 2; scrapbooks, 1897–1901, "Current Comment" and foreign affairs, reels 94–98, McKinley Papers.
94. Hilderbrand, *Power and the People,* 21–29; Ponder, *Managing the Press,* 4–10.
95. Chauncey Depew, "Memorial Address on President McKinley at the Lincoln

Anniversary Banquet of the Republican Club of New York, February 12, 1902," qtd. in Linderman, *Mirror of War,* 27.

96. Henry S. Pritchett, "Some Recollections of President McKinley and the Cuban Intervention," *North American Review* 189 (March 1909): 401.
97. For examples see Cortelyou diary, March 26 and April 16, 22, and 23, 1898, Cortelyou Papers.
98. Hamilton, *Journalism's Roving Eye,* 100–102; Hamilton, Coleman, Grable, and Cole, "An Enabling Environment," 78–93.
99. Offner, *Unwanted War,* 233–34. On the extensive role of the press in late nineteenth-century politics, see Blondheim, *News over the Wires*; Gleijeses, "1898"; Smythe, *Gilded Age Press*; Kilmer, "The Press in Industrial America"; Campbell, "Age of New Journalism"; Schwarzlose, *The Nation's Newsbrokers,* 109–212; and Brands, *Reckless Decade,* 298–335. For thoughtful assessments of the causes of the Spanish American War, see Louis Pérez, "The Meaning of the *Maine*: Causation and the Historiography of the Spanish-American War," *Pacific Historical Review* 58.3 (1989): 293–322, and Louis Pérez, *The War of 1898: The United States and Cuba in History and Historiography* (Chapel Hill: University of North Carolina Press, 1998).
100. *New York Sun* of April 27, 1898, qtd. in Wisan, *Cuban Crisis,* 447. See also *New York Tribune,* March 6, 1898, 5; *San Francisco Chronicle,* March 9, 1898, 6; and *New York Times,* March 13, 1898, 21. Gleijeses found six business newspapers that were "steadfast and vocal in their opposition to war." Five others were less consistent or noncommittal ("1898," 711–12).

CHAPTER ELEVEN

1. Lubow, *Reporter,* 171–80; Evan Thomas, *The War Lovers: Roosevelt, Lodge, Hearst and the Rush to Empire, 1898* (New York: Little, Brown, 2010), 301–15; Dale Walker, *The Boys of '98: Theodore Roosevelt and the Rough Riders* (New York: Tom Doherty, 1998), 169–88. Davis had quit the *New York Journal* because of his embarrassment concerning misleading illustrations that accompanied his article on the *Olivette* affair in February 1898. See Lubow, *Reporter,* 143–44.
2. United States and Haiti opened for business after the settlement of a lawsuit in federal court in New York City. See *Electrical World,* October 10, 1897, 413–14, November 21, 1897, 620, and December 12, 1897, 716.
3. Arthur Brisbane, "The Modern Newspaper in War Time," *Cosmopolitan* 20 (September 1898): 553–54; West India and Panama, directors' report, December, 14, 1898, 2–3, May 20, 1899, 2–3, Cable and Wireless Archive, Porthcurno Telegraph Museum, Cornwall, UK; West India and Panama, board minute books, April 13 and 21, May 11, June 8 and 29, and July 25, 1898,

Cable and Wireless Archive. West India and Panama frequently routed its U.S.-bound telegrams through the lines of Cuba Submarine. See Cuba Submarine, directors' report, April 26, 1899, 2–3, directors' report, October 24, 1899, 2–3, and board minute books, April 3, April 23, April 24, and September 14, 1898, Cable and Wireless Archive. See also "The West Indian Cables," *Electrical World*, July 2, 1898, 10–11; and Thomas Grier, "The Use of Electricity in the Spanish-American War," *Electrical World*, October 15, 1898, 390–92.

4. There are several accounts of the Spanish-American War, including David Trask, *The War with Spain in 1898* (Lincoln: University of Nebraska Press, 1981); Joseph Smith, *The Spanish-American War: Conflict in the Caribbean and the Pacific 1895–1902* (London: Longman, 1994); Musicant, *Empire by Default*; and Albert Nofi, *The Spanish-American War, 1898* (Conshoken, PA: Combined Books, 1996).
5. Howard Giddings, *Exploits of the Signal Corps in the War with Spain* (Kansas City, MO: Hudson-Kimberly Publishing, 1900), 113–14. Perhaps as many as 155 reporters reached the war zone. The dispatch boats made it possible for those without credentials to reach Cuba. See Brown, *Correspondents' War*, 446.
6. Brisbane, "Modern Newspaper," 542.
7. Ibid., 552; "The Newspaper Correspondents in the War," *Review of Reviews* 28 (November 1898): 540; *New York Times*, June 5, 1898, 1.
8. "Newspaper Correspondents," 540; Brown, *Correspondents' War*, 445–46. For the perspective of a magazine editor, see "McClure's Magazine in War Times," *McClure's Magazine* 11 (June 1898): 206; and "McClure's and the War," *McClure's Magazine* 11 (July 1898): 496.
9. Brown, *Correspondents' War*, 168–69. For a description of other dispatch boat races, see Spears, "Afloat for News," 503. On these and other problems in reporting the war, see Mary S. Mander "Pen and Sword: Problems of Reporting the Spanish-American War," *Journalism History* 9.1 (1982): 2–9, 28.
10. *Atlanta Constitution*, June 27, 1898, 1. For general descriptions of the use of dispatch boats, see George Kennan, *Campaigning in Cuba* (New York: Century, 1899), 37–38; Brisbane, "Modern Newspaper," 553–54; and Baker, "How the News of the War Is Reported."
11. *New York Times*, June 27, 1898, 1.
12. Kennan, *Campaigning*, 38; Spears, "Afloat for News," 501–4.
13. Spears, "Afloat for News," 503–4. See also Baker, "How the News of the War Is Reported," 492; Richard Harding Davis, "Our War Correspondents in Cuba and Puerto Rico," *Harper's Magazine* 98 (May 1899): 946.
14. "Newspaper Correspondents," 540.
15. Brown, *Correspondents' War*, 388.
16. Spears, "Afloat for News," 504.

17. *Atlanta Constitution*, May 10, 1898, 1; *Washington Post*, May 10, 1898, 1.
18. *Chicago Tribune*, May 11, 1898, 2.
19. *Atlanta Constitution*, May 16, 1898, 1; Brown, *Correspondents' War*, 212–20.
20. Milton, *Yellow Kids*, 280–83.
21. Ibid., 278–79; Mander, "Pen and Sword," 6–8; *Atlanta Constitution*, June 4, 1898, 1.
22. Adolphus W. Greely to Wesley Merritt, April 28, 1898, Letterbooks, Adolphus W. Greely Papers, 1753–1959, Manuscript Division, Library of Congress, Washington, D.C.; Brown, *Correspondents' War*, 225.
23. Greely to Merritt, April 28, 1898; Adolphus W. Greely, "The Signal Corps in War-Time," *Century Magazine* 66 (September 1903): 813–15; Adolphus W. Greely, "Report of the Chief Signal Officer, U.S. Army, 1898," in *Military Signal Communications*, ed. Paul J. Scheips (New York: Arno Press, 1950), 891–95. See also West India and Panama, minute books, April 27, 1898, Cable and Wireless Archive; and Robbins Raines, *Getting the Message Through*, 88–89.
24. Qtd. in Brown, *Correspondents' War*, 225.
25. Poultney Bigelow, "In Camp at Tampa with the Troops under Canvas" (dateline May 22, 1898), *Harper's Weekly*, June 4, 1898, 550; Brown, *Correspondents' War*, 229–34.
26. Adolphus W. Greely to J. W. Brady, June 7, 1898, Greely Papers. See also Greely, "Report of the Chief Signal Officer," 891–95; and Greely, "Signal Corps in War-Time," 813–15.
27. Grant Squires, "Experiences of a War Censor," *Atlantic Monthly* 83 (March 1899): 426.
28. Adolphus W. Greely to William McKinley, May 24, May 25, and May 27, 1898, Greely Papers; Giddings, *Exploits of the Signal Corps*, 114.
29. Trask, *The War with Spain*, 108–224; Smith, *The Spanish-American War*, 48–129; Musicant, *Empire*, 235–368.
30. Ibid., 134–36.
31. *Chicago Tribune*, June 5, 1898, 1.
32. *New York Times*, June 5, 1898, 1. The article is datelined "Port Antonio, Jamaica, June 3."
33. Brown, *Correspondents' War*, 293–94.
34. *San Francisco Chronicle*, June 5, 1898, 1; *San Francisco Examiner*, June 5, 1898, 1.
35. *Atlanta Constitution*, June 6, 1898, 1.
36. *New York Times*, June 5, 1898, 1.
37. Trask, *War with Spain*, 288. For Richard Harding Davis's eyewitness account of the Hobson exchange see his *The Cuban and Porto Rican Campaigns* (New York: Charles Scribner's Sons, 1898), 262–73.
38. Brown, *Correspondents' War*, 296.

39. *New York Times*, June 15, 1898, 1; Trask, *War with Spain*, 138.
40. *New York Times*, June 15, 1898, 1.
41. *San Francisco Examiner*, June 16, 1898, 1.
42. *San Francisco Chronicle*, June 15, 1898, 4.
43. *New York Times*, June 15, 1898, 1
44. *Washington Post*, June 15, 1898, 1.
45. Trask, *War with Spain*, 138, 186–88. For Victor Blue's report, see William T. Sampson to John D. Long, June 13, 1898 (telegram quoted in Long to Blue, June 20, 1898), Blue to Daniel Delehanty, June 27, 1898, and Delehanty to Sampson, June 29, 1898, box 1, folder 4, Victor Blue Papers, Southern Historical Collection, Wilson Library, University of North Carolina, Chapel Hill, NC.
46. Musicant, *Empire*, 377–89.
47. Walker, *Boys of '98*, 184; Smith, *Spanish-American War*, 127.
48. Musicant, *Empire*, 382; Walker, *Boys of '98*, 177–78.
49. Lubow, *Reporter*, 173–74.
50. Brown, *Correspondents' War*, 312–14.
51. Walker, *Boys of '98*, 175.
52. Lubow, *Reporter*, 178.
53. The *San Francisco Chronicle* and the *San Francisco Examiner*, both of which went to press a few hours after the East Coast newspapers, named Wood and Roosevelt as leaders of the charge (*San Francisco Chronicle*, June 25, 1898, 1). The *Examiner* placed a portrait of Roosevelt on the front page.
54. *New York Times*, June 26, 1898, 1. See also *San Francisco Examiner*, June 25, 1898, 1; and *Philadelphia Inquirer*, June 26, 1898, 1. The other major big-city papers covered the story as well; see *San Francisco Chronicle* for June 25, 26, and 27, 1898; and *Atlanta Constitution*, *Washington Post*, *Boston Globe*, *Chicago Tribune*, and *New York Tribune* for June 26 and 27, 1898.
55. *Chicago Tribune*, June 26, 1898, 1, 2.
56. *New York Tribune*, June 26, 1898, 1, 2.
57. *New York Times*, June 27, 1898, 1. On the same day the *Washington Post* carried essentially the same story on its front page.
58. Musicant, *Empire*, 377–89; Walker, *Boys of '98*, 169–205; Trask, *War with Spain*, 217–24; Nofi, *Spanish-American War*, 123–32.
59. *San Francisco Chronicle*, July 4, 1898, 3. See similar coverage in *San Francisco Examiner*, July 4, 1898, 3.
60. *Boston Globe*, July 3, 1898, 1, and July 4, 1898, 1, 4; *Philadelphia Inquirer*, July 3, 1898, 1, and July 4, 1898, 1.
61. *New York Times*, July 5, 1898, 6.
62. Ibid., July 9, 1898, 3.
63. Ibid., July 4, 1898, 1; *Chicago Tribune*, July 4, 1898, 1; *San Francisco*

Examiner, July 4, 1898, 1. The *San Francisco Examiner,* July 5, 1898, 2, placed Lieutenant Colonel Allen on its honor roll as one of five "heroes who fought in the cause of humanity at Santiago," explaining that he was the signal officer who sent "the authentic news" of Sampson's victory.

64. *Atlanta Constitution,* July 4, 1898, 1–2.
65. Brown, *Correspondents' War,* 377–86.
66. *New York Times,* July 5, 1898, 1.
67. *New York Tribune,* July 5, 1898, 1.
68. *San Francisco Examiner,* July 6, 1898, 2.
69. *Washington Post,* July 5, 1898, 1.
70. *New York Journal,* qtd. in Brown, *Correspondents' War,* 389. A similar version of this dateline appeared in the *San Francisco Examiner,* July 6, 1898, 1. See also Santos, *1898,* 123–30.
71. Lubow, *Reporter,* 187. See also Lubow, *Reporter,* 180–88; and Brands, *TR,* 333–57.
72. Davis, *The Cuban and Porto Rican Campaigns,* esp. chaps. 5 and 6.
73. Qtd. in Walker, *Boys of '98,* 216. On the historical judgments that have been passed on the role played by Roosevelt and the Rough Riders in the battle, see Walker, *Boys of '98,* 206–40; Musicant, *Empire,* 390–31; Smith, *Spanish-American War,* 129–50; and Trask, *War with Spain,* 242–44.
74. Richard Harding Davis to Theodore Roosevelt, April 21, 1899, reel 2, Theodore Roosevelt Papers, 1759–1919, reel 2, Manuscript Division, Library of Congress, Washington, D.C. See also Theodore Roosevelt to Richard Harding Davis, February 2, 1899, reel 319, Richard Harding Davis to Theodore Roosevelt, August 24, 1899, reel 3, Richard Harding Davis to Theodore Roosevelt, August 19, 1900, reel 6, and Richard Harding Davis to Theodore Roosevelt, November 14, 1900, reel 6, Roosevelt Papers. Davis commented extensively on Stephen Bonsal's book *The Fight for Santiago: the Story of the Soldier in the Cuban Campaign from Tampa to the Surrender* (New York: Doubleday and McClure, 1899). See Richard Harding Davis to Theodore Roosevelt, April 21, 1899, reel 2, Roosevelt Papers.
75. Charles Johnson Post, *The Little War of Private Post* (Boston: Little, Brown, 1960), 203–4. Post also described Davis as "a brilliant and superficial writer to whom glamour was always gold" and as "an intense worshiper of Colonel Roosevelt" (118).
76. John Black Atkins, *The War in Cuba: The Experience of an Englishman with the United States Army* (London: Smith, Elder, and co., 1899), 96–97.
77. Davis, "Our War Correspondents," 944.
78. Philip S. Foner, *The Spanish-Cuban-American War and the Birth of*

American Imperialism, 2 vols. (New York: Monthly Review Press, 1972), 2:360–61. See also Foner, *Spanish-Cuban American War*, 2:355–61, 2:394–95. For an extensive analysis for U.S. prejudice toward Cubans, see Peréz, *Cuba in the American Imagination*, 95–174. Lars Schoultz sets the prejudice against Cubans in the larger context of U.S.–Latin American cultural relations in his *Beneath the United States*, esp. 124–51. See also *Literary Digest*, July 30, 1898, 121–25; and Kennan, *Campaigning*, 100, 128, 143–44.

79. *New York Times*, July 23, 1898, 1.
80. Pérez, *The War of 1898*, 97. For more on the Garcia-Shafter dispute, see *Literary Digest*, July 30, 1898, 121–22.
81. *New York Times*, July 18, 1898, along with front-page coverage in all ten metropolitan newspapers on July 18 and for several days thereafter. See also *Literary Digest*, July 23, 1898, 93–96.
82. *New York Tribune*, August 7, 1898, 5.
83. Quoted in Foner, *Spanish-Cuban-American War*, 2:395.
84. *New York Times*, July 29, 1898, 4, qtd. in Pérez, *United States and Cuba*, 85.
85. Fitzhugh Lee and Joseph Wheeler, *Cuba's Struggle against Spain* (New York: American Historical Press, 1899), 617; see also 608–20.
86. Quesada and Northrop, *The War in Cuba*.
87. J. Rankin Young and J. Hampton Moore, *Reminiscences and Thrilling Stories of the War by Returned Heroes* (Chicago: Providence Publishing, 1899).
88. Lubow, *Reporter*, 155–96; Milton, *Yellow Kids*, xix, 319–20, 346–61.
89. Hilderbrand, *Power and the People*, 21; Ponder, *Managing the Press*, 8–15; scrapbooks, 1897–1901, "Current Comments" and foreign affairs, reels 94–98, William McKinley Papers, ca. 1847–1945, Manuscript Division, Library of Congress, Washington, D.C.
90. Hilderbrand, *Power and the People*, 21–29.
91. Ibid., 31–32.
92. Ida Tarbell, "President McKinley in War Times," *McClure's* 11 (July 1898): 208–24.
93. For examples, see George Cortelyou diary, March 26, April 16, 22, and 23, and July 20, 1898, and February 17, 1899, in George L. Cortelyou Papers, 1871–1948, Manuscript Division, Library of Congress.
94. Leech, *In the Days of McKinley*, 230–31.
95. Campbell, *Yellow Journalism*, assembles evidence that indicates that such a telegram never existed.
96. *Literary Digest*, October 15, 1898, 455.

CHAPTER TWELVE

1. McCullough, *Path Between the Seas*, 329–60; Philippe Bunau-Varilla, *Panama: Its Creation, Destruction, and Resurrection* (London: Constable, 1913), 288–93; Anguizola, *Philippe Bunau-Varilla*, 157–297. On the Prestán revolt, see chapter 5.
2. Bunau-Varilla, *Panama*, 304–19; McCullough, *Path*, 329–86; Miles DuVal, *Cadiz to Cathay: The Story of the Long Diplomatic Struggle for the Panama Canal* (Stanford, CA: Stanford University Press, 1940), 273–316; Richard Collin, *Theodore Roosevelt's Caribbean: The Panama Canal, the Monroe Doctrine, and the Latin American Context* (Baton Rouge: Louisiana State University Press, 1990), 215–68.
3. The code is reproduced in McCullough, *Path*, 343–45, and in U.S. Congress, *The Story of Panama: Hearings of the Rainey Resolution before the Committee on Foreign Affairs of the House of Representatives* (Washington, D.C.: Government Printing Office, 1913), 358–59. See also Díaz Espino, *How Wall Street Created a Nation*, 76–81.
4. McCullough, *Path*, 387–402; Bunau-Varilla, *Panama*, 320–97.
5. The unusually rapid diplomatic recognition of the Panamanian government made possible the quick signing and ratification of the treaty that gave the United States authority over what became the Panama Canal Zone. Precedent for U.S. recognition of new governments in Latin America dated back to the early decades of the nineteenth century; in these cases, the pace had been more gradual, allowing for policy discussion and negotiation. Secretary of State John Quincy Adams, Speaker of the House Henry Clay, and other officials debated the issue from 1819 to 1822 before deciding to recognize the new governments.

   In addition to the issue of recognition, the Panama episode created the potential for a lengthy crisis in U.S.-Colombian relations. Through the rapid recognition of Panama, the United States, in effect, endorsed the independence of the new nation and Colombia's loss of its isthmian province. This intrusive action by the United States with its broad territorial and geopolitical ramifications created international tensions comparable to those generated by the *Virginius* affair of 1873, the Chilean-U.S. confrontation of 1891–1892, also known as the *Baltimore* affair, and the Venezuela–British Guiana boundary dispute of 1895–1896. Resolving all of these crises required time-consuming diplomatic exchanges over several months. In the Panama case, however, this potentially lengthy crisis found resolution in a matter of days through the abrupt actions of the Roosevelt administration.

   On U.S. recognition of the Latin American governments in the 1820s, see Arthur P. Whitaker, *The United States and the Independence of Latin*

America (New York: Norton, 1964); Wesley P. Newton, "Origins of United States-Latin American Relations," in *United States-Latin American Relations, 1800–1850: The Formative Generation,* ed. T. Ray Shurbutt (Tuscaloosa: University of Alabama Press, 1991), 1–24; Dent, *The Legacy of the Monroe Doctrine;* William Earl Weeks, *John Quincy Adams and American Global Empire* (Lexington: University Press of Kentucky, 1992); and Perkins, *Monroe Doctrine.* The role of diplomatic recognition in international relations is examined in Stephen D. Krasner, *Sovereignty: Organized Hypocrisy* (Princeton, NJ: Princeton University Press, 1999), 3–42; and M. J. Peterson, *Recognition of Governments: Legal Doctrine and State Practice* (New York: St. Martin's Press, 1997). For the *Virginius* dispute, see Bradford, Virginius *Affair;* and Nevins, *Hamilton Fish,* 2:607–87. The Chilean-U.S. crisis is discussed in Goldberg, *The* Baltimore *Affair;* and Britton, "Confusion Provoked." On the Venezuela–British Guiana boundary controversy, see Bertram, *Birth;* and Harris, "Venezuela."

6. Collin, *Theodore Roosevelt's Caribbean,* 173–218.
7. George Grafton Wilson, *Submarine Telegraph Cables in their International Relations* (Washington, D.C.: Government Printing Office, 1901), 3–5, 35–38.
8. West India and Panama, directors' report, December 31, 1902, 3, Cable and Wireless Archive, Porthcurno Telegraph Museum, Cornwall, UK; Ahvenainen, *History of the Caribbean Telegraphs,* 166–67; Gordon Thomas and Max Morgan Witts, *The Day the World Ended* (New York: Stein and Day, 1969).
9. Mitchell, *Danger of Dreams,* 64–107.
10. Richard Challener, *Admirals, Generals, and American Foreign Policy, 1898–1914* (Princeton, NJ: Princeton University Press, 1973), 16–17, 28–29, 32–35, 66–67, 111–19; Anguizola, *Bunau-Varilla,* 233–34.
11. Anguizola, *Bunau-Varilla,* 183–87; McCullough, *Path,* 270–38.
12. Davis, *Three Gringos,* 198. This chapter first appeared in serial form in *Harper's Weekly.*
13. Davis, *Three Gringos,* 211; see also 205–11.
14. Bunau-Varilla, *Panama,* 226–53.
15. McCullough, *Path,* 318–28.
16. Collin, *Theodore Roosevelt's Caribbean,* 163–236; Delpar, "Colombia."
17. Anguizola, *Bunau-Varilla,* 229–31; Bunau-Varilla, *Panama,* 286–88.
18. *New York Times,* August 31, 1903, 1.
19. Ibid., September 14, 1903, 7.
20. DuVal, *Cadiz to Cathay,* 277.
21. Thomas M. Leonard, *Panama, the Canal, and the United States: Guide to Issues and References* (Claremont, CA: Regina Books, 1993), 7–15; Collin, *Theodore Roosevelt's Caribbean,* 266–68; Delpar, "Colombia," 58–80.

22. Nickles, *Under the Wire*, 169–73.
23. Díaz Espino, *How Wall Street Created a Nation*, 77–81; McCullough, *Path*, 341–42.
24. McCullough, *Path*, 345–46.
25. Ibid., 346–47; Díaz Espino, *How Wall Street Created a Nation*, 61.
26. Bunau-Varilla, *Panama*, 289–92; McCullough, *Path*, 349–50.
27. Bunau-Varilla, *Panama*, 312–16; McCullough, *Path*, 352–53.
28. Bunau-Varilla, *Panama*, 294–97.
29. Ibid., 312.
30. Ibid., 304–12; McCullough, *Path*, 350–52; Anguizola, *Bunau-Varilla*, 235–38.
31. Bunau-Varilla, *Panama*, 312–16, 320–24. For the authorizing telegram, see Bunau-Varilla, *Panama*, 324. See also McCullough, *Path*, 356–58.
32. Bunau-Varilla, *Panama*, 316–19; McCullough, *Path*, 354–55.
33. Bunau-Varilla, *Panama*, 324–26; DuVal, *Cadiz to Cathay*, 311; McCullough, *Path*, 358.
34. U.S. Congress, *Story of Panama*, 380; McCullough, *Path*, 358; Díaz Espino, *How Wall Street Created a Nation*, 88.
35. U.S. Congress, *Story of Panama*, 381.
36. Ibid.
37. Ibid., 381–82; Collin, *Theodore Roosevelt's Caribbean*, 261–62. Collin views the U.S. role as "largely accidental" and "symbolic," but telegraphic orders for and coordination of U.S. ship movements before the beginning of the revolution point to a more direct and deliberate involvement on the part of the United States.
38. Richard W. Turk, "The United States Navy and the 'Taking' of Panama, 1901–1903," *Military Affairs* 38.3 (1974): 92–96; Collin, *Theodore Roosevelt's Caribbean*, 258.
39. U.S. Congress, *Story of Panama*, 381.
40. Turk, "United States Navy"; U.S. Congress, *Story of Panama*, 382–83.
41. Thomas Schoonover, "Max Farrand's Memorandum on the U.S. Role in the Panamanian Revolution of 1903," *Diplomatic History* 12.4 (1988): 504.
42. Bunau-Varilla, *Panama*, 333–34.
43. U.S. Congress, *Story of Panama*, 384.
44. McCullough, *Path*, 367; see also 639. For a brief account of Hubbard's situation, see U.S. Congress, *Story of Panama*, 387–89.
45. U.S. Congress, *Story of Panama*, 442; McCullough, *Path*, 366–67; Díaz Espino, *How Wall Street Created a Nation*, 101–2; DuVal, *Cadiz to Cathay*, 331.
46. Turk, "United States Navy," 94.
47. Ibid., 93–96.
48. U.S. Congress, *Story of Panama*, 443.

49. McCullough, *Path,* 368; see also 369–77.
50. Bunau-Varilla, *Panama,* 335.
51. U.S. Congress, *Story of Panama,* 395; McCullough, *Path,* 371. There was one death when the Colombian gunboat *Bogotá* shelled Panama City. Shore batteries returned fire, and the *Bogotá* withdrew (McCullough, *Path,* 371).
52. See the telegrams between Arthur Matthias Beaupré and John Hay, November 6, 7, 9, 11, 12, 14, and 18, 1903, in *Papers Relating to Foreign Relations of the United States, 1903* (Washington, D.C., Government Printing Office, 1904), 225–33; DuVal, *Cadiz to Cathay,* 341–42. Bunau-Varilla claimed that Hay telegraphed notification of U.S. recognition of Panama to Bogotá on November 5. See Bunau-Varilla, *Panama,* 350–51.
53. Qtd. in Anguizola, *Bunau-Varilla,* 254.
54. Bunau-Varilla, *Panama,* 351–408.
55. John Major, "Who Wrote the Hay-Bunau-Varilla Convention?," *Diplomatic History* 8.2 (1984): 115–23.
56. Bunau-Varilla, *Panama,* 378; McCullough, *Path,* 395.
57. Bunau-Varilla, *Panama,* 379–86, 398–408; McCullough, *Path,* 394–97.
58. McCullough, *Path,* 398; Collin, *Theodore Roosevelt's Caribbean,* 305.
59. For examples of prerevolt coverage of unrest, see *New York Times,* August 31, 1903, 1; *Washington Post,* September 1, 1903, 1; and Díaz Espino, *How Wall Street Created a Nation,* 58. See also Díaz Espino, *How Wall Street Created a Nation,* 50–58; and McCullough, *Path,* 341–71.
60. *Washington Post,* November 9, 1903, 6.
61. *Chicago Tribune,* November 7, 1903, 1, 2, and November 8, 1903, 1, 2, 5.
62. *New York Tribune,* qtd. in *Literary Digest,* November 14, 1903, 650–51.
63. *San Francisco Chronicle,* November 11, 1903, 6; *Boston Globe,* November 4–7, 1903, 1, and November 9, 1903, 2; *Philadelphia Inquirer,* November 4–7, 1903, 1, and November 5, 1903, 8 (editorial); *Atlanta Constitution,* November 9, 1903, 4; *Literary Digest,* November 14, 1903, 649–51, and November 21, 1903, 689–92.
64. *San Francisco Examiner,* November 7, 1903, 16, and November 9, 1903, 14.
65. *Literary Digest,* November 14, 1903, 649–51, and November 21, 1903, 689–92.
66. *New York Times,* November 6, 1903, 8.
67. *Nation,* February 4, 1904, 81.
68. J. P. Gordy, "Ethics in the Panama Case," *Forum* 36 (July 1904): 115–16, 124.
69. Theodore Roosevelt to George Brinton McClellan Harvey, December 19, 1903, in *The Letters of Theodore Roosevelt,* 8 vols., ed. Elting E. Morrison (Cambridge, MA: Harvard University Press, 1951), 3:673–74. See also *Harper's Weekly,* November 28, 1903, 1892.

70. Theodore Roosevelt to James Bryce, December 19, 1903, *Letters of Theodore Roosevelt*, 3: 671–72.
71. Brands, *TR*, 367. See also McCullough, *Path*, 379–86.
72. For a succinct account, see Schoultz, *Beneath the United States*, 164–75.
73. *New York Times*, November 6, 1903, 8.
74. *Chicago Tribune*, November 8, 1903, 18.

CHAPTER THIRTEEN

1. Hoganson, *American Manhood*, esp. 110–18 and 200–208.
2. *San Francisco Examiner*, July 4, 1898, 1.
3. *New York Times*, July 18, 1898, 1.
4. *Literary Digest*, July 23, 1898, 95–99.
5. J. Hampton Moore and James Rankin Young, *Thrilling Stories of the War by Returning Heroes* (Chicago: Providence Publishing Company, 1899), iv; A. C. Parkerson, *How Uncle Sam Fights; or, Modern Warfare—How Conducted* (Baltimore, MD: R. H. Woodward Company, 1898). Parkerson had prepared most of his volume before the fighting started. The text on the war itself made up only about one-eighth of the book. See also Theodore Roosevelt, *The Rough Riders* (New York: Charles Scribner's Sons, 1899); and Brands, *TR*, 372–76.
6. Davis, *The Cuban and Porto Rican Campaigns*; Lubow, *Reporter*, 171–72.
7. *Literary Digest*, October 15, 1898, 455.
8. McCartney, *Power and Progress*, esp. 258–28; Robert Beisner, *Twelve Against Empire: The Anti-Imperialists* (Chicago: University of Chicago Press, 1985); Schoonover, *Uncle Sam's War.*
9. Trumbull White, *Our New Possessions* (Chicago: Henry Publishing Company, 1901), 17. White also commented on the Filipino armed uprising against U.S. control of the islands; see 222–30 and 291–305. Murat Halstead wrote a similar book under the title *Our New Possessions: Natural Riches, Industrial Resources of Cuba, Porto Rico, Hawaii, the Ladrones and Philippine Islands with Episodes of Their Early History* (Chicago: Dominion Company, 1898).
10. White, *Our New Possessions*, 535, 550.
11. J. Michael Hogan, *The Panama Canal and American Politics* (Carbondale: Southern Illinois University Press, 1986), 45–46; Willis J. Abbot, *Panama and the Canal in Pictures and Prose: A Complete Story of Panama as well as the History, Purpose, and Promise of Its World-Famous Canal—the Most Gigantic Engineering Undertaking since the Dawn of Time* (New York: Syndicate Publishing, 1913).
12. Hoganson, *American Manhood*, esp. 1–42.

13. Creelman, *On the Great Highway*, 340, 343; see also 338–56.
14. Ibid., 344.
15. Lee and Wheeler, *Cuba's Struggle against Spain*.
16. Much of the content of these three volumes originally appeared as essays in *Harper's New Monthly Magazine* (*Interests of America and Problem of Asia*) and *McClure's Magazine* (*Lessons of War*).
17. Robert Seager II, *Alfred Thayer Mahan: The Man and His Letters* (Annapolis, MD: Naval Institute Press, 1977), 349.
18. Seager, *Alfred Thayer Mahan*, 462.
19. Alfred Thayer Mahan, *Lessons of the War with Spain* (Boston: Little, Brown, 1899), 277–82; Alfred Thayer Mahan, *The Problem of Asia and Its Effect upon International Policies* (Boston: Little, Brown, 1900), 129–131.
20. *New York Times*, December 1, 1898, 5, April 7, 1899, 5, February 25, 1900, 3, and March 12, 1900, 3.
21. Alfred Thayer Mahan, *The Interest of America in Sea Power Present and Future* (Boston: Little, Brown, 1897).
22. Mahan, *Lessons*, 286.
23. *New York Times*, March 12, 1900, 3.
24. Ibid., March 12, 1900, 3.
25. Mahan, *Problem of Asia*, 13–17, 179–86, 201–2.
26. Ibid., esp. 125–31, 155–58, and 184–97.
27. Brands, *TR*, 333–433.
28. *New York Times*, October 6, 1898, 1, 2.
29. Ibid., October 6, 1898, 2.
30. Ibid.
31. Ibid
32. Ibid.
33. Ibid.
34. Ibid.
35. Kennan, *Campaigning in Cuba*, 130–91, 213–69. For critical press commentary on the administration of the War Department, see *Literary Digest*, August 20, 1898, 212–17.
36. Mahan, *Lessons*, 161, 168–70.
37. Walter Millis, *The Martial Spirit* (Boston: Houghton Mifflin, 1931).
38. Challener, *Admirals, Generals, and American Foreign Policy*; Schoonover, *Uncle Sam's War*, 102–22; Steven T. Ross, *American War Plans, 1890–1939* (London: Frank Cass, 2002); James Chace and Caleb Carr, *American Invulnerable: The Quest for Absolute Security from 1812 to Star Wars* (New York: Summit Books, 1988), 109–44; Brian McAllister Linn, *Guardians of Empire: The U.S. Army and the Pacific, 1902–1940* (Chapel Hill: University of North

Carolina Press, 1997), 5–113; Jeffery M. Dorwart, *The Office of Naval Intelligence: The Birth of America's First Intelligence Agency, 1865–1918* (Annapolis, MD: Naval Institute Press, 1979), 55–95.

39. Henry Cabot Lodge to Theodore Roosevelt, July 6, 1898, in *Selections from the Correspondence of Theodore Roosevelt and Henry Cabot Lodge, 1884–1918*, 2 vols., ed. Henry Cabot Lodge and Charles Redmond (1925; repr., New York: Da Capa Press, 1979), 1:318; see also 1:318–21 and 1:323–25.
40. For examples, see *San Francisco Chronicle*, July 1, 1898, 1; and *New York Times*, July 1, 1898, 1.
41. *New York Times*, July 2, 1898, 1; *Boston Globe*, July 3, 1898, 1. Most of the ten metropolitan newspapers placed this telegram at the head of their coverage on July 2 or 3. The *Boston Globe* and *Philadelphia Inquirer* emphasized casualties on these dates.
42. *San Francisco Examiner*, July 4, 1898, 1.
43. *New York Times*, July 4, 1898, 1.
44. *Atlanta Constitution*, July 4, 1898, 1; *Philadelphia Inquirer*, July 4, 1898, 1; *Washington Post*, July 4, 1898, 1.
45. *New York Times*, July 4, 1898, 1. The other nine metropolitan newspapers published the cable on the same date on page 1.
46. *New York Times*, July 5, 1898, 1. The other nine metropolitan newspapers printed this telegram on the same date on page 1 or 2.
47. "Cable Cutting in Warfare," *Literary Digest*, July 16, 1898, 75; "Rapid Cable-Laying in War," *Literary Digest*, July 30, 1898, 135; "West India Cables," *Electrical World*, July 2, 1898, 10–11; Thomas G. Grier, "The Use of Electricity in the Spanish American War, *Electrical World*, October 15, 1898, 390–91.
48. *New York Times*, April 10, 1898, 5, May 1, 1898, 2, 11, May 4, 1898, 7, May 11, 1898, 1, May 21, 1898, 1, May 24, 1898, 2 (editorial).
49. Wilson, *Submarine Telegraph Cables*.
50. Lodge, *The War with Spain*, 33.
51. Duncan, *Whitelaw Reid*, 197–217. McKinley was also an effective advocate for public acceptance of the new U.S. empire. See Herring, "Imperial Tutor."
52. Daniel Margolies, *Henry Watterson and the New South: The Politics of Empire, Free Trade, and Globalization* (Lexington: University of Kentucky Press, 2006), 177–243.
53. Elbert Hubbard, *A Message to Garcia* (East Aurora, NY: Roycrofters, 1899); O'Leary, *To Die For*, 190–91; Katherine Isabel Bemis, ed., *The Patriotic Reader for Seventh and Eighth Grades and Junior High Schools* (Boston: Houghton Mifflin, 1917), 115–19.
54. *New York Times*, July 3, 1898, illustrated magazine supplement, 12, Squier's emphasis.

55. George O. Squier, "The Influence of Submarine Cables upon Military and Naval Supremacy," *National Geographic* 12 (January 1901): 2.
56. Squier, "Influence," 7.
57. Howard S. Abramson, *National Geographic: Behind America's Lens on the World* (New York: Crown, 1987), 27–57.
58. Greely, "The Signal Corps in War-Time,"; Henry MacFarland, "The Signal Corps of the Army in the War," *Review of Reviews* 29 (February 1899): 183–88.
59. Robbins Raines, *Getting the Message Through,* 98–110.
60. Headrick, *Invisible Weapon,* 93–137.
61. O'Leary, *To Die For,* 150–93. See also Kammen, *Mystic Chords of Memory,* 93–296; and Ruth Miller Elson, *Guardians of Tradition: American Schoolbooks of the Nineteenth Century* (Lincoln: University of Nebraska Press, 1964), esp. 186–218.
62. On the Monroe Doctrine, see note 4 in chapter 1.
63. Mitchell, *Danger of Dreams.*
64. See the illustrations in Britton, "Confusion Provoked."
65. *Washington Post,* December 18, 1895, 1; *San Francisco Chronicle,* December 28, 1895, 1.
66. *Chicago Tribune,* December 20, 1895, 1.
67. Ibid., April 9, 1898, 1.
68. Hoganson, *American Manhood,* 43–67; Kaplan, *Anarchy of Empire,* 92–145; Hunt, *Ideology and U.S. Foreign Policy,* 46–91; Pérez, *Cuba in the American Imagination,* 95–174.
69. *Philadelphia Inquirer,* February 11, 1898, 1; *Chicago Tribune,* February 11, 1898, 1.
70. *Philadelphia Inquirer,* July 6, 1898, 1.
71. *Boston Globe,* July 5, 1898, 6.
72. *Chicago Tribune,* October 20, 1895, 1, and December 22, 1895, 25.
73. *San Francisco Chronicle,* December 15, 1895, 16.
74. *New York Journal,* February 17, 1898, 1; and *New York World,* February 17, 1898, 1, reproduced in Sloan, *Media in America,* 241.
75. *Chicago Tribune,* February 16, 1898, 3.
76. Ibid., April 14, 1898, 2.
77. *San Francisco Examiner,* June 6, 1898, 1.
78. *Atlanta Constitution,* July 4, 1898, 1.
79. Charles Lewis Bartholomew, *Cartoons of the Spanish-American War* (Minneapolis, MN: Journal Printing Press, 1899).
80. Douglas Allen, *Frederick Remington and the Spanish-American War* (New York: Crown, 1971).
81. For a sampling of *Harper's, Collier's,* and *Frank Leslie's* illustrations, see Frank Freidel, *The Splendid Little War* (Boston: Little, Brown, 1958).

82. Joseph Pennell, *Joseph Pennell's Pictures of the Panama Canal: Reproductions of a Series of Lithographs made by Him on the Isthmus of Panama, January–March, 1912* (Philadelphia, PA: Lippincott, 1913).
83. Abbot, *Panama and the Canal*, esp. 125 and 130.
84. The surge of Latin American anti-imperialism in the 1890s and early twentieth century and the fact that a part of the U.S. information system eventually began to report on this trend deserves further discussion. For helpful pioneering efforts, see Schoultz, *Beneath the United States*; Pérez, *Cuba and the United States*; Frank Ninkovich, *The United States and Imperialism* (Malden, MA: Blackwell, 2001); E. Berkeley Tompkins, *Anti-Imperialism in the United States: The Great Debate, 1890–1920* (Philadelphia: University of Pennsylvania Press, 1970); Tomás Pérez Vejo, "La guerra hispano-estadosunidense del '98 en la prensa mexicana," *Historia mexicana* 50.2 (2000): 271–308; Rafael Rojas, "Retóricos de la raza: Intelectuales mexicanos ante la guerra del 98," *Historia mexicana* 49.4 (2000): 593–629; and María Elena Rodríguez Ozán, "La guerra hispano-america en la prensa mexicana (1898–1899)," in *1898 ¿Desastre o reconciliación?*, ed. Leopoldo Zea and Mario Magallón (Mexico City: Instituto panamericano de geografía e historia, 1998), 51–75, along with other articles in this volume. For examples of anti-imperialist writing from the time period, see José Enrique Rodó, *Ariel* (1900; repr., Cambridge: Cambridge University Press, 1967); Manuel Ugarte, *El destino de un continente* (Madrid: Editorial mundo latina, 1923); Jose Martí, *Inside the Monster: Writings on the United States and American Imperialism*, ed. Philip Foner (New York: Monthly Review Press, 1975), esp. "The Truth about the United States," 49–54; Samuel Guy Inman, *Intervention in Mexico* (New York: Doran, 1919); Carleton Beals, *Mexico: An Interpretation* (New York: B. W. Huebesch, 1923); and Lamar T. Beman, ed., *Intervention: Selected Articles on Intervention in Latin America* (New York: H. W. Wilson, 1928). For biographical perspectives, see Kirk, *José Martí*; John A. Britton, *Carleton Beals: A Radical Journalist in Latin America* (Albuquerque: University of New Mexico Press, 1986); and Norberto Galasso, *Manuel Ugarte y la lucha para la unidad latinoamericana* (Buenos Aires: Corregidor, 2001).

CONCLUSION

1. Cuba Submarine, directors' reports, 1890–99, Cable and Wireless Archive, Porthcurno Telegraph Museum, Cornwall, UK; Brazilian Submarine, directors' reports, 1890–98, Cable and Wireless Archive.
2. West India and Panama, directors' reports, 1890–98, Cable and Wireless Archive.
3. "Report of the Central and South American Telegraph Company for 1891,"

enclosed in a letter from Charles Flint to James G. Blaine, April 20, 1892, James G. Blaine Papers, Manuscript Division, Library of Congress, Washington, D.C.

4. Winseck and Pike, *Communication and Empire,* 43–91; Headrick, *Invisible Weapon,* 11–92.
5. Western Union, annual report, October 12, 1904, 7, Western Union Telegraph Company Records, Lemelson Center, National Museum of American History, Washington, D.C.
6. Smyth, *Gilded Age Press,* 203–14; Campbell, "Age of New Journalism"; Baldasty, *Commercialization of News,* 81–112.
7. On the Latin American responses to U.S. imperialism, see Ugarte, *Destiny of a Continent*; Galasso, *Manuel Ugarte*; Zea and Magallón, eds., *1898: ?Desastre o reconciliación?*; Pérez Vejo, "La guerra hispano- estadounidense"; and Rojas, "Retóricas de la raza." For analyses of Latin American and U.S. reactions to U.S. imperialism, see Joseph, LeGrand, and Salvatore, eds., *Close Encounters of Empire*; and Topik and Wells, eds., *Second Conquest of Latin America.* For broad interpretive studies, see Langley, *America and the Americas*; Lester Langley, *The Banana Wars: United States Intervention in the Caribbean, 1898–1934* (Chicago: Dorsey, 1985); Schoonover, *Uncle Sam's War*; and Schoultz, *Beneath the United States.* On the Philippine issue, see Beisner, *Twelve Against Empire*; and Richard Welch, *Response to Imperialism: The United States and the Philippine-American War, 1899–1902* (Chapel Hill: University of North Carolina Press, 1979).
8. John G. Walker to Bowman McCalla, April 5, 1885, John G. Walker Papers, Manuscript Division, Library of Congress, Washington, D.C.
9. *New York Tribune,* April 29, 1885, 1.
10. *Congressional Record,* 52nd Congress, 1st sess., vol. 23 (Washington, D.C.: U.S. Government Printing Office 1893), 549.
11. *Nation,* March 24, 1892, 222–23.
12. *Times* (London), November 11, 1891, 5.
13. Stafford Northcote to Lord Salisbury, March 28, 1895, and April 8, 1896, A/139, Salisbury Papers, Hatfield House, Hertfordshire, UK.
14. Stafford Northcote to Lord Salisbury, March 22, 1896, A/139, Salisbury Papers.
15. See Richard Olney to James Storrow, June 12, 1896, box 55, reel 20, Richard Olney Papers, 1830–1928, Manuscript Division, Library of Congress, Washington, D.C.
16. Whitelaw Reid to Donald Nicholson, March 12, 1898, reel 72, Reid Family Papers, Manuscript Division, Library of Congress, Washington, D.C.
17. C. A. Bayly, *The Birth of the Modern World, 1780–1914* (Oxford, UK: Blackwell, 2003), 451–87.

18. Edwards, *New Spirits*, 37–59.
19. Nugent, *Habits of Empire*, 237–304.
20. Tom Lutz, *American Nervousness, 1903: An Anecdotal History* (Ithaca, NY: Cornell University Press, 1991), 7–13.
21. Philip M. Taylor, *Munitions of the Mind: A History of Propaganda from the Ancient World to the Present Day* (Manchester, UK: Manchester University Press, 2003); Garth S. Jowett and Victoria O'Donnell, *Propaganda and Persuasion* (Thousand Oaks, CA: Sage Publications, 1999); Jacques Ellul, *Propaganda: The Formation of Men's Minds* (New York: Vintage, 1965); Stewart Ewen, *PR! A Social History of Spin* (New York: Basic Books, 1996); Benjamin Ginsberg, *The Captive Public: How Mass Opinion Promotes State Power* (New York: Basic Books, 1986); Paddock, ed., *A Call to Arms*. Some perceptive contemporary analyses of the new information system can be found in Gustav Le Bon, *The Crowd: A Study of the Popular Mind* (New York: Penguin, 1960), first published in French in 1895 and in English in 1896; Ferdinand Tonnies, *Community and Society* (New York: Dover, 2002), originally published in German in 1887; Walter Lippmann, *Public Opinion* (New York: Free Press, 1997), originally published in 1922; and Edward Bernays, *Propaganda* (New York: Liveright, 1928).
22. Hilderbrand, *Power and the People*, 30–71.
23. Mitchell, *Danger of Dreams*, 64–107.
24. Brands, *TR*, 469–70.
25. Theodore Roosevelt to John Hay, March 13, 1903, reel 33, Theodore Roosevelt Papers, 1759–1919, Manuscript Division, Library of Congress, Washington, D.C.
26. Mitchell, *Danger of Dreams*, 98–107; James R. Holmes, *Theodore Roosevelt and World Power in International Relations* (Washington, D.C.: Potomac Books, 2006), 168–74.
27. For example, see Susan A. Brewer, *Why America Fights: Patriotism and War Propaganda from the Philippines to Iraq* (New York: Oxford University Press, 2009); Knightley, *The First Casualty*' Jowett and O'Donnell, *Propaganda and Persuasion*; Taylor, *Munitions of the Mind*; Ewen, *PR!*; Ginsberg, *Captive Public*; Ellul, *Propaganda*; and Anthony Pratkanis and Elliot Aronson, *Age of Propaganda: The Everyday Use and Abuse of Persuasion* (New York: Henry Holt, 2001). On World War I, see Stephen Vaughan, *Holding Fast the Inner Lines: Democracy, Nationalism, and the Committee on Public Information* (Chapel Hill: University of North Carolina Press, 1980); Ward, *Mass Communications*, 21–109; and Gary Messinger, *British Propaganda and the State in the First World War* (Manchester, UK: University of Manchester Press, 1992). Of special help in this study was Paddock, ed., *A Call to Arms*,

especially Paddock's "Introduction: Newspapers, Public Opinion, and Propaganda," 1–14. There is a large literature on particular cases, including Judith Brown, *Gandhi: Prisoner of Hope* (New Haven, CT: Yale University Press, 1989); John A. Britton, *Revolution and Ideology: Images of the Mexican Revolution in the United States* (Lexington: University of Kentucky Press, 1995); and David Welch, *The Third Reich: Politics and Propaganda* (London: Routledge, 2002).

28. Chapman, *Comparative Media History*, 238–65; Briggs and Burke, *Social History of the Media*, 267–333; Tim Wu, *The Master Switch: The Rise and Fall of Information Empires* (New York: Knopf, 2011); Neil Harvey, *The Chiapas Rebellion* (Durham, NC: Duke University Press, 1998); Subcomandante Marcos, *Our Word Is Our Weapon*, ed. Juana Ponce de León (New York: Seven Stories Press, 2002); Osgood and Frank, eds., *Selling War in a Media Age*; Eva Bellin, "Reconsidering the Robustness of Authoritarianism in the Middle East: Lessons from the Arab Spring," *Comparative Politics* 44.2 (2012): 127–49; Naila Hamdy and Ehab H. Gomaa, "Framing the Egyptian Uprising in Arabic Language Newspapers and Social Media," *Journal of Communication* 62.2 (2012): 195–211; Merlyna Lim, "Clicks, Cabs, and Coffee Houses: Social Media and Oppositional Movements in Egypt, 2004–2011," *Journal of Communication* 62.2 (2012): 231–48; Zizi Papacharissi and Maria de Fatima Oliveira, "Affective News and Networked Publics: The Rhythms of News Storytelling on Egypt," *Journal of Communication* 62.2 (2012): 266–82; Zeynep Tufekci and Christopher Wilson, "Social Media and the Decision to Participate in Political Protest: Observations from Tahrir Square," *Journal of Communication* 62.2 (2012): 363–79.

Bibliography

PRIMARY SOURCES

Cable and Wireless Archive, Porthcurno, Cornwall, UK

COMPANY RECORDS

Brazilian Submarine Telegraph Company

Cuba Submarine Telegraph Company

River Plate Telegraph Company

West Coast of America Telegraph Company

Western and Brazilian Telegraph Company

West India and Panama Telegraph Company

UNPUBLISHED MANUSCRIPTS

Anonymous. "West India and Panama Telegraph Company, Ltd."

Tarver, F. H. C. "The Associated Companies in South America." April 20, 1934.

Lemelson Center, Museum of American History, Smithsonian Institution, Washington, D.C.

Western Union Archive

Manuscripts Division, Library of Congress, Washington, D.C.

Adolphus Greely Papers

Benjamin Harrison Papers

Benjamin Tracy Papers

Bowman McCalla Collection

George Cortelyou Papers
Grover Cleveland Papers
James G. Blaine Papers
John G. Walker Papers
Reid Family Papers (Whitelaw Reid)
Richard Olney Papers
Theodore Roosevelt Papers
William McKinley Papers
William Scruggs Papers

Archives and Records of the London Times, *London, UK*
Archives and Records of the London *Times*

Public Records Office, London, UK
British Foreign Office

ING Barings Bank, London, UK
Records of Barings Bank

National Archives of the United States, Washington, D.C.
Department of State. Despatches from U.S. Ministers to Spain, 1792–1906.
Navy Department, General Orders, 1884–1896, Microcopy 984.

Hatfield House, Hatfield, Hertfordshire, UK
Salisbury Papers

Southern Historical Collection, Wilson Library, University of North Carolina, Chapel Hill, NC
Victor Blue Papers

Newspapers
THE TEN METROPOLITAN DAILIES
Atlanta Constitution
Boston Globe
Chicago Tribune
New Orleans Picayune
New York Times
New York Tribune

Philadelphia Inquirer
San Francisco Chronicle
San Francisco Examiner
Washington Post

OTHER NEWSPAPERS

Journal of Commerce
La Nación (Buenos Aires)
La Prensa (Buenos Aires)
Times (London)
South American Journal

Contemporary Commentary on the Press and Information Flow

Atkins, John Black. *The War in Cuba: The Experience of an Englishman with the United States Army*. London: Smith, Elder, and Co., 1899.

Baker, Ray Stannard. "How the News of the War is Reported." *McClure's Magazine* 11 (September 1898): 491–95.

Bartholomew, Charles Lewis. *Cartoons of the Spanish-American War*. Minneapolis, MN: Journal Printing Press, 1899.

Bigelow, Poultney. "In Camp at Tampa with the Troops under Canvas." *Harper's Weekly*, June 4, 1898, 550.

Brisbane, Arthur. "The Modern Newspaper in War Time." *Cosmopolitan* 20 (September 1898): 541–57.

Bryce, James. "British Feeling on the Venezuelan Question." *North American Review* 161 (February 1896): 145–53.

Creelman, James. *On the Great Highway*. Boston: Lothrop, 1901.

Davis, Richard Harding. "$300,000 Reward." *Harper's Weekly*, January 23, 1892, 91–93.

———. *The Cuban and Porto Rican Campaigns*. New York: Charles Scribner's Sons, 1898.

———. "The Garza Revolution." *Harper's Weekly*, January 16, 1892, 51–53.

———. "Our War Correspondents in Cuba and Puerto Rico." *Harper's Magazine* 98 (May 1899): 938–48.

———. "The Rough Riders' Fight at Guasimas." *Scribners' Magazine* 24 (September 1898): 261–73.

———. *The West from a Car-Window*. New York: Harper and Brothers, 1892.

Dicey, Edward. "Common Sense and Venezuela." *Nineteenth Century* 39 (January 1896): 7–15.

Editorial (no title). *Times* (London), November 30, 1872, 9.
Edwards, E. J. "Charles H. Taylor of the Boston 'Globe.'" *McClure's Magazine* (July 1896): 144–47.
"England and America—The Use and Abuse of Ocean Telegraphy." *New York Times*, December 14, 1872, 2.
Godkin, E. L. "Diplomacy and the Newspaper." *North American Review* 160 (May 1895): 570–79.
———. "The Growth and Expansion of Public Opinion." *Atlantic Monthly* 81 (January 1898): 1–15.
———. "The Secret of Warriorism." *Nation* (March 24, 1892): 222–23.
———. "The Ultimatum." *Nation* (January 28, 1892): 64–66.
Gossip, G. D. H. "England in Nicaragua and Venezuela from the American Point of View." *Fortnightly Review* 64 (December 1895): 829–42.
———. "Venezuela Before Europe and America from an American Point of View." *Fortnightly Review* 65 (January 1896): 397–411.
Hadfield, William .*Brazil and the River Plate*. London: Edward Stanford, 1877.
Kennan, George. *Campaigning in Cuba*. New York: Century, 1899.
"McClure's and the War." *McClure's Magazine* 11 (July 1898): 496.
"McClure's Magazine in War Times." *McClure's Magazine* 11 (June 1898): 206.
"Newspaper Men in the War." *Literary Digest* 17 (October 15, 1898): 455–57.
Norton, Charles Eliot. "Some Aspects of Civilization in America." *Forum* 20 (February 1896): 641–51.
Olney, Richard. "International Isolation of the United States." *Atlantic Monthly* 81 (May 1898): 577–88.
Post, Charles Johnson. *The Little War of Private Post*. Boston: Little Brown, 1960.
Rea, George Bronson. *Facts and Fakes about Cuba*. New York: George Munro's Sons, 1897.
Spears, John R. "Afloat for the News in War Times." *Scribners' Magazine* 24 (October 1898): 501–4.
———. "The Chase for Cervera." *Scribners' Magazine* 24 (August 1898): 144–52."
Stanley, Henry M. "The Issue Between Great Britain and America." *Nineteenth Century* 39 (January 1896): 1–6.
Stead, W. T. "Jingoism in America." *Comtemporary Review* 67 (September 1895): 334–47.
Tarbell, Ida. "President McKinley in War Times." *McClure's* 11 (July 1898): 208–24.
"United States and Chili." *Times* (London), November 11, 1891, 8.

Selected Contemporary Publications in Propaganda, Public Relations, and Persuasion

Abbot, Willis J. *Panama and the Canal in Pictures and Prose: A Complete Story of Panama as well as the History, Purpose, and Promise of Its World-Famous Canal—the Most Gigantic Engineering Undertaking since the Dawn of Time*. New York: Syndicate Publishing, 1913.

Allen, H. W. "President Díaz and the Mexico of Today." *Review of Reviews* 61 (January 1893): 676–96.

Beals, Carleton. *Mexico: An Interpretation*. New York: B.W. Huebesch, 1923.

Beman, Lamar T., ed. *Intervention: Selected Articles on Intervention in Latin America*. New York: H. W. Wilson, 1928.

Bemis, Katherine Isabel, ed. *The Patriotic Reader for Seventh and Eighth Grades and Junior High Schools*. Boston: Houghton Mifflin, 1917.

Bernays, Edward. *Propaganda*. New York: Liveright, 1928.

Bourke, John G. "An American Congo." *Scribners* (May 1894): 590–610.

Bunau-Varilla, Philippe. *Panama: Its Creation, Destruction, and Resurrection*. London: Constable, 1913.

Central and South American Cable Company, *Via Galveston*. New York: n.p., 1895.

Cisneros, Evangelina. *The Story of Evangelina Cisneros Told by Herself*. New York: Continental Publishing Company, 1898.

Curtis, William Eleroy. *Between the Andes and the Ocean*. Chicago: H. S. Stone and Company, 1900.

———. *The Capitals of Spanish America*. New York: Harper and Brothers, 1888.

———. *Handbook of the American Republics*. Vol. 3. Washington, D.C.: Bureau of the American Republics, 1893.

———. *The Relics of Columbus: A Illustrated Description of the Historical Collection in the Monastery of La Rabida*. Washington, D.C.: Lowdermilk, 1893.

———. *Trade and Transportation Between the United States and Spanish America*. Washington, D.C.: U.S. Government Printing Office, 1889.

———. *The United States and the Foreign Powers*. Meadville, PA: Flood and Vincent, 1892.

———. *Venezuela: A Land Where It's Always Summer*. New York: Harper & Brothers, 1896.

Davis, Richard Harding. *Soldiers of Fortune*. New York: Scribner's, 1897.

———. *Three Gringos in Venezuela and Central America*. New York: Harper and Brothers, 1896.

———. "United States to Dominate Hemisphere." www.learner.org/channel/workshops/primarysources.

de Quesada, Gonzalo, and Henry Davenport Northrop. *The War in Cuba, Being a Full Account of Her Struggle for Freedom.* New York: Liberty Publishing Company, 1896.

Duclós Salinas, Adolfo. *The Riches of Mexico and Its Institutions.* St. Louis, MO: Nixon-Jones Printing Company, 1893.

Flint, Grover. *Marching with Gomez.* New York: Lamson and Wolfe, 1898.

Gossip, G. D. H. "England in Nicaragua and Venezuela from the American Point of View." *Fortnightly Review* 64 (December 1895): 838.

———. "Venezuela before Europe and America from an American Point of View." *Fortnightly Review* 65 (January 1896): 403–4.

Halstead, Murat. *Our New Possessions: Natural Riches, Industrial Resources of Cuba, Porto Rico, Hawaii, the Ladrones and Philippine Islands with Episodes of their Early History.* Chicago: Dominion Company, 1898.

Hubbard, Elbert. "A Message to Garcia." East Aurora, NY: Roycrofters, 1899.

Inman, Samuel Guy. *Intervention in Mexico.* New York: Doran, 1919.

Kasson, John. "The Monroe Doctrine in 1881." *North American Review* 133 (December, 1881): 525–33.

Le Bon, Gustav. *The Crowd: A Study of the Popular Mind.* New York: Penguin, 1960.

Lee, Fitzhugh, and Joseph Wheeler. *Cuba's Struggle against Spain.* New York: American Historical Press, 1899.

Lippmann, Walter. *Public Opinion.* New York: Free Press, 1997 (originally published 1922).

Lodge, Henry Cabot. "England, Venezuela, and the Monroe Doctrine." *North American Review* 160 (June, 1895): 651–58.

———. *The War with Spain.* New York: Harper and Brothers, 1899.

Mahan, Alfred Thayer. *The Interest of America in Sea Power Present and Future.* Boston: Little, Brown, 1897.

———. *Lessons of the War with Spain.* Boston: Little, Brown, 1899.

———. *The Problem of Asia and Its Effect upon International Policies.* Boston: Little, Brown, 1900.

Markham, Clements R. *The War Between Peru and Chile, 1879–1882.* London: Sampson Low, Marston, Searle and Rivington, 1882.

Martí, José. "The Truth about the United States." In *Inside the Monster by José Martí: Writings on the United States and American Imperialism*, edited by Philip Foner, 49–54. New York: Monthly Review Press, 1975.

Martínez, Marcial. *Cuestión Chileno-Peruana.* Santiago: La ilustración, 1914.

———. *General Considerations in Relation to the Difficulty Between Chile, on the One Hand, and Peru and Bolivia on the Other.* Philadelphia: Times Printing House, 1881.

———. *Memorandum delivered by the Minister of Chile to the Secretary of State, stating the actual condition of affairs on the Pacific Coast.* Washington, D.C.: n.p., 1881.

———. *Opusculos sobre asuntos internacionales.* Vol. 3 of *Obras completas.* Santiago: La ilustracíon, 1919.

Matte, Augusto, and Agustín Ross. *Memoria presentada a la excma junta de gobierno.* Paris: Imprimerie et Librarie Administratives et Critiques Paul Dupont, 1892.

Maurtua, Victor. *The Question of the Pacific.* Translated by F.A. Pezet. Philadelphia: Lasher, 1901.

Moore, J. Hampton, and James Rankin Young. *Thrilling Stories of the War by Returning Heroes.* Chicago: Providence Publishing Company, 1899.

Parkerson, A. C. *How Uncle Sam Fights; or, Modern Warfare—How Conducted.* Baltimore, MD: R. H. Woodward Company, 1898.

Pennell, Joseph. *Joseph Pennell's Pictures of the Panama Canal: Reproductions of a Series of Lithographs made by Him on the Isthmus of Panama, January–March, 1912.* Philadelphia, PA: Lippincott, 1913.

Rodó, José Enrique. *Ariel.* 1900. Reprinted Cambridge: Cambridge University Press, 1967.

Romero, Matías. *Articulos sobre México publicado en los Estados Unidos de América por Matías Romero en 1891–1892.* México: Oficina impresor de estampillas, 1892.

———. "The Garza Raid and Its Lessons." *North American Review* 155 (September 1892): 324–37.

———. "Mexican Trade." *North American Review* 154 (June 1892): 626–27,

———. *Mexico and the United States: A Study of Subjects Affecting their Political, Commercial, and Social Relations Made with a View toward their Promotion.* New York: G. P. Putnam's Sons, 1898.

———. "The Pan-American Conference." Part 1. *North American Review* 151 (September 1890): 354–69.

———. "The Pan-American Conference." Part 2. *North American Review* 151 (October 1890): 407–21.

———. "Wages in Mexico." *North American Review* 154 (January 1892): 33–49.

Roosevelt, Theodore. *The Rough Riders.* New York: Scribner's, 1899.

Scruggs, William L. *The Venezuelan Question: British Aggressions in Venezuela; or, the Monroe Doctrine on Trial.* Atlanta, GA: Franklin Printing Company, 1896.

Smalley, George W. *Anglo-American Memories.* New York: Putnam's, 1911.

Squires, Grant. "Experiences of a War Censor." *Atlantic Monthly* 83 (March 1899): 426.

Taylor, Hannis. "A Review of the Cuban Question in Its Economic, Political, and Diplomatic Aspects." *North American Review* 165 (November 1897): 610–35.

Tonnies, Ferdinand. *Community and Society*. New York: Dover, 2002.

Ugarte, Manucl. *El destino de un continente*. Madrid: Editorial Mundo Latina, 1923. Translated and reprinted as *The Destiny of a Continent*, New York: Knopf, 1925.

Walker, Jeanie Mort. *Life of Capt. Joseph Frye: The Cuban Martyr*. Hartford, CT: J.B. Burr, 1875.

White, Trumbull. *Our New Possessions*. Chicago: Henry Publishing Company, 1901.

Young, J. Rankin, and J. Hampton Moore. *Reminiscences and Thrilling Stories of the War by Returned Heroes*. Chicago: Providence Publishing, 1899.

Contemporary Publications on Technology and Business

Amores, Emilio M. "Trade and Industry of South America." *Engineering Magazine* 10 (February 1896): 811–20.

Bright, Charles. "The Extension of Submarine Telegraphy in a Quarter-Century." *Engineering Magazine* 16 (1899): 417–28.

———. *Submarine Telegraphs: Their History, Construction, and Working*. London: Crosby Lockwood and Son, 1898.

"Cable Cutting in Warfare." *Literary Digest* 17 (July 16, 1898): 75

Curtis, William Eleroy. *Trade and Transportation Between the United States and Spanish America*. Washington, D.C.: Government Printing Office, 1889.

Delany, Patrick. "The Future of the Electric Telegraph." *Engineering Magazine* 10 (October 1895): 39–48.

Giddings, Howard. *Exploits of the Signal Corps in the War with Spain*. Kansas City, MO: Hudson-Kimberly Publishing, 1900.

Greely, Adolphus W. "Report of the Chief Signal Officer, U.S. Army, 1898." In *Report of the Secretary of War, House of Representatives, 56th Congress, 1st Session*. Reprinted in *Military Signal Communications*, edited by Paul J. Scheips, 891–95. New York: Arno Press, 1950.

———. "The Signal Corps in War-Time." *Century Magazine* 66 (September 1903): 811–26.Grier, Thomas G. "The Use of Electricity in the Spanish-American War." *Electrical World*, October 15, 1898, 390–92.

Hurd, Percy A. "Our Telegraphic Isolation." *Contemporary Review* 69 (June 1896): 899–908.

Lambert, Francis. "On Grapnels for Raising Submarine Cables in Deep Water." *Society of Telegraph Engineers and Electricians*, November 13, 1878, 417–33.

MacFarland, Henry. "The Signal Corps of the Army in the War." *Review of Reviews* 29 (February 1899): 183–188.

Maver, William J., Jr. "Ocean Telegraphy." *Electrical World*, May 10, May 17, and June 14, 1884.

Muir, Henry. "The Telegraph Systems of the World." *McClure's Magazine* 2 (July 1895): 98–112.

"Ocean Cables in War Times." *Boston Globe*, July 3, 1898, 16.

"Our News Service." *Wall Street Journal*, July 19, 1889, 3.

Pearce, Arthur L. "Gold and Silver in South America." *Engineering Magazine* 7 (May 1894): 182–92.

"Prince Bismarck's Lament." *Electrical World* 5 (January 1885): 1.

"Progress of Telegraphy." *The Electrician and Electrical Engineer* 2 (April 1883): 98–99.

"Rapid Cable-Laying in War." *Literary Digest* 18 (July 1898): 135.

Scrymser, James. *Personal Reminiscences of James Scrymser in Times of Peace and War*. Easton, PA: Eschenbach Printing Company, 1915.

Squire, George O. "The Influence of Submarine Cables upon Military and Naval Supremacy." *National Geographic* 12 (January 1901): 1–12.

Squires, Grant. "Experiences of a War Censor." *Atlantic Monthly* 83 (March 1899): 425–32.

"Submarine Cables." *New York Times*, August 25, 1870, 2.

"Submarine Telegraphs." *Banker's Magazine* 5 (July 1870): 9–12.

"Telegraphic Errors." *The Electrician and Electrical Engineer* 4 (August 1885): 283.

"Telegraphing Around the Globe." *Scientific American* 74 (May 1895): 347.

Trott, Samuel, and Frederic Adams Hamilton. "Submarine Telegraph Cables: Their Decay and Renewal." *Society of Telegraph Engineers and Electricians* 12 (November 1883): 495–504.

Webb, Herbert Laws. "A Boom in Submarine Cables." *Electrical World*, June 29, 1889.

———. "A New Chilean Cable." *Electrical World*, June 22, 1889.

———. "Puck's Girdle Round the Earth." *Electrical World*, March 1, 1890.

———. "With a Cable Expedition." *Scribner's Magazine* 7 (October 1890): 399–415.

Weiller, Lazare. "The Annihilation of Distance." *The Living Age* (October 15, 1898): 163–78.

Wilson, George Grafton. *Submarine Telegraph Cables in their International Relations*. Washington, D.C.: Government Printing Office, 1901.

Winslow, Cameron McR. "Cable-Cutting at Cienfuegos." *Century Magazine* 57 (March 1899): 708–17.

Secondary Works

Abramson, Howard S. *National Geographic: Behind America's Lens on the World*. New York: Crown Publishers, 1987.

Adas, Michael. *Dominance by Design: Technological Imperatives and America's Civilizing Mission*. Cambridge: Harvard University Press, 2006.

Aguilar, Alonso. *Pan-Americanism from Monroe to the Present: A View from the Other Side*. New York: Monthly Review Press, 1968.

Ahvenainen, Jorma. *The European Cable Companies in South America before the First World War*. Helsinki: Finnish Academy of Science and Letters, 2004.

———. *The Far Eastern Telegraphs: The History of Telegraphic Communications between the Far East, Europe, and America before the First World War*. Helsinki: Suomalainen Tiedeakatemia, 1981.

———. *The History of the Caribbean Telegraphs before the First World War*. Helsinki: Finnish Academy of Science and Letters, 1996.

Alcorn, Paul A. *Social Issues in Technology: A Format for Investigation*. Upper Saddle River, NJ: Prentice Hall, 2003.

Allen, Douglas. *Frederick Remington and the Spanish-American War*. New York: Crown Publishers, 1971.

Alonso, Paula, ed. *Construcciones impresas: Panfletos, diarios y revistas en la formación de los estados nacionales en América Latina, 1820–1920*. Buenos Aires: Fondo de Cultura Económica, 2003.

Altschull, J. Herbert. *From Milton to McLuhan: The Ideas Behind American Journalism*. White Plains, NY: Longman, 1990.

Anderson, Benedict. *Imagined Communities: Reflections on the Origin and Spread of Nationalism*. Second edition. London: Verso, 1991.

Andrews, J. Cutler. *The North Reports the Civil War*. Pittsburgh: University of Pittsburgh Press, 1955.

Anguizola, Gustave. *Philippe Bunau-Varilla: The Man behind the Panama Canal*. Chicago: Nelson Hall, 1980.

Arceniegas, German. *Latin America: A Cultural History*. London: Barrie and Rockliff, 1969.

Armstrong, William M. *E. L. Godkin: A Biography*. Albany: State University of New York Press, 1978.

———. *E. L. Godkin and American Foreign Policy, 1865–1900*. New York: Bookman, 1957.

———, ed. *The Gilded Age Letters of E. L. Godkin*. Albany: State University of New York Press, 1974.

Arnold, A. J., and Robert Greenhill. "Business, Government, and Profit: The Management and Regulation of Overseas Mail Contracts, 1836–90." *Accounting, Business and Financial History* 5.3 (1995): 334–59.

———. "Mail Contracts, Subsidies, and Profits: The Royal Mail Steam Packet Company, 1842–1905." In *Management, Finance and Industrial Relations in the Maritime Industry*, edited by Simon P. Ville and David M. Williams, 47–60.

St. John's, Newfoundland: International Maritime Economic History Association, 1994.

Auxier, George. "Middle Western Newspapers and the Spanish-American War, 1895–1898." *Mississippi Valley Historical Review* 26.4 (1939): 523–34.

———. "The Propaganda Activities of the Cuban Junta in Precipitating the Spanish-American War, 1895–1898." *Hispanic American Historical Review* 19.3 (1939): 288–305.

Bailyn, Bernard. *Atlantic History: Concept and Contours.* Cambridge, MA: Harvard University Press, 2005.

Baldasty, Gerald. *The Commercialization of News in the Nineteenth Century.* Madison: University of Wisconsin Press, 1992.

Balfour, Sebastian. *The End of the Spanish Empire.* Oxford, UK: Clarendon Press, 1997.

Barbance, Martha. *Vie commercial de la route de Cap Horn au xixe siecle: L'armament A. D. Bordes et fils.* Paris: SEVPEN, 1969.

Barber, Nancy Nichols. *The French Experience in Mexico, 1821–1861: A History of Constant Misunderstanding.* Chapel Hill: University of North Carolina Press, 1979.

Barnhurst, Kevin, and John Nerone. *The Form of the News: A History.* New York: Guilford Press, 2001.

Barros, Mario. *Historia diplomatica de Chile, 1541–1930.* Barcelona: Ediciones Ariel, 1970.

Barros Franco, José Miguel. *El Caso del* Baltimore*: Apuntes para la historia diplomatica de Chile.* Santiago: Universidad de Chile, 1950.

Barty-King, Hugh. *Girdle Round the Earth*: The Story of Cable and Wireless. London: Heineman, 1979.

Basadre, Jorge. *Reflexiones en Torno a la Guerra de 1879.* Lima: F. Campodonico, 1979.

Bastert, Russell. "Diplomatic Reversal: Frelinghuysen's Opposition to Blaine's Pan-American Policy." *Mississippi Valley Historical Review* 42.4 (1956): 653–70.

Bayly, C. A. *The Birth of the Modern World, 1780–1914.* Oxford, UK: Blackwell, 2003.

Beatty, Jack. *The Age of Betrayal: the Triumph of Money in America, 1865–1900.* New York: Vintage, 2007.

Beauchamp, Ken. *A History of Telegraphy: Its History and Technology.* New Brunswick, NJ: Institute of Engineering and Technology, 2001.

Beisner, Robert. *Twelve Against Empire: The Anti-Imperialists.* Chicago: University of Chicago Press, 1985.

Bellin, Eva. "Reconsidering the Robustness of Authoritarianism in the Middle East: Lessons from the Arab Spring." *Comparative Politics* 44.2 (2012): 127–49.

Beniger, James R. *The Control Revolution: Technological Change and Economic Origins of the Information Society*. Cambridge, MA: Harvard University Press, 1986.

Bernecker, Walther L. "Las relaciones entre Europa y Latinoamerica durante el siglo XIX: Ofensivas comerciales e intereses economicos." *Hispania* 53.183 (1993): 177–212.

Bernstein, Harry. *Matías Romero, 1837–1898*. Mexico: Fondo de Cultura Económica, 1973.

Bertram, Marshall. *The Birth of Anglo-American Friendship: The Prime Facet of the Venezuelan Boundary Dispute. A Study of the Interaction of Diplomacy and Public Opinion*. Lanham, MD: University Press of America, 1992.

Bethel, Leslie. *The Cambridge History of Latin America*. 11 vols. Cambridge: Cambridge University Press, 1986.

Blakemore, Harold. *British Nitrates and Chilean Politics, 1886–1896: Balmaceda and North*. London: Athlone Press, 1974.

Blaufarb, Rafe. "The Western Question: The Geopolitics of Latin American Independence." *American Historical Review* 112 (June 2007): 742–62.

Bledstein, Burton. *The Culture of Professionalism*. New York: Norton, 1976.

Blondheim, Menahem. *News over the Wires: The Telegraph and the Flow of Public Information in America, 1844–1897*. Cambridge: Harvard University Press, 1994.

Bonilla, Heraclio. *Gran Bretana y el Peru*. Lima: Instituto de Estudios Peruanos, 1997.

Bose, Walter B. L. "Historia de las comunicaciones." In *Historia Argentina contemporanea, 1862–1930*, vol. 3 of *Historia Economica*, by Horacio Juan Cuccorese et al., 582–603. Buenos Aires: Libreria "El Ateno" Editorial, 1966.

Botana, Natalio R. "1880: La federalización de Buenos Aires." In *La Argentina del ochenta al centenario*, edited by Gustavo Ferrari and Ezequiel Gallo, 107–122. Buenos Aires: Editorial Sudamerica, 1989.

Botero, Rodrigo. *Ambivalent Embrace: America's Troubled Relationship with Spain from the Revolutionary War to the Cold War*. Westport, CT: Greenwood, 2001.

Boyd-Barrett, Oliver. *The International News Agencies*. London: Constable, 1980.

———, and Michael Palmer. *Le trafique de nouveles: les agencies mondiales d'information*. Paris: Alain Moreau, 1981.

Bradford, James C., ed. *Makers of the American Naval Tradition*. Annapolis: Naval Institute Press, 1990.

Bradford, Richard. *The* Virginius *Affair*. Boulder: Colorado Associated University Press, 1980.

Brands, H. W. *The Reckless Decade: America in the 1890s*. New York: St. Martin's, 1995.

———. *TR: The Last Romantic*. New York: Basic Books, 1997.

Brewer, Susan A. *Why America Fights: Patriotism and War Propaganda from the Philippines to Iraq*. New York: Oxford University Press, 2009.

Brian, Denis. *Pulitzer: A Life*. New York: Wiley, 2001.

Briggs, Asa, and Peter Burke. *A Social History of the Media: From Gutenberg to the Internet*. Cambridge, England: Polity Press, 2002.

Britton, John A. *Carleton Beals: A Radical Journalist in Latin America*. Albuquerque: University of New Mexico Press, 1986.

———. "'The Confusion Provoked by Instantaneous Discussion': The New International Communications Network and the Chilean Crisis of 1891–92 in the United States." *Technology and Culture: The International Quarterly of the Society for the History of Technology* 48.2 (2007): 729–57.

———. "International Communications and the Anglo-Argentine Crises of the 1890s." Paper presented at the annual meeting of SECOLAS, March 3, 2001, Veracruz, Mexico.

———. *Revolution and Ideology: Images of the Mexican Revolution in the United States*. Lexington: University of Kentucky Press, 1995.

———, and Jorma Ahvenainen. "Showdown in South America: James Scrymser, John Pender, and the United States–British Cable Competition." *Business History Review* 78.1 (2004): 1–27.

Brown, Clarence. *The Correspondents' War*. New York: Scribner's, 1967.

Brown, Judith. *Gandhi: Prisoner of Hope*. New Haven, CT: Yale University Press, 1989.

Brown, Matthew, ed. *Informal Empire in Latin America: Culture, Commerce, and Capital*. Malden, MA: Blackwell, 2008.

Brown, Richard D. *Knowledge is Power: The Diffusion of Information in Early America, 1700–1865*. New York: Oxford University Press, 1989.

Buchenau, Jürgen. *In the Shadow of the Giant: The Making of Mexico's Central American Policy, 1876–1930*. Tuscaloosa: University of Alabama Press, 1996.

Bulmer-Thomas, Victor. *The Economic History of Latin America since Independence*. Cambridge: Cambridge University Press, 1994.

Bulnes, Gonzalo. *La Guerra del Pacífico*. 3 vols. Valparaíso: Sociedad Imprenta Litografía Universo, 1911–1919. Reprinted by Santiago: Editorial del Pacifico, 1955.

Burgess, Paul. *Justo Rufino Barrios: A Biography*. Quetzaltenango, Guatemala: El noticiero evangelico, 1946.

Burr, Robert N. *By Reason or Force: Chile and the Balancing of Power in South America, 1830–1905*. Berkeley: University of California Press, 1967.

Bushnell, David, and Neill Macauley. *The Emergence of Latin America in the Nineteenth Century*. New York: Oxford University Press, 1994.

Cain, Robert J. "Telegraphs in the British Empire, 1850–1900." PhD diss., Duke University, 1971.

Calhoun, Charles W. *Benjamin Harrison*. New York: Times Books, 2005.

Campbell, W. Joseph. "The Age of New Journalism, 1883–1890." In *The Media in America: A History*, edited by William David Sloan, 223–48. Newport, AL: Vision Press, 2005.

———. *Yellow Journalism: Puncturing the Myths, Defining the Legacies*. Westport, CT: Praeger, 2003.

Cardona, Gabriel, and Juan Carlos Losada. *Weyler, nuestro hombre en la Habana*. Barcelona: Editorial Planeta, 1997.

Carmangani, Marcello. *El otro occidente: América Latina desde la invasión europa hasta la globalización*. Mexico City: El Colegio de México, 2004.

Castro-Klarén, Sara, and John Charles Chasteen, eds. *Beyond Imagined Communities: Reading and Writing the Nation in Nineteenth Century Latin America*. Baltimore: Johns Hopkins University Press, 2003.

Caudill, Edward. "E. L. Godkin and His (Special and Influential) View of 19th Century Journalism." *Journalism Quarterly* 69.4 (1992): 1039–49.

———, ed. *Constitucionalismo y orden liberal América Latina, 1850–1920*. Torino, Italy: Otto Editore, 2000.

———, ed. *Federalismos latinamericanos: México/Brasil/Argentina*. Mexico City: El Colegio de México, 1993.

Chace, James, and Caleb Carr. *American Invulnerable: The Quest for Absolute Security from 1812 to Star Wars*. New York: Summit Books, 1988.

Challener, Richard. *Admirals, Generals, and American Foreign Policy, 1898–1914*. Princeton: Princeton University Press, 1973.

Chandler, Alfred D., Jr., and James Cortada, eds. *A Nation Transformed by Information: How Information Has Shaped the United States from Colonial Times to the Present*. Cambridge: Harvard University Press, 2000.

Chapman, Jane. *Comparative Media History*. Malden, MA: Polity, 2005.

Chen Daley, Mercedes. "The Watermelon Riot: Cultural Encounters in Panama City, April 15, 1856." *Hispanic American Historical Review* 70.1 (1990): 85–108.

Cisneros, Andrés, Carlos Escudé, et al. *Historia general de las relaciones exteriores de la República Argentina*. (2000): www.argentina-rree.com.

Clayton, Larry. *Grace: W. R. Grace and Company, 1850–1930*. Ottowa, IL: Jameson Books, 1985.

Coats, Vary, and Bernard Finn. *Peru and the United States*. Athens: University of Georgia Press, 1999.

———. *A Retrospective Technology Assessment: Submarine Telegraphy, the Transatlantic Cable of 1866*. San Francisco: San Francisco Press, 1979.

Coatsworth, John, and Alan M. Taylor, eds. *Latin America and the World Economy since 1800*. Cambridge, MA: Rockefeller Center for Latin American Studies, Harvard University, 1998.

Coclanis, Peter. "*Drang Nach Osten*. Bernard Bailyn, the World-Island, and the Idea of Atlantic history." *Journal of World History* 13.1 (2002): 169–82.

Coerver, Don M. "Mexico: Conflicting Self-Interests." In *United States-Latin American Relations, 1850–1903*, edited by Thomas M. Leonard, 22–27. Tuscaloosa: University of Alabama Press, 1999.

Coletta, Paolo E. *Bowman Hendry McCalla: A Fighting Sailor*. Washington, D.C.: University Press of America, 1979.

Collier, Simon, and William F. Sater. *A History of Chile, 1808–1994*. Cambridge: Cambridge University Press, 1996.

Collin, Richard. *Theodore Roosevelt's Caribbean: The Panama Canal, the Monroe Doctrine, and the Latin American Context*. Baton Rouge: Louisiana State University Press, 1990.

Cooling, B. Franklin. *Benjamin Franklin Tracy: Father of the Modern American Fighting Navy*. Hamden, CT: Archon, 1973.

Coontz, Robert. *From the Mississippi to the Sea*. Philadelphia: Dorrance, 1930.

Cosio Villegas, Daniel, ed. *Historia Moderna de Mexico*. Mexico City: Editorial Hermes, 1974.

Couch, Carl J. *Information Technologies and Social Orders*. New York: Aldine de Gruyter, 1996.

Crapol, Edward. *James G. Blaine: Architect of Empire*. Wilmington, DE: Scholarly Resources, 2000.

Czitrom, Daniel. *Media and the American Mind*. Chapel Hill: University of North Carolina Press, 1982.

Davis, Harold E. "Henry W. Grady, Master of the Atlanta Ring—1880–1886." *Georgia Historical Quarterly* 69.1 (1988): 1–16.

Davis, William C. *The Last Conquistadores*. Athens: University of Georgia Press, 1950.

De Fleur, Melvin L., and Sandra Ball-Rokeach. *Theories of Mass Communications*. Boston: Allyn & Bacon, 1998.

De La Pedraja, Rene. *Oil and Coffee: Latin American Merchant Shipping from the Imperial Era to the 1950s*. Westport: Greenwood Press, 1998.

Della Paolera, Gerardo, and Alan Taylor. *Straining at the Anchor: The Argentine*

Currency Board and the Search for Macroeconomic Stability, 1880–1935. Chicago: University of Chicago Press, 2001.

Delpar, Helen. "Colombia: Troubled Friendship." In *United States-Latin American Relations, 1850–1903*, edited by Thomas M. Leonard, 58–80. Tuscaloosa: University of Alabama Press, 1999.

Dent, David W. *The Legacy of the Monroe Doctrine*. Westport, Conn.: Greenwood Press, 1999.

de Quesada, Gonzalo. *Paginas escogidas*. Havana: Ediciones Politicas, 1968.

Desbordes, Rhoda."Transnationales de l'information au Nouveau Monde: Quelques aspects de l'implementation de agences internationales de information en Amerique du Sud, 1874–1919." PhD diss., Sorbonne Nouvelle, 2005.

Díaz Alejandro, Carlos F. "La economía argentina durante el período 1880–1913." In *La Argentina del ochenta al centenario*, edited by Gustavo Ferrari and Ezequiel Gallo, 369–76. Buenos Aires: Editorial Sudamerica, 1989.

Díaz Espino, Ovidio. *How Wall Street Created a Nation: J. P. Morgan, Teddy Roosevelt, and the Panama Canal*. New York: Four Walls Eight Windows, 2001.

D'Olwer, Luis Nicolau. "Las inversiones extranjeras." In *Historia moderna de Mexico*, edited by Daniel Cosio Villegas, 973–1185. Mexico City: Editorial Hermes, 1974.

Dorwart, Jeffery M. *The Office of Naval Intelligence: The Birth of America's First Intelligence Agency, 1865–1918*. Annapolis, MD: Naval Institute Press, 1979.

Doyle, Don, and Marco Antonio Pamplona, eds. *Nationalism in the New World*. Athens: University of Georgia Press, 2006.

DuBoff, Richard. "Business Demand and the Development of the Telegraph in the United States." *Business History Review* 54.4 (1986): 459–79.

Duncan, Bingham. *Whitelaw Reid: Journalist, Politician, Diplomat*. Athens: University of Georgia Press, 1975.

Duncan, Robert H. "Political Legitimization and Maximilian's Second Empire in Mexico, 1864–1867." *Mexican Studies* 12.1 (1996): 27–66.

DuVal, Miles. *Cadiz to Cathay: The Story of the Long Diplomatic Struggle for the Panama Canal*. Stanford, CA: Stanford University Press, 1940.

Eakin, Marshall. *British Enterprise in Brazil: The St. John d'el Rey Mining Company and the Morro Velho Gold Mine, 1830–1960*. Durham, NC: Duke University Press, 1989.

Edwards, Rebecca. *New Spirits: Americans in the Gilded Age, 1865–1925*. New York: Oxford University Press, 2006.

Eggert, Gerald G. "Our Man in Havana: Fitzhugh Lee." *Hispanic American Historical Review* 47.4 (1967): 463–85.

———. *Richard Olney: Evolution of a Statesman.* University Park: Penn State University Press, 1974.

Elena, Alberto, and Javier Ordóñez. "Science, Technology, and the Spanish Colonial Experience in the Nineteenth Century." *Osiris* 15, s.2 (2000): 70–82.

Ellul, Jacques. *Propaganda: The Formation of Men's Minds.* New York: Vintage, 1965.

———. *The Technological Society.* New York: Vintage, 1964.

Elson, Ruth Miller. *Guardians of Tradition: American Schoolbooks of the Nineteenth Century.* Lincoln: University of Nebraska Press, 1964.

Ewell, Judith. *Venezuela and the United States: From Monroe's Hemisphere to Petroleum's Empire.* Athens: University of Georgia Press, 1996.

Ewen, Stewart. *PR! A Social History of Spin.* New York: Basic Books, 1996.

Farcau, Bruce W. *The Ten Cents War: Chile, Peru, and Bolivia in the War of the Pacific, 1879–1884.* Westport, CT: Praeger, 2000.

Fernández, Carmen Adams. *La América distorsionada: una visión española de finales del siglo XIXi.* Oviedo, Spain: Universidad de Oviedo, 1999.

Fernández Almagro, Melchor. *Historia política de la España contempóranea.* 2 vols. Madrid: Alianza Editorial, 1968–1969.

Ferns, H. S. "The Baring Crisis Revisited." *Journal of Latin American Studies* 24.2 (1992).

———. *Britain and Argentina in the Nineteenth Century.* Oxford: Clarendon Press, 1960. 241–73.

Ferrara, Orestes. *The Last Spanish War: Revelations in Diplomacy.* New York: Paisley Press, 1937.

Ferrari, Gustavo, and Ezequiel Gallo, eds. *La Argentina del Ochenta al Centenario.* Buenos Aires: Editorial Sudamerica, 1989.

Field, Alexander James Field. "The Magnetic Telegraph, Price and Quantity Data, and the New Management of Capital." *Journal of Economic History* 52.2 (1992): 401–13.

Flayhart. William Henry. *The American Line (1871–1902).* New York: Norton, 2000.

Folkerts, Jean, and Dwight L. Teeter. *Voices of a Nation: A History of Mass Media in the United States.* Boston: Allyn and Bacon, 2002.

Foner, Philip S. *The Spanish-Cuban-American War and the Birth of American Imperialism.* 2 vols. New York: Monthly Review Press, 1972.

———, ed. *Inside the Monster by José Martí: Writings on the United States and American Imperialism.* New York: Monthly Review Press, 1975.

Forbes, Ian. "German Informal Imperialism in South America before 1914." *Economic History Review* 31.3 (1978): 384–98.

Ford, A. G. "Comercio exterior y inversiones extranjeras." In *La Argentina del ochenta al centenario*, edited by Gustavo Ferrari and Ezequiel Gallo, 497–512. Buenos Aires: Editorial Sudamerica, 1989.

Frankel, Benjamin Frankel. *Venezuela y los Estados Unidos*. Caracas: Fundación Boulton, 1977.

Fraser, Steve. *Every Man a Speculator: A History of Wall Street in American Life*. New York; Harper Collins, 2005.

Freidel, Frank. *The Splendid Little War*. Boston: Little, Brown, 1958.

Fuenzalida Bade, Rodrigo. *La armada de Chile*. Valparaíso: Talleres Empresa Periodista, 1978.

Galasso, Norberto. *Manuel Ugarte y la lucha por la unidad latinoamericana*. Buenos Aires: Corregidor, 2001.

Gallo, Ezequiel. "Argentina: Society and Politics, 1880–1916." In *The Cambridge History of Latin America*, vol. 5, *c. 1870–1916*, edited by Leslie Bethel, 359–63. Cambridge: Cambridge University Press, 1986.

Games, Alison. "Atlantic History: Definitions, Challenges, and Opportunities." *American Historical Review* 111.3 (2006): 741–57.

Garner, Paul. *Porfirio Díaz*. London: Longman, 2001.

Gibb, Paul. "Unmasterly Inactivity? Sir Julian Pauncefote, Lord Salisbury, and the Venezuelan Boundary Dispute." *Diplomacy and Statecraft*, 16.1 (2005): 23–55.

Gibson, George H. "Attitudes in North Carolina Regarding the Independence of Cuba, 1868–1898." *North Carolina Historical Review* 43.1 (1966): 43–65.

Gillis, John R. *Islands of the Mind: How the Human Imagination Created the Atlantic World*. New York: Palgrave Macmillan, 2004.

Ginsberg, Benjamin. *The Captive Public: How Mass Opinion Promotes State Power*. New York: Basic Books, 1986.

Gleijeses, Piero. "1898: The Opposition to the Spanish-American War." *Journal of Latin American Studies* 35.4 (2003): 681–719.

Goldberg, Joyce. *The* Baltimore *Affair*. Lincoln: University of Nebraska Press, 1986.

González, Jorge Enrique, ed. *Nación y nacionalismo en América Latina*. Buenos Aires: Consejo Latinoamericano de Ciencias Sociales, 2007.

Gordon, John Steele. *The Scarlet Woman of Wall Street*. New York: Weidenfeld and Nicholson, 1988.

Gordy, J. P. "Ethics in the Panama Case." *Forum* 36 (July 1904): 115–24.

Gould, Lewis L. *The Spanish-American War and President McKinley*. Lawrence: University Press of Kansas, 1980.

Greenhill, Robert. "The Royal Mail Steam Packet Company and the Development of Steamship Links with Latin America." *Maritime History* 3.1 (April 1973): 67–91.

Grenville, John A. S. *Lord Salisbury and Foreign Policy : The Close of the Nineteenth Century*. London: Athlone Press, 1964.
———, and George Berkeley Young. *Politics, Strategy, and American Diplomacy, 1873–1917*. New Haven: Yale University Press, 1966.
Gwynn, Stephen. *The Letters and Friendships of Sir Cecil Spring Rice: A Record*. 2 vols. Boston: Houghton Mifflin, 1929.
Haber, Stephen. *Industry and Underdevelopment: The Industrialization of Mexico, 1890–1940*. Stanford, CA: Stanford University Press, 1989.
Hadfield, William. *Brazil, the River Plate, and the Falkland Islands*. London: Longman, Brown, Green, and Longmans, 1854.
———. *Brazil and the River Plate*. London: Edward Stanford, 1877.
Hagan, Kenneth. *American Gunboat Diplomacy and the Old Navy, 1877–1889*. Westport, CT: Greenwood, 1973.
Hale, Oron J. *Publicity and Diplomacy with Special Reference to England and Germany, 1890–1914*. Charlottesville: University of Virginia Institute for Research in the Social Sciences, 1940.
Hamdy, Naila, and Ehab H. Gomaa. "Framing the Egyptian Uprising in Arabic Language Newspapers and Social Media." *Journal of Communication* 62.2 (2012): 195–211.
Hamilton, John Maxwell. *Journalism's Roving Eye: A History of American Foreign Reporting*. Baton Rouge: Louisiana State University Press, 2009.
———, Renita Coleman, Bettye Grable, and Jaci Cole. "An Enabling Environment: A Reconsideration of the Press and the Spanish-American War." *Journalism Studies* 7.1 (2006): 78–93.
Hamilton, Richard F. *President McKinley, War, and Empire*. 2 vols. New Brunswick, NJ: Tansaction Publishers, 2006.
Hammett, Brian. *Juárez*. London: Longman, 1994.
Handy, Jim. *Gift of the Devil: A History of Guatemala*. Boston: South End Press, 1984.
Hanna, Alfred Jackson, and Kathryn Hanna. *Napoleon III and Mexico*. Chapel Hill: University of North Carolina Press, 1971.
Hannigan, Robert. *New World Power: American Foreign Policy, 1898–1917*. Philadelphia: University of Pennsylvania Press, 2002.
Hardy, Osgood. "The *Itata* Incident." *Hispanic American Historical Review* 5.2 (1922): 204–12.
Harris, William L. "Venezuela: Wars, Claims, and the Cry for a Stronger Monroe Doctrine." In *United States-Latin American Relations, 1850–1903*, edited by Thomas M. Leonard, 107–22. Tuscaloosa: University of Alabama Press, 1999.

Harvey, Neil. *The Chiapas Rebellion*. Durham, NC: Duke University Press, 1998.

Headrick, Daniel. *The Invisible Weapon: Telecommunications and International Politics, 1851–1945*. New York: Oxford University Press, 1991.

———. *The Tentacles of Progress: Technology Transfer in the Age of Imperialism, 1850–1940*. New York: Oxford University Press, 1988.

———. *The Tools of Empire: Technology and European Imperialism in the Nineteenth Century*. New York: Oxford University Press, 1981.

Healy, David. *James G. Blaine and Latin America*. Columbia: University of Missouri Press, 2001.

Henkin, David M. *The Postal Age: The Emergence of Modern Communications in Nineteenth-Century America*. Chicago: University of Chicago Press, 2007.

Herman, Christian. *La politique de la France en Amérique Latine, 1826–1850*. Bordeaux: Maison des Pays Iberique, 1996.

Herrick, Walter R. *The American Naval Revolution*. Baton Rouge: Louisiana State University Press, 1966.

Herring, George. "Imperial Tutor: William McKinley, the War of 1898, and the New Empire, 1898–1902." In *Selling War in a Media Age: The Presidency and Public Opinion in the American Century*, edited by Kenneth Osgood and Andrew K. Frank, 18–47. Gainesville: University Press of Florida, 2010.

Herweg, Holger. *Germany's Vision of Empire in Venezuela, 1871–1914*. Princeton, NJ: Princeton University Press, 1986.

Heuman, Gad. "The British West Indies." In *The Oxford History of the British Empire: The Nineteenth Century*, edited by Andrew Porter, 470–94. New York: Oxford University Press, 1999.

Hietala, Thomas. *Manifest Design: Anxious Aggrandizement in Late Jacksonian America*. Ithaca, NY: Cornell University Press, 1985.

Hilderbrand, Robert C. *Power and the People: Executive Management of Public Opinion in Foreign Affairs, 1897–1921*. Chapel Hill: University of North Carolina Press, 1981.

Hills, Jill. *The Struggle for the Control of Global Communications: The Formative Century*. Urbana: University of Illinois Press, 2002.

Hirsch, Mark D. *William C. Whitney: Modern Warwick*. New York: Dodd, Mead, 1948.

Hixson, Walter. *The Myth of American Diplomacy: National Identity and U.S. Foreign Policy*. New Haven: Yale University Press, 2008.

Hodge, John E. "The Role of the Telegraph in the Consolidation of the Argentine Republic." *The Americas* 41.1 (1984): 59–80.

Hogan, J. Michael. *The Panama Canal and American Politics*. Carbondale: Southern Illinois University Press, 1986.

Hoganson, Karen. *Fighting for American Manhood: How Gender Politics Provoked the Spanish-American War and the Philippine-American War.* New Haven: Yale University Press, 1998.

Hohenberg, John. *Foreign Correspondence: The Great Reporters and Their Times.* New York: Columbia University Press, 1964.

Holmes, James R. *Theodore Roosevelt and World Power in International Relations.* Washington, D.C.: Potomac Books, 2006.

Hoogenboom, Ari. *Rutherford B. Hayes, Warrior and President.* Lawrence: University Press of Kansas, 1995.

Hopkins, A. G. "Informal Empire in Argentina: An Alternative View." *Journal of Latin American Studies* 26.2 (1994): 469–84.

Hubbard, Walter, and Richard F. Winter. *North Atlantic Mail Sailings, 1840–75.* Canton, OH: U.S. Philatelic Society, 1988.

Hughes, Thomas P. *American Genesis: A Century of Invention and Technological Enthusiasm.* New York: Viking, 1989.

Hugill, Peter. *Global Communications since 1844: Geopolitics and Technology.* Baltimore: Johns Hopkins University Press, 1999.

Humphreys, R. A. "Anglo-American Rivalries and the Venezuela Crisis of 1895." In *Tradition and Revolt in Latin America and Other Essays,* edited by R. A. Humphreys, 186–215. London: Weidenfeld and Nicholson, 1969.

———, ed. *Tradition and Revolt in Latin America and Other Essays.* London: Weidenfeld and Nicholson, 1969.

Hunt, Michael. *Ideology and U.S. Foreign Policy.* New Haven, CT: Yale University Press, 1987.

Jacobson, Nils. "Liberalismo tropical: Como explicar el auge de una doctrina económica europea en America Latina." *Historia Critica* 34 (Julio–Diciembre 2007): 118–47.

James, William. *Richard Olney and His Public Service.* Boston: Houghton Mifflin, 1923.

Jervy, Theodore D. "William Lindsay Scruggs—A Forgotten Diplomat." *South Atlantic Quarterly* 27.3 (1928): 292–309.

Johannsen, Robert. *To the Halls of Montezuma: The Mexican War in the American Imagination.* New York: Oxford University Press, 1985.

John, Richard. "Recasting the Information Infrastructure for the Industrial Age." In *A Nation Transformed by Information,* edited by Alfred D.Chandler Jr. and James Cortada, 55–105. Cambridge: Harvard University Press, 2000.

Johnson, John J. *Latin America in Caricature.* Austin: University of Texas Press, 1980.

———. *Pioneer Telegraphy in Chile, 1852–1876*. Stanford CA: Stanford University Press, 1948.

Jones, Charles. *El Reino Unido y América: Inversiones y influencia económica*. Madrid: Mapfre, 1992.

Joseph, Gilbert M. "Close Encounters: Toward a Cultural History of U.S.–Latin American Relations." In *Close Encounters of Empire: Writing the Cultural History of U.S.–Latin American Relations*, edited by Gilbert M. Joseph, Catherine C. LeGrand, and Ricardo D. Salvatore, 3–46. Durham, NC: Duke University Press, 1998.

———, Catherine Leogrand, and Ricardo Salvatore, eds. *Close Encounters of Empire: Writing the Cultural History of U.S.–Latin American Relations*. Durham, NC: Duke University Press, 1998.

Jowett, Garth S., and Victoria O'Donnell. *Propaganda and Persuasion*. Thousand Oaks, CA: Sage Publications, 1999.

Kagan, Richard L. "Prescott's Paradigm: American Historical Scholarship and the Decline of Spain." *American Historical Review* 101.2 (1996): 423–46.

Kammen, Michael. *Mystic Chords of Memory: The Transformation of Tradition in American Culture*. New York: Vintage, 1991.

Kaplan, Amy. *The Anarchy of Empire in the Making of U.S. Culture*. Cambridge, MA: Harvard University Press, 2002.

Karnes, Thomas. *The Failure of Union: Central America, 1824–1960*. Chapel Hill: University of North Carolina Press, 1961.

Katz, Frederick, and Jane-Dale Lloyd, eds. *Porfirio Díaz frente al descontento popular regional (1891–1893)*. Mexico: Universidad Iberoamericana, 1986.

Kemble, John H. *The Panama Route*. New York: Da Capa, 1972.

Kendall, George Wilkins. *Dispatches from the Mexican War*. Norman: University of Oklahoma Press, 1999.

Kennedy Paul. *The Samoan Tangle: A Study in Anglo-German-American Relations, 1878–1900*. New York: Harper and Row, 1974.

Kilmer, Paulette. "The Press in Industrial America, 1865–1883." In *The Media in America: A History*, edited by William David Sloan, 199–222. Newport, AL: Vision Press, 2005.

Kirk, John M. *José Martí: Mentor of the Cuban Nation*. Gainesville: University Presses of Florida, 1983.

Knight, Alan. "Britain and Latin America." In *The Oxford History of the British Empire: The Nineteenth Century*, edited by Andrew Porter, 122–45. New York: Oxford University Press, 1999.

———. "U.S. Imperialism/Hegemony and Latin American Resistance." In *Empire and Dissent: The United States and Latin America*, edited by Fred Rosen, 23–52.

Durham, NC: Duke University Press, 2008.

Knightley, Phillip. *The First Casualty: From the Crimea to Vietnam: The War Correspondent as Hero, Propagandist, and Myth Maker.* New York: Harcourt Brace Jovanovich, 1975.

Krasner, Stephen D. *Sovereignty: Organized Hypocrisy.* Princeton, NJ: Princeton University Press, 1999.

Kroeger, Brooke. *Nellie Bly: Daredevil, Reporter, Feminist.* New York: Three Rivers Press, 1995.

Kubicek, Robert. "British Expansion, Empire, and Technological Change." In *The Oxford History of the British Empire: The Nineteenth Century,* edited by Andrew Porter, 247–69. New York: Oxford University Press, 1999.

LaFeber, Walter. *The New Empire: An Interpretation of American Expansion, 1860–1898.* Ithaca: Cornell University Press, 1963.

Landes, David. *The Unbound Prometheus: Technological Change and Industrial Development in Western Europe from 1750 to the Present.* Cambridge: Cambridge University Press, 1972.

———. *The Wealth and Poverty of Nations.* New York: Norton, 1998.

Langer, William. *The Diplomacy of Imperialism.* Second edition. New York: Knopf, 1960.

Langley, Lester. *America and the Americas: The United States in the Western Hemisphere.* Athens: University of Georgia Press, 2010.

———. *The Americas in the Age of Revolution, 1750–1850.* Princeton, NJ: Princeton University Press, 1996.

———. *The Banana Wars: United States Intervention in the Caribbean, 1898–1934.* Chicago: Dorsey, 1985.

———. *Struggle for the American Mediterranean: United States–European Rivalry in the Gulf Caribbean, 1776–1904.* Athens: University of Georgia Press, 1982.

Leech, Margaret. *In the Days of McKinley.* New York: Harper, 1959.

Leonard, Thomas M. "Central America: The Search for Economic Development." In *United States-Latin American Relations, 1850–1903: Establishing a Relationship,* edited by Thomas M. Leonard, 93–95. Tuscaloosa: University of Alabama Press, 1999.

———. *Panama, the Canal, and the United States: Guide to Issues and References.* Claremont, CA: Regina Books, 1993.

———, ed. *United States-Latin American Relations, 1850–1903: Establishing a Relationship.* Tuscaloosa: University of Alabama Press, 1999.

Liehr, Reinhard, ed. *América Latina en la época de Simón Bolívar.* Berlin: Colloquium Verlag, 1989.

Lim, Merlyna. "Clicks, Cabs, and Coffee Houses: Social Media and Oppositional Movements in Egypt, 2004–2011." *Journal of Communication* 62.2 (2012): 231–48.

Linderman, Gerald F. *The Mirror of War: American Society and the Spanish-American War.* Ann Arbor: University of Michigan Press, 1974.

Linn, Brian McAllister. *Guardians of Empire: The U.S. Army and the Pacific, 1902–1940.* Chapel Hill: University of North Carolina Press, 1997.

Livezey, William E. *Mahan on Sea Power.* Norman: University of Oklahoma Press, 1947.

Lodge, Henry Cabot, and Charles Redmond, eds. *Selections from the Correspondence of Theodore Roosevelt and Henry Cabot Lodge, 1884–1918.* 2 vols. 1925. Reprinted New York: Da Capa Press, 1979.

López-Ocón, Leoncio, Jean-Pierre Chaumeil, and Ana Verde Casanova, eds. *Los Americanistas del siglo XIX: La construcción de una comunidad cienífica internacional.* Madrid: Iberoamericana, 2005.

Lubow, Arthur. *The Reporter Who Would Be King: A Biography of Richard Harding Davis.* New York: Scribner's, 1992.

Lutz, Tom. *American Nervousness, 1903: An Anecdotal History.* Ithaca, NY: Cornell University Press, 1991.

Mack, Gerstel. *The Land Divided: A History of the Panama Canal and Other Isthmian Projects.* New York: Knopf, 1944. Reprinted New York: Octagon, 1974.

Mahan, Alfred Thayer. *The Influence of Seapower upon History.* New York: Sagamore Press, 1957.

———. *The Letters and Papers of Alfred Thayer Mahan.* 3 vols. Edited by Robert Seager II and Doris Maguire. Annapolis, MD: Naval Institute Press, 1975.

Major, John. "Who Wrote the Hay-Bunau-Varilla Convention?" *Diplomatic History* 8.2 (1984): 115–23.

Mander, Mary S. "Pen and Sword: Problems of Reporting the Spanish-American War." *Journalism History* 9.1 (1982): 2–9, 28.

Marcos, Subcomandante. *Our Word Is Our Weapon.* Edited by Juana Ponce de León. New York: Seven Stories Press, 2002.

Marcus, Alan J., and Howard P. Segal. *Technology in America: A Brief history.* Belmont, CA: Wadsworth Thomson, 1999.

Margolies, Daniel. *Henry Watterson and the New South: The Politics of Empire, Free Trade, and Globalization.* Lexington: University of Kentucky Press, 2006.

Marichel, Carlos. *A Century of Debt Crises in Latin America: From Independence to the Great Depression, 1820–1930.* Princeton, NJ: Princeton University Press, 1989.

Marin Vicuña, Santiago. *Los hermanos Clark.* Santiago: Balcells, 1929.

Martin, Michéle. *Images at War: Illustrated Periodicals and Constructed Nations.* Toronto: University of Toronto Press, 2006.

Martín Jiménez, Hilario. *Valeriano Weyler: Capitán-General, Duque de Rubi, Marqués de Tenerife y Grande de España: de su vida y personalidad, 1838–1930.* Santa Cruz de Tenerife: Ediciones del Umbral, 1998.

Matthews, Joseph. "Informal Diplomacy in the Venezuelan Crisis of 1896." *Mississippi Valley Historical Review* 50.2 (1963): 195–212.

Maurtua, Victor M. *The Question of the Pacific.* Translated by F. A. Pezet. Philadelphia: Lasher, 1901.

May, Ernest R. *Imperial Democracy: The Emergence of America as a Great Power.* New York: Harper and Row, 1961.

McCartney, Paul T. *Power and Progress: American National Identity, the War of 1898, and the Rise of American Imperialism.* Baton Rouge: Louisiana State University Press, 2006.

McCreery, David. *Development and the State in Reforma Guatemala.* Athens, OH: Ohio University, Center for International Studies, 1983.

McCullough, David. *The Path Between the Seas: The Creation of the Panama Canal, 1870–1914.* New York: Simon and Schuster, 1977.

McEvoy, Carmen, and Ana María Stuven, eds. *La república peregrina: hombres de armas y letras en América del Sur, 1800–1884.* Lima: Instituto Francés de Estudios Andinos, 2007.

McGann, Thomas F. *Argentina, the United States, and the Inter-American System, 1880–1914.* Cambridge: Harvard University Press, 1957.

McGuiness, Aims. *Path of Empire: Panama and the California Gold Rush.* Ithaca, NY: Cornell University Press, 2009.

Meneses Ciuffardi, Emilio. *El Factor naval en las relaciones entre Chile y los Estados Unidos (1885–1951).* Santiago: Ediciones pedagogicas chileanas, 1989.

Messinger, Garry. *British Propaganda and the State in the First World War.* Manchester, UK: University of Manchester Press, 1992.

Mettelart, Armand. *The Invention of Communication.* Minneapolis: University of Minnesota Press, 1994.

———. *Networking the World, 1974–2000.* Minneapolis: University of Minnesota Press, 2000.

Michelson, Charles. *The Ghost Talks.* New York: Putnam's, 1944.

Miller, Robert R. *Arms Across the Border: United States Intervention in Mexico.* Philadelphia: American Philosophical Society, 1973.

Miller, Rory. *Britain and Latin America in the Nineteenth and Twentieth Centuries.* London: Longman, 1993.

Millington, Herbert. *American Diplomacy and the War of the Pacific.* New York: Columbia University Press, 1948.

Millis, Walter. *The Martial Spirit.* Boston: Houghton Mifflin, 1931.

Milton, Joyce. *The Yellow Kids: Foreign Correspondents in the Heyday of Yellow Journalism.* New York: Harper and Row, 1989.

Mitchell, Nancy. *The Danger of Dreams: German and American Imperialism in Latin America.* Chapel Hill: University of North Carolina Press, 1999.

Monaghan, Jay. *Chile, Peru, and the California Gold Rush.* Berkeley: University of California Press, 1973.

Morgan, H. Wayne. *William McKinley and His America.* Syracuse, NY: Syracuse University Press, 1963.

Morrison, Elting E., ed. *The Letters of Theodore Roosevelt.* 8 vols. Cambridge, MA: Harvard University Press, 1951.

Morrogh, Earl. *Information Architecture: An Emerging 21st Century Profession.* Upper Saddle River, NJ: Prentice Hall, 2003.

Mowat, R. B. *The Life of Lord Pauncefote.* Boston: Houghton Mifflin, 1929.

Murphey, Gretchen. *Hemispheric Imaginings: The Monroe Doctrine and Narratives of US Empire.* Durham, N.C.: Duke University Press, 2005.

Musicant, Ivan. *The Banana Wars: A History of United States Military Intervention in Latin America from the Spanish-American War to the Invasion of Panama.* New York: Macmillan, 1990.

———. *Empire by Default: The Spanish-American War and the Dawn of the American Century.* New York: Henry Holt, 1998.

Nalbach, Alex. "'Poisoned at the Source'? Telegraphic News Services and Big Business in the Nineteenth Century." *Business History Review* 77 1 (2003): 577 610.

Nasaw, David. *The Chief: The Life of William Randolph Hearst.* Boston: Houghton Mifflin, 2000.

Navarro Burciaga, José Luis. "Catarino Garza, periodista opositor a Porfirio Díaz en Tamaulipas." In *Porfirio Díaz frente al descontento popular regional (1891–1893)*, edited by Frederick Katz and Jane-Dale Lloyd, 59–78. Mexico: Universidad Iberoamericana, 1986.

Neale, R. G. *Britain and American Imperialism.* Brisbane: University of Queensland Press, 1965.

Nevins, Allan. *Hamilton Fish: The Inner History of the Grant Administration.* 2 vols. New York: Unger, 1957.

Newton, Wesley P. "Origins of United States-Latin American Relations." In *United States-Latin American Relations, 1800–1850: The Formative Generation*, edited by T. Ray Shurbutt, 1–24. Tuscaloosa: University of Alabama Press, 1991.

Nickles, David Paul. "Telegraph Diplomats: The United States' Relations with France in 1848 and 1870." *Technology and Culture* 40.1 (1999): 1–25.

———. *Under the Wire: How the Telegraph Changed Diplomacy*. Cambridge: Harvard University Press, 2002.

Ninkovich, Frank. *The United States and Imperialism*. Malden, MA: Blackwell, 2001.

Nofi, Albert. *The Spanish-American War, 1898*. Conshoken, PA: Combined Books, 1996.

Norman, Henry. "Anglo-American Arbitration." *Review of Reviews* 15 (January 1897): 75–76.

Nugent, Walter. *Habits of Empire: A History of American Empire*. New York: Knopf, 2008.

Núñez, Enrique Bernardo. *Tres momentos en la controversia de limites de Guyana: el incidente de Yuruán*. Caracas: Imprenta Nacional, 1962.

O'Brien, Thomas. *The Century of U.S. Capitalism in Latin America*. Albuquerque: University of New Mexico Press, 1999.

———. *Making the Americas: The United States and Latin America from the Age of Revolutions to the Era of Globalization*. Albuquerque: University of New Mexico Press, 2007.

Offner, John. "United States Politics and the 1898 War Over Cuba." In *The Crisis of 1898: Colonial Redistribution and Nationalist Mobilization*, edited by Angel Smith and Emma Davila-Cox, 18–44. London: Macmillan, 1999.

———. *An Unwanted War: The Diplomacy of the United States and Spain over Cuba, 1895–1898*. Chapel Hill: University of North Carolina Press, 1992.

Olasky, Marvin. "Hawks or Doves? Texas Press and the Spanish American War." *Journalism Quarterly* 64.1 (1987), 205–8.

O'Leary, Cecilia Elizabeth. *To Die For: The Paradox of American Patriotism*. Princeton, NJ: Princeton University Press, 1999.

Ortiz, David. *Paper Liberals: Press and Politics in Restoration Spain*. Westport, CT: Greenwood, 2000.

Osgood, Kenneth, and Andrew K. Frank, eds. *Selling War in a Media Age: The Presidency and Public Opinion in the American Century*. Gainesville: University Press of Florida, 2010.

O'Toole, G. J. A. *The Spanish War: An American Epic*. New York: Norton, 1984.

Ovalle Castillo, Francisco Javier. *La personalidad de don Marcial Martínez*. Santiago: Imprenta universitaria, 1918.

Paddock, Troy R. E., ed. "Introduction: Newspapers, Public Opinion, and Propaganda." In *A Call to Arms: Propaganda, Public Opinion, and Newspapers in the Great War*, edited by Troy R. E. Paddock, 1–13. Westport, CT: Praeger, 2004.

———. *A Call to Arms: Propaganda, Public Opinion, and Newspapers in the Great War*. Westport, CT: Praeger, 2004.

Pakenham, Thomas. *The Boer War*. New York: Random House, 1979.

———. *The Scramble for Africa*. New York: Basic Books, 1991.

Palacis, Guillermo, ed. *Ensayos sobre la nueva historia política de América Latina, siglo XIX* .Mexico City: Colegio de México, 2007.

Papacharissi, Zizi, and Maria de Fatima Oliveira. "Affective News and Networked Publics: The Rhythms of News Storytelling on Egypt." *Journal of Communication* 62.2 (2012): 266–82.

Parks, E. Taylor. *Colombia and the United States, 1765–1934*. Durham: Duke University Press, 1935.

Pauly, John J. "The Great Chicago Fire as a National Event." *American Quarterly* 36.5 (1984): 677.

Pérez, Louis A. *Cuba and the United States: Ties of Singular Intimacy*. Athens: University of Georgia Press, 1990.

———. *Cuba between Empires, 1878–1902*. Pittsburgh: University of Pittsburgh Press, 1983.

———. *Cuba in the American Imagination*. Chapel Hill: University of North Carolina Press, 2008.

———. "The Meaning of the *Maine*: Causation and the Historiography of the Spanish-American War." *Pacific Historical Review* 58.3 (1989): 293–322.

———. *On Becoming Cuban: Identity, Nationality, and Culture*. Chapel Hill: University of North Carolina Press, 1999.

———. *The War of 1898: The United States and Cuba in History and Historiography*. Chapel Hill: University of North Carolina Press, 1998.

Pérez Guzmán, Francisco. *Herida profunda*. Havana: Unión de Escritores y Artistas de Cuba, 1998.

Pérez Vejo, Tomás. "La guerra hispano-estadounidense del '98 en la prensa mexicana." *Historia Mexicana* 50.2 (2000): 271–308.

Perkins, Bradford. *The Great Rapprochement*. New York: Atheneum, 1968.

Perkins, Dexter. *A History of the Monroe Doctrine*. Boston: Little, Brown, 1963.

Perry, Richard O. "Argentina and Chile: the Struggle for Patagonia, 1843–1881." *The Americas* 36.3 (1980): 347–63.

Peterson, Harold. *Argentina and the United States, 1810–1960*. New York: State University of New York, 1964.

Peterson, M. J. *Recognition of Governments: Legal Doctrine and State Practice*. New York: St. Martin's Press, 1997.

Picón-Salas, Mario, Augusto Mijares, et al., *Venezuela Independiente*. Caracas: Fundación Eugenio Mendoza, 1962.

Pike, Frederick. *Chile and the United States, 1880–1962*. Notre Dame, IN: University of Notre Dame Press, 1963.

———. *The United States and Latin America: Myths and Stereotypes of Civilization and Nature*. Austin: University of Texas Press, 1992.

Pletcher, David. *The Awkward Years: American Foreign Relations Under Garfield and Arthur*. Columbia: University of Missouri Press, 1961.

———. *The Diplomacy of Trade and Investment. American Economic Expansion in the Hemisphere, 1865–1900*. Columbia: University of Missouri Press, 1998.

———. *Rails, Mines, and Progress: Seven American Promoters in Mexico, 1867-1911*. Ithaca, NY: Cornell University Press, 1958.

Ponder, Stephen. *Managing the Press: Origins of the Media Presidency*. New York: St. Martin's Press, 1999.

Porter, Andrew, ed. *The Oxford History of the British Empire: The Nineteenth Century*. New York: Oxford University Press, 1999.

Potter, Simon. *News and the British World: The Emergence of an Imperial Press System*. Oxford: Clarendon Press, 2003.

Poyo, Gerald E. *"With All, and for the Good of All": The Emergence of Popular Nationalism in the Cuban Communities of the United States, 1848–1898*. Durham, NC: Duke University Press, 1989.

Pratkanis, Anthony, and Elliot Aronson. *Age of Propaganda: The Everyday Use and Abuse of Persuasion*. New York: Henry Holt, 2001.

Pritchett, Henry S. "Some Recollections of President McKinley and the Cuban Intervention." *North American Review* 189 (March 1909): 401.

Querejazu-Calvo, Roberto. *Guano, salitre, y sangre*. La Paz: Editorial Los Amigos del Libro, 1979.

Quijada, Monica. "Sobre el origin y difusión del nombre América Latina (O una variación heterodoxa en torno el tema de la construción social de la verdad)." *Revista Indiana* 58.214 (1998): 595–616.

Randall, Stephen J. *Colombia and the United States: Hegemony and Interdependence*. Athens: University of Georgia Press, 1992.

Rauch, George. *Conflict in the Southern Cone: The Argentine Military and the Boundary Dispute with Chile, 1870–1902*. Westport, CT: Praeger, 1999.

Raymond, Dora Neill. *British Policy and Opinion during the Franco-Prussian War*. New York: AMS Press, 1967.

Read, Donald. *The Power of News: The History of Reuters*. New York: Oxford University Press, 1992.

Reggini, Horacio. *Sarmiento y las telecomunicaciones: La obsesión del hilo*. Buenos Aires: Ediciones Galápago, 1996.

Rénique, José Luis. "Benjamín Vicuña Mackenna: exilio, historia y nación." In *La república peregrina: hombres de armas y letras en América del Sur, 1800–1884*, edited by Carmen McEvoy and Ana María Stuven, 487–529. Lima: Instituto Francés de Estudios Andinos, 2007.

Robbins Raines, Rebecca. *Getting the Message Through: A Branch History of the U.S. Army Signal Corps*. Washington, D.C.: Center for Military History, United States Army, 1996.

Rodríguez Ozán, María Elena. "La Guerra Hispano-America en la prensa mexicana (1898–1899)." In *1898 ¿desastre o reconciliación?*, edited by Leopoldo Zea and Mario Magallón. Mexico: Instituto Panamericano de Geografía e Historia, 1998.

Roehl, Charlotte. "Porfirio Díaz in the Press of the United States." PhD diss., University of Chicago, 1953.

Roggenkamp, Karen. *Narrating the News: New Journalism and Literary Genre in Late Nineteenth Century American Newspapers and Fiction*. Kent, OH: Kent State University Press, 2005.

Roig de Leuchsenring, Emilio. *Weyler en Cuba: un precusor de la barbarie fascista*. Havana: Páginas, 1947.

Rojas, Armando. *Historia de las relaciones diplomaticas entre Venezuela y los Estados Unidos*. Vol. 1, *1810–1899*. Caracas: Ediciones de la Presidencia de la República, 1979.

———. *Las misiones diplomaticas de Guzmán Blanco*. Caracas: Monte Avila Editores, 1971.

Rojas, Rafael. "Retóricas de la raza: Intelectuales mexicanos ante la Guerra del 98." *Historia Mexicana* 49.4 (2000): 593–629.

Rosen, Fred, ed. *Empire and Dissent: The United States and Latin America*. Durham, NC: Duke University Press, 2008.

Rosenberg, Emily. *Financial Missionaries to the World: The Politics and Culture of Dollar Diplomacy, 1900–1930*. Durham: Duke University Press, 2003.

Rosenberg, Martin M., and Thomas P. Ruff. *Indiana and the Coming of the Spanish-American War*. Muncie, IN: Ball State University, 1976.

Rosenzweig, Fernando. "El comercio exterior." In *Historia moderna de Mexico*, edited by Daniel Cosio Villegas, 635–729. Mexico City: Editorial Hermes, 1974.

Rosewater, Victor. *History of Coöperative News-Gathering in the United States*. New York: Appleton, 1930.

Ross, Steven T. *American War Plans, 1890–1939*. London: Frank Cass, 2002.

Samuels, Peggy, and Harold Samuels. *Remembering the* Maine. Washington, D.C.: Smithsonian Institution Press, 1995.

San Román, María Cristina, and Guillermo Horacio Gasió. "La presidencia de

Avellaneda." In *La Argentina del ochenta al centenario*, edited by Gustavo Ferrari and Ezequiel Gallo, 35–44. Buenos Aires: Editorial Sudamerica, 1989.

Santos, Felix. *1898: La Prensa y la guerra de Cuba*. Bilbao: Asociación Julián Zugazagoita, 1998.

Sarabia, Nydia. *Noticias confidenciales sobre Cuba, 1870–1895*. Havana: Editorial Politica, 1985.

Sater, William F. "Chile: Clash of Global Visions II." In *United States-Latin American Relations, 1850–1903: Establishing a Relationship*, edited by Thomas M. Leonard, 169–96. Tuscaloosa: University of Alabama Press, 1999.

———. *Chile and the United States: Empires in Conflict*. Athens: University of Georgia Press, 1990.

———. *Chile and the War of the Pacific*. Lincoln: University of Nebraska Press, 1986.

Scheina, Robert L. *Latin America's Wars: The Age of the Caudillo, 1791–1899*. Washington, D.C.: Brassey's, 2003.

Schneider, Jurgen. "Trade Relations Between France and Latin America, 1810–1850." In *América Latina en la época de Simón Bolívar*, edited by Reinhard Liehr, 423–37. Berlin: Colloquium Verlag, 1989.

Schoonover, Thomas. *Dollars over Dominion: The Triumph of Liberalism in Mexican–United States Relations, 1861–1867*. Baton Rouge: Louisiana State University Press, 1978.

———. *The French in Central America: Culture and Commerce, 1820–1930*. Wilmington, DE: Scholarly Resources, 2000.

———. *Germany in Central America: Competitive Imperialism, 1829–1929*. Tuscaloosa: University of Alabama Press, 1998.

———. "Max Farrand's Memorandum on the U.S. Role in the Panamanian Revolution of 1903." *Diplomatic History* 12.4 (1988): 504.

———. *Uncle Sam's War of 1898 and the Origins of Globalization*. Lexington: University Press of Kentucky, 2003.

———. *The United States in Central America, 1860–1911: Episodes of Social Imperialism and Imperial Rivalry in the World System, 1860–1911*. Durham: Duke University Press, 1991.

Schott, Joseph. *Rails Across Panama*. New York: Bobbs Merrill, 1967.

Schoultz, Lars. *Beneath the United States: A History of U.S. Policy toward Latin America*. Cambridge, MA: Harvard University Press, 1998.

Schroeder, John H. *Mr. Polk's War: American Opposition and Dissent, 1846–1848*. Madison: University of Wisconsin Press, 1973.

Schwartz Cowan, Ruth. *A Social History of American Technology* . New York: Oxford University Press, 1997.

Schwarzlose, Richard. *The Nation's Newsbrokers*. 2 vols. Evanston, IL: Northwestern University Press, 1989–1990.

Seager, Robert II. "Alfred Thayer Mahan: Christian Expansionist, Navalist, Historian." In *Makers of the American Naval Tradition*, edited by James C. Bradford, 24–72. Annapolis: Naval Institute Press, 1990.

———. *Alfred Thayer Mahan: The Man and His Letters*. Annapolis, MD: Naval Institute Press, 1977.

Sexton, Jay. *The Monroe Doctrine: Empire and Nation in Nineteenth Century America*. New York: Hill and Wang, 2011.

———. "The United States, the Cuban Rebellion, and the Multilateral Initiative of 1875." *Diplomatic History* 30.3 (2006): 335–50.

Shurbutt, T. Ray, ed. *United States-Latin American Relations, 1800–1850: The Formative Generation*. Tuscaloosa: University of Alabama Press, 1991.

Simon, Maron J. *The Panama Affair*. New York: Scribner's, 1971.

Sloan, William David, ed. *The Media in America: A History*. Newport, AL: Vision Press, 2005.

Smalley, George W. *Anglo-American Reminiscences*. New York: Putnam, 1911.

Smith, Angel. "The People and the Nation: Nationalist Mobilization and the Crisis of 1898 in Spain." In *The Crisis of 1898: Colonial Redistribution and Nationalist Mobilization*, edited by Angel Smith and Emma Davila-Cox, 152–79. London: Macmillan, 1999.

———, and Emma Davila-Cox. *The Crisis of 1898: Colonial Redistribution and Nationalist Mobilization*. London: Macmillan, 1999.

Smith, Geoffrey S. "The Role of José M. Balmaceda in Preserving Argentine Neutrality in the War of the Pacific." *Hispanic American Historical Review* 49.2 (1969): 254–67.

Smith, Joseph. *Illusions of Conflict: Anglo-American Diplomacy Toward Latin America, 1865–1896*. Pittsburgh: University of Pittsburgh Press, 1979.

———. "The Latin American Trade Commission of 1884–5." *Inter-American Economic Affairs* 24.4 (1971): 3–24.

———. *The Spanish-American War: Conflict in the Caribbean and the Pacific 1895–1902*. London: Longman, 1994.

Smythe, Ted Curtis. *The Gilded Age Press*. Westport, CT: Praeger, 2003.

Spencer, David R. *The Yellow Journalism: The Press and America's Emergence as a World Power*. Evanston, IL: Northwestern University Press, 2007.

Standage, Tom. *The Victorian Internet*. New York: Walker, 1998.

Starr Paul. *The Creation of the Media: Political Origins of Modern Communications*. New York: Basic Books, 2004.

Stevens, John D. *Sensationalism and the New York Press.* New York: Columbia University Press, 1991.

Summers, Mark Wahlgren. *The Gilded Age; or, The Hazards of New Functions.* Upper Saddle River, NJ: Prentice Hall, 1997.

Swanberg, W. A. *Citizen Hearst.* New York: Scribners, 1961.

———. *Pulitzer.* New York: Scribners, 1967.

Swann, Leonard. *John Roach: Maritime Entrepreneur.* Annapolis, MD: U.S. Naval Institute Press, 1965.

Tansill, Charles C. *The Congressional Career of Thomas Francis Bayard, 1869–85.* New York: Fordham University Press, 1940.

———. *The Foreign Policy of Thomas F. Bayard.* New York: Fordham University Press, 1940.

Taylor, Philip M. *Munitions of the Mind: A History of Propaganda from the Ancient World to the Present Day.* Manchester, UK: Manchester University Press, 2003.

Tenorio-Trillo, Mauricio. *Mexico at the World's Fairs: Crafting a Modern Nation.* Berkeley: University of California Press, 1996.

Thomas, Evan. *The War Lovers: Roosevelt, Lodge, Hearst and the Rush to Empire, 1898.* New York: Little, Brown, 2010.

Thomas, Gordon, and Max Morgan Witts. *The Day the World Ended.* New York: Stein and Day, 1969.

Thompson, Andrew. "Informal Empire? An Exploration in the History of Anglo-Argentine Relations, 1810–1914." *Journal of Latin American Studies* 24.2 (1992): 419–36.

Tompkins, E. Berkeley. *Anti-Imperialism in the United States: The Great Debate, 1890–1920.* Philadelphia: University of Pennsylvania Press, 1970.

Topik, Steven. *Trade and Gunboats: The United States and Brazil in the Age of Empire.* Stanford, CA: Stanford University Press, 1996.

———, and Allen Wells, eds. *The Second Conquest of Latin America: Coffee, Henequen, and Oil during the Export Boom, 1850–193.* Austin: University of Texas Press, 1998.

Trask, David. *The War with Spain in 1898.* Lincoln: University of Nebraska Press, 1981.

Tufekci, Zeynep, and Christopher Wilson. "Social Media and the Decision to Participate in Political Protest: Observations from Tahrir Square." *Journal of Communication* 62 (2012): 363–79.

Tulchin, Joseph. "Argentina: Clash of Global Visions I." In *United States-Latin American Relations, 1850–1903: Establishing a Relationship,* edited by Thomas M. Leonard, 147–68. Tuscaloosa: University of Alabama Press, 1999.

Turk, Richard W. "The United States Navy and the 'Taking' of Panama, 1901–1903." *Military Affairs* 38.3 (1974): 92–96.

Valdivieso, Bravo, Bulnes Serrano, and Vial Correa. *Balmaceda y la guerra civil*. Santiago: Editorial fundación, 1991.

Vaughan, Stephen. *Holding Fast the Inner Lines: Democracy, Nationalism, and the Committee on Public Information*. Chapel Hill: University of North Carolina Press, 1980.

Veliz, Claudio. *Historia de la marina mercante de Chile*. Santiago: Universidad de Chile, 1961.

Villarías Robles, Juan J. R. "El Peruanismo de Sir Clements Markham (1830–1916)." In *Los Americanistas del siglo XIX: La construcción de una comunidad científica internacional*, edited by Leoncio López-Ocón, Jean-Pierre Chaumeil, and Ana Verde Casanova, 111–43. Madrid: Iberoamericana, 2005.

Ville, Simon P., and David M. Williams, eds. *Management, Finance and Industrial Relations in the Maritime Industry*. St. John's, Newfoundland: International Maritime Economic History Association, 1994.

Voss, Stuart. *Latin America in the Middle Period, 1750–1929*.Wilmington, DE: Scholarly Resources, 2002.

Walker, Dale. *The Boys of '98: Theodore Roosevelt and the Rough Riders*. New York: Tom Doherty, 1998.

Ward, Ken. *Mass Communications and the Modern World*. Chicago: Dorsey Press, 1989.

Wawro, Geoffrey. *The Franco-Prussian War: The German Conquest of France*. Cambridge: Cambridge University Press, 2003.

Weber, Ralph E. *United States Diplomatic Codes and Ciphers*. Chicago: Precedent Publishing, 1979.

Weeks, William Earl. *John Quincy Adams and American Global Empire*. Lexington: University Press of Kentucky, 1992.

Welch, David. *The Third Reich: Politics and Propaganda*. London: Routledge, 2002.

Welch, Richard. *Response to Imperialism: The United States and the Philippine-American War, 1899–1902*. Chapel Hill: University of North Carolina Press, 1979.

———. *The Presidencies of Grover Cleveland*. Lawrence: University Press of Kansas, 1988.

Welter, Mark M. "The 1895–1898 Crisis in Minnesota Newspapers: Testing the 'Yellow Journalism' Theory." *Journalism Quarterly* 47.4 (1970): 719–24.

Whates, Harry R. *The Third Salisbury Administration, 1895–1900*. London: Westminster Press, 1900.

Whitaker, Arthur P. *The United States and the Independence of Latin America*. New York: Norton, 1964.

Wicks, Daniel H. "Dress Rehearsal: United States Intervention on the Isthmus of Panama, 1885." *Pacific Historical Review* 49.4 (1980): 581–605.

———. "New Navy and New Empire: The Life and Times of John Grimes Walker." PhD diss., University of California, Berkeley, 1979.

Wiebe, Robert. *The Search for Order, 1877–1920*. New York: Hill and Wang, 1967.

Wilkerson, Marcus. *Public Opinion and the Spanish American War: A Study in War Propaganda*. New York: Russell and Russell, 1967. Reprint of 1932 edition.

Williams, R. Hal. *Years of Decision: American Politics in the 1890s*. Prospect Heights, IL: Waveland Press, 1993.

Winseck, Dwayne, and Robert M. Pike, *Communication and Empire: Media, Markets, and Globalization, 1860–1930*. Durham, NC: Duke University Press, 2007.

Wisan, Joseph. *The Cuban Crisis as Reflected in the New York Press, (1895–1898)*. New York: Columbia University Press, 1934.

Woodward, Ralph Lee. *Central America: A Nation Divided*. New York: Oxford University Press, 1999.

Wu, Tim. *The Master Switch: The Rise and Fall of Information Empires*. New York: Knopf, 2011.

Yates, JoAnne. "Business Use of Information and Technology during the Industrial Age." In *A Nation Transformed by Information*, edited by Alfred D. Chandler Jr. and James Cortada, 107–35. Cambridge: Harvard University Press, 2000.

———. *Control Through Communication: The Rise of System in American Management*. Baltimore: Johns Hopkins University Press, 1989.

Young, Elliot. *Catarino Garza's Revolution on the Texas-Mexico Border*. Durham, NC: Duke University Press, 2004.

Yrarrázaval Larrain, José Miguel. *El Presidente Balmaceda*. 2 vols. Santiago: Editorial Nascimiento, 1940.

Zanetti, Oscar, and Alejandro Garcia. *Sugar and Railroads: A Cuban History*. Translated by Franklin Knight and Mary Todd. Chapel Hill: University of North Carolina Press, 1998.

Zea, Leopoldo, and Mario Magallón. *1898 ¿desastre o reconciliación?* Mexico: Instituto Panamericano de Geografía e Historia, 1998.

Ziegler, Philip. *The Sixth Great Power: A History of One of the Greatest of All Banking Families: The House of Barings, 1762–1929*. New York: Knopf, 1988.

Zimmermann, Warren. *First Great Triumph*. New York: Farrar, Straus and Giroux, 2002.

Index

Page numbers in italic text indicate illustrations.